OPEC in a Shale Oil World

Mohamed Ramady • Wael Mahdi

OPEC in a Shale Oil World

Where to Next?

Springer

Mohamed Ramady
Department of Finance and Economics
  Visiting Associate Professor
King Fahd University of Petroleum
  and Minerals
Dhahran, Saudi Arabia

Wael Mahdi
Regus, Building 12, Level 4 Trust Tower
OPEC & Middle East Energy
  Correspondent Bloomberg News
Manama, Bahrain

ISBN 978-3-319-22370-4   ISBN 978-3-319-22371-1   (eBook)
DOI 10.1007/978-3-319-22371-1

Library of Congress Control Number: 2015946616

Springer Cham Heidelberg New York Dordrecht London
© Springer International Publishing Switzerland 2015
This work is subject to copyright. All rights are reserved by the Publisher, whether the whole or part of the material is concerned, specifically the rights of translation, reprinting, reuse of illustrations, recitation, broadcasting, reproduction on microfilms or in any other physical way, and transmission or information storage and retrieval, electronic adaptation, computer software, or by similar or dissimilar methodology now known or hereafter developed.
The use of general descriptive names, registered names, trademarks, service marks, etc. in this publication does not imply, even in the absence of a specific statement, that such names are exempt from the relevant protective laws and regulations and therefore free for general use.
The publisher, the authors and the editors are safe to assume that the advice and information in this book are believed to be true and accurate at the date of publication. Neither the publisher nor the authors or the editors give a warranty, express or implied, with respect to the material contained herein or for any errors or omissions that may have been made.

Printed on acid-free paper

Springer International Publishing AG Switzerland is part of Springer Science+Business Media (www.springer.com)

*Dedicated to our wives, Fatina and Muna,
Thank you for your love, patience,
and support*

Foreword

Ever since its inception in 1960, OPEC has been the focus of deep-seated insecurity, if not outright hostility, from a wide range of Western oil companies and governments. At that time, the hostility was not surprising given OPEC's years of acrimonious negotiations with Western oil companies and its goal of unifying the policies of its member states with the intent of increasing their oil income and limiting the financial returns to foreign investors.

Against this history of contentious relations in an era of resource nationalism, the creation of OPEC was viewed by the international oil companies as a revolutionary challenge to their economic interests and their dominance in the global oil markets. OPEC countries, on the other hand, saw it as the vehicle through which they would finally achieve full political and economic independence from Western interests and gain effective sovereignty over their national resources.

In fact it took OPEC a full decade before it could actually impose its aspirations in the oil industry and advance the national interests of its member states. This began in 1970 when it succeeded in moving Arabian Light oil prices off their historic low of $1.80 per barrel to clear $2.20. This was followed by more aggressive OPEC campaigning, including Libya's renegotiation of the terms of its oil concessions in 1970 and Iraq's nationalization of its industry in 1972. Additional oil price adjustments followed driven by political events, including the brief Saudi oil embargo of 1973, and by 1974 Arabian Light prices hit $11.60 per barrel, equivalent to $55.60 in 2014 dollars!

This rapid escalation in oil prices was to surge even further as a result of the Iranian revolution of 1979 and the breakout of military hostilities between Iraq and Iran the following year. By 1981, Iran's oil output had dropped from 5.3 million barrels per day (mbpd) in 1978 to 1.3 mbpd, and deep-seated concerns regarding the security of oil supplies permeated global markets. Saudi Arabian efforts to moderate the concurrent surge in OPEC prices were not successful, and oil prices more than doubled from 1979 through 1985, peaking at $36.80 in 1980, the equivalent of $106 in 2014 dollars.

Having witnessed oil prices rise to such unsustainable levels, it became clear that OPEC was in fact unable or unwilling to stabilize oil markets which had become dominated by near-term commercial thinking and subordinated to geopolitical conflicts.

This reality became even more evident when oil demand collapsed in the mid-1980s following the price surge of previous years. OPEC's efforts to starve the flooded oil markets by reducing supplies and thereby defend prices had the perverse effect of cutting its production from 30 to 16 mbpd by 1985, while new capacity from Mexico, Alaska, the North Sea, China, and Russia took over its original market share. Nor did OPEC's strategic planning improve during the following decade. Although its production increased gradually into the early 1990s, oil prices remained volatile due to widespread quota violations by virtually all the OPEC member states. A new oil price collapse became inevitable in the mid-1990s when OPEC made an ill-advised effort to raise prices in the midst of the Asian debt crisis of 1996. This led to an extended oil price slump which saw some crude grades selling below $10 per barrel by 1998 for the first time since the 1970s. The crisis was not resolved until the direct intervention of key heads of states from within and outside of OPEC and the collective reduction in OPEC's production by over 1 mbpd.

More crises were to follow, however, starting in 2003 with the growth in Chinese demand for oil in an already overstretched global market. In spite of the many OPEC claims of spare capacity and production capabilities, neither OPEC nor any of the other oil producers could easily provide the increases in conventional oil supplies that China required. It was inevitable that oil markets would tighten again and oil prices surged from $29 per barrel in 2004 to over $111 in 2007, peaking at over $147 per barrel of Brent in July 2008.

This rapid escalation in oil prices was again driven by events outside of OPEC's control and confirmed the reality that OPEC's massive oil reserves were irrelevant in volatile markets where rapid increases in oil demand required equally rapid adjustments in oil prices and supplies.

As the financial crisis of 2008 took hold and OPEC prices collapsed from $112 in August to $32 per barrel in December, OPEC members met and cut back supplies from what had become a severely flooded oil market. This resulted in a production rollback of 3 mbpd and succeeded in lifting oil prices back up over $70 per barrel within mid-year 2009. This reduction in supplies, however, had the unintended effect of signaling that OPEC was again prepared to return to its role of swing producer and to defend oil prices any time they dropped below a floor now deemed to be $70 per barrel.

The conviction of the existence of such a presumed floor price created a new confidence among marginal oil producers and allowed a widespread surge in costly oil developments from the deep offshore of Brazil and West Africa, and bitumens from Canada, as well as the financing of unconventional oil supplies from the oil-rich shale basins of the USA. Inevitably, these new realities raised the prospects of a future flood of additional oil production including the spread of unconventional oil supplies from vast accumulations of shale oil resources from across the world.

If this were to occur, the result would be millions of barrels of costly new oil supplies gradually flooding conventional oil markets and marginalizing OPEC's near-term ability to balance oil supply and demand and attempt to regulate oil prices.

Although the prospects for an outlook of this nature remain exceptionally slim for a variety of fundamental factors including offshore capital costs, onshore well productivities, severe shale decline rates, and environmental issues, the very influx of these and other new technologies will continue to challenge conventional resources and low-cost oil production.

Although it remains to be seen whether these and other challenging scenarios are likely to materialize in the next few years, it is inevitable that such challenges will continue to unfold, and OPEC must reexamine its role in seeking to balance oil supplies and demand in markets where technology is suppressing the significance of its oil reserves. Should OPEC discard its self-appointed mission of defending oil prices and define for itself a more effective strategy that increases its relevance and broadens its engagement with the wider spectrum of energy-based industries?

The assessments of this and other fundamental questions are the core issues that are raised and explored at length with great expertise by Mohamed Ramady and Wael Mahdi, the two authors of *OPEC in a Shale Oil World: Where to Next?* They have delved in detail into OPEC's history of successes, failures, motivation, and challenges and documented the key milestones that have shaped OPEC configuration over its five decades of history. Their book presents a detailed understanding of OPEC's capabilities and limitations as they have evolved with time and is a vital resource for anyone wishing to consider OPEC's and the oil industry's future options and opportunities.

Although the book concludes with many insightful recommendations for revitalizing critical sectors of OPEC as an institution, some of these will remain highly improbable given the loss of national sovereignty that they would entail. On the other hand, one thing is for certain. As the authors do emphasize, OPEC today has no choice but to accept the reality that its institutions are less than functional and its role as an oil cartel was never realistic. Its self-appointed mission to moderate oil supply volatility and manage oil price is not realistic and is not in its interest in any case.

Whether shale oil will grow to challenge OPEC's oil reserves and take more market share or not, the continuous evolution in energy markets and the environmental pressures to reduce the consumption of fossil fuels are sufficient factors in and of themselves to require a profound internal reassessment of OPEC's mission and its own options in any case.

Unless OPEC finds the will and leadership to recast itself as a relevant modern institution, endowed with intellectual and organizational assets that go beyond its massive oil reserves, it may ultimately find itself relegated, one way or another, to the same marginalized role that its members once held prior to its establishment in 1960. The call for action by Mohamed Ramady and Wael Mahdi in this book makes it clear that time, and not oil, is the precious commodity that is running out fast on OPEC's side.

Sadad Al Husseini
Former Executive Vice President
Saudi Aramco

About the Authors

Mohamed Ramady is Visiting Associate Professor in the Faculty of Finance and Economics, King Fahd University of Petroleum and Minerals, Dhahran, Saudi Arabia. He obtained both his B.A. and his Ph.D. from Leicester University, UK, and a postgraduate degree from the University of Glasgow. He is a Fellow of the Chartered Institute of Bankers (UK). He specializes on regional economics, energy, and the Saudi economy, as well as on money and banking. Among his publications are *The Saudi Arabian Economy: Policies, Achievements and Challenges* (Springer, New York, 2010), *The GCC Economies: Stepping Up To Future Challenges* (Springer, 2012), and *Political, Economic and Financial Country Risk: Analysis of the Gulf Cooperation Council* (Springer, 2014). Prior to his academic career, Dr. Ramady had over 25 years of experience working at senior level positions in the banking, finance, and investment sectors in the Middle East and Europe with Citibank, Chase Manhattan, First City Texas Bancorp, and Qatar National Bank. He also served as Senior Advisor to the Chairman of Qatar International Islamic Bank.

Wael Mahdi is Bloomberg OPEC Energy Correspondent and has over 10 years of experience in financial, economic, and energy journalism, including 5 years as senior energy reporter dedicated to covering OPEC annual and extraordinary meetings and the meetings of the Organization of Arab Petroleum Exporting Countries (OAPEC), as well as GCC ministerial energy meetings. He covers the energy sector in Saudi Arabia and Bahrain for Bloomberg News. He enjoys strong personal ties with Saudi oil policy makers in Riyadh and other key OPEC officials in Qatar, Libya, Kuwait, and the UAE. Besides Bloomberg, he has worked with various publications in the Gulf, including *The National, Arab News, Arabian Business, Al Watan*—a newspaper in Saudi Arabia, and *Asharq Al-Awsat*—a leading Pan-Arab newspaper. He holds a B.A. in Political Economy from American University in Cairo with a minor in Economics and is also a member of the International Association of Energy Economics. He has also participated in many financial and business reporting training in London with Thomson Reuters and Bloomberg.

Acknowledgments

Numerous colleagues, mentors, and friends have been instrumental in encouraging us to write this book, and many have shared their unique insights as well as their rich and varied experiences on the evolving geopolitics of the regional and international energy markets. Without such encouragement and enthusiasm, this book would have been much more difficult to complete.

With apologies to those omitted, the authors wish to thank the following for their encouragement, input, frank insights, and comments on drafts: HRH Prince Abdulaziz bin Salman bin Abdulaziz, Dr. Khaled Albinali, Sultan Albinali, Ali Aissaoui, Samer Al Ashgar Omar Attefah, Khalid Abuleif, Abdulsamad Al Awadhi, Maher Chmaytelli, HE Khaled Al Falih, HE Dr. Abdulaziz Al Furaih, Nawal Al Fuzaie, Hasan Hafidh, Riad Hamade, Kamel Al Harami, Dr. Awad Al Harithy, Sadad Al Husseini, Dr. Anas Al Hajji, Abdullah Juma'a, Othman Al Khowaiter, Nabil Al Khowaiter, Julian Lee, Dr. Mohamed Al Mady, Abdelkareem Al Marzooq, Dr. Faisal Mirza, Faisal Al Mudhaf, HE Dr. Majed Al Muneef, Hassan Qabazard, Dr. Mohammed Al Qahtani, Dr. Ayed Al Qahtani, Jamal Al Rammah, Nayla Razzouk, Abdullah Al Roumi, Innas Saied, Sheila Al Rowaily, Prof. Mohammed Al Sahlawi, Mohammed Sergie, Mohammed Al Shatti, HE Dr. Mohamed Al Suwaiyel, Samir Al Tubbayeb, Robert Tuttle, Steve Voss, Stuart Wallace, and Ali Al Yabhouni.

Last, but by no means least, a special word of appreciation for the tireless efforts of Junaid Akhtar for his secretarial work. Any shortcomings are the sole responsibility of the authors.

Dr. Mohamed A. Ramady
Wael Mahdi

Contents

Part I	**A History of Mistrust and Struggle**	1
1	**A New Paradigm: Protecting Market Share?**	3
	Introduction	3
	OPEC's 166th Meeting: A New Remedy for an Old Illness	4
	2014 Price Crisis: Another 1986 or Another 2008?	7
	A 1986-Like Event?	8
	A 2008-Like Crisis?	10
	A New Paradigm or a New Low-Price Cycle?	12
	The "Happy Meeting"	14
	No "Swing Producer" to the Rescue	17
	A New "Oil Order": The Search for a New Swing Producer	21
	How Did Saudi Arabia Transform OPEC?	23
	Why Did Saudi Arabia Relinquish Its OPEC Role?	24
2	**Oil and Geopolitics in the 1970s and 1980s:**	
	OPEC's Boom and Bust Eras	33
	Introduction	33
	The Rise and Fall of the Major Oil Companies	33
	The OPEC Era: Gaining Market Power	37
	The Pre-embargo Oil Order	40
	The Post-embargo Period: The Rise and Fall of OPEC	42
	The Price War of the 1980s	45
	A New Era After the Price War	50
	The Rebirth of the IOC's and Producer Cooperation	51
Part II	**OPEC and the New Reality**	55
3	**A False Dawn: Myths and Realities of OPEC's Power**	57
	OPEC's New Reality	57
	Monopoly Profits and Market Power	58

OPEC-Non-OPEC Cooperative Limits	60
Brazil: One More OPEC Headache	62
Quotas: Did They Ever Work?	64
OPEC's Quota Setting Formalities	66
Setting Quotas: Theory and Realities	68
Price Signaling, Compliance, and Cheating	71
Oil Market Crude Pricing	72
Not All Oil Is the Same	74
Futures, Hedgers, and Speculators	75
Equilibrium Price of Oil: In Search of the Holy Grail	76
4 Non-OPEC Producers, the Ever Fading "Peak Oil," and the Rise of the USA	**79**
Energy "Independence": Weaning Consumers Off OPEC	79
"Peak Oil": The End of the Oil Era?	81
"Peak Oil" on the Back Burner?	81
Was Hubbert Wrong?	82
Are International Reserves Overinflated?	84
Technology to the Rescue	85
Barriers to Renewable Energy Technology Deployment	86
What Is Unconventional Oil and Gas?	88
How Much Global Energy Resources Are Available?	90
The New Unconventional Energy "Stars"	92
The US Shale Revolution: *Drill, Baby, Drill…*	94
The USA in the Driver's Seat: A New "North American OPEC"?	97
US Shale Revolution: Key Success Drivers	99
Success Comes at a Price…	101
North American Shale "Plays": Location, Location, Location	102
A Heady Cocktail of Productivity, New Finds, and Price	103
The November 2014 Oil Price Shock: What Future for the Shale Oil Boom?	110
OPEC and the Shale Producers: Who Blinks First?	114
Shale Oil and OPEC: A Threat or a Stabilizer?	119
Enter the Dragon But Beware of the Year of the Goat	121
China's Nonconventional Gas Reserves: Large But Difficult to Exploit	123
Maintaining Producer's Market Share: Discount and More Discounts	125
The Russians Are Coming: The Bear Awakens	126
Russia and OPEC	129
5 Facing Realities: OPEC Fiscal Stress and Break-Even Pricing	**131**
Introduction	131
OPEC: A Phantom Menace?	131
Fiscal Challenges: Addressing Subsidies First	132
Economic Cost of Subsidies	134

	Break-Even Prices: Different Pain Thresholds	135
	Uneven Pain and Gain	136
	What Happens to OPEC Energy Investments?	141
	Who Blinks First?	144
6	**Environmental Obligations and Climate Change Politics**	147
	Introduction: Looming Challenges	147
	What Future for Fossils? Oil Producers' Fears	148
	The Scientific Debate	149
	The Long and Winding Road from Rio to Lima (Through Kyoto): Political Pressure and Shifting Alliances	154
	Does Kyoto Still Have Life?	157
	US-China Breakthrough	158
	India's Conversion from "Climate Agnostic" to Possible "Climate Evangelical?"	159
	From Fringe to Mainstream: Enter the "Greens"	161
	Global Climate Change: Who Pays?	162
	Paris 2015: A Half Glass Full Better Than None	167
	OPEC and the Environmental Bandwagon	168
	OPEC's Policy Challenges: Choices, Choices	173
7	**Charting a New OPEC Role: Avoiding the Coming Perfect Storm**	179
	Oil: A Blessing or a Curse?	179
	OPEC: Can It Ensure a "Fair" Market Price?	180
	Predicting Future Oil Price Curves	185
	Future Oil Price Curves: An Alphabet Soup of V, L, W, U, and M	188
	OPEC's Quandary: Dealing with the Elephants in the Room: Iraq and Iran	194
	The Long and Winding Road: Iran's "Historic" Nuclear Standstill Agreement	196
	Saudi Arabia and OPEC: End of the Hegemony?	199
	Saudi Arabia's Policy Options	202
	To Invest or Not to Invest in Spare Capacity?	210
8	**OPEC Reinvented: A New OPEC for the Twenty-First Century**	215
	Introduction	215
	Let's Not Call It a Cartel	215
	Consumer-Producer Cooperation and Supplier's Security to the Center Stage	220
	Why Cooperate?	221
	Suppliers' Energy Security: An Unattainable Goal?	223
	OPEC: Holding Together?	225
	OPEC Members: Some More Equal Than Others	226
	OPEC's Founding Charter: A Time for a Revisit?	232
	OPEC: Birth of a New Club	237

OPEC: Re-branding the Institutional Survivor 240
OPEC's "Re-branding": Some Recommendations 240
Conclusion 242
Closing Afterthought: OPEC—Has Its Strategy Succeeded?..................... 243

References 247

Index 263

Abbreviations

ADNOC	Abu Dhabi National Oil Company
APEC	Asia Pacific Economic Cooperation
APICORP	Arab Petroleum Investment Corporation
Bcfd	Billion cubic feet per day
Bcm	Billion cubic meters
BRICS	Brazil, Russia, India, China, and South Africa
Btu	British thermal unit
CCGT	Combined cycle gas turbine
CCS	Carbon capture and sequestration
CFD	Contracts for differences
CNPC	China National Petroleum Company
CSR	Corporate social responsibility
DUC	Drilled uncompleted wells
EFP	Exchange of futures for physicals
EIA	US Energy Information Administration
EOR	Enhanced oil recovery
EU	European Union
FCCC	Framework Convention on Climate Change
FDI	Foreign direct investment
GCC	Gulf Cooperation Council
GCF	Green Climate Fund
GDP	Gross domestic product
GW	Gigawatt
IAEA	International Atomic Energy Agency
IEA	International Energy Agency
IMF	International Monetary Fund
INDC	Intended Nationally Determined Contributions
IOC	International oil company
IOSCO	International Organization of Security Commission
IPCC	Intergovernmental Panel on Climate Change

IPO	Initial public offering
IPP	Independent power producer
IRENA	International Renewable Energy Agency
JODI	Joint Organizations Data Initiative
k	Thousand
KA-CARE	King Abdullah City for Atomic and Renewable Energy
KACST	King Abdul-Aziz City for Science and Technology
KAPSARC	King Abdullah Petroleum Studies and Research
KPC	Kuwait Petroleum Company
kW	Kilowatt
LTO	Light tight oil
mD	Millidarcy
mm	Million
MOC	Market-on-Close
NOPEC	No-OPEC Act
NPT	Nuclear Non-proliferation Treaty
O&M	Operation and maintenance
OAPEC	Organization of Arab Petroleum Exporting Countries
OECD	Organization for Economic Cooperation and Development
OFID	OPEC Fund for International Development
OLADE	Latin American Energy Organization
OPEC	Organization of the Petroleum Exporting Countries
OSP	Official selling prices
OTC	Over the counter
OWEM	OPEC World Energy Model
p.a.	Per annum
PPP	Public-private partnership
PV	Photovoltaic
QP	Qatar Petroleum
R&D	Research and development
SABIC	Saudi Arabian Basic Industries Corporation
SAMA	Saudi Arabian Monetary Agency
SOE	State-owned enterprise
SOEC	Statistical Office of the European Communities
SPR	Strategic petroleum reserves
SWF	Sovereign wealth fund
t/CO_2	Ton of CO_2
t/CO_2e	Ton of CO_2 equivalent
tcf	Trillion cubic feet
tcm	Trillion cubic meters
UN	United Nations
UNCTAD	United Nations Conference on Trade and Development
UNFCC	United Nations Framework Convention on Climate Change
UNSD	United Nations Statistics Division

URR	Ultimate recoverable reserves
US	United States
USCAP	United States Climate Action Partnership
USGS	United States Geological Survey
WTI	West Texas Intermediate

Introduction and Overview

Everything that happens happens as it should, and if you observe carefully you will find this to be so.

Marcus Aurelius

"OPEC is dead. Long live OPEC." There seems to be a bit of truth in both statements. The expression "mid-life crisis" is associated with that time of life when individuals reach their 40s and ask themselves many searching questions on what they have done and, more importantly, what they should now be doing with whatever existence they think they may have left. And so with OPEC at 54 years, where does it go next? For the organization, like individuals, it is a time of crisis, reflection, and sometimes profound changes with OPEC wondering what new role to play in a fast-changing, shale oil-led world. It is a time for reflection and not recrimination, and one that has to assess in fundamental terms what it is that the organization wishes to achieve to reach its 100 years and continue contributing to the world's energy needs. As a senior OPEC official commented, "if OPEC did not exist, it would be necessary to invent another organization much like it." The book will highlight OPEC policy options by drawing on the authors' firsthand knowledge of how key OPEC decision makers operate and are likely to act to meet such challenges.

Since OPEC was established in those heady days in Baghdad in 1960, petroleum crude is still viewed as a strategic and critical commodity for the world economy, highlighted by the many territorial disputes over possession of this essential commodity. However, the international petroleum industry has changed substantially since 1960, with a bewildering array of alternative energy sources and a mix of public and private players, along with an added element of environmental pressure groups that was not prevalent in 1960. When OPEC came to life, production, marketing, and prices were controlled unilaterally by the international oil companies (IOCs), and there was little or no concern for the environmental and the social impacts of the industry such as global warming and ozone depletion. Today things have fundamentally changed, and at their peril, neither OPEC nor the IOCs can ignore these issues.

The key is *what type* of organization and role should OPEC play that will make it acceptable to consumers and reduce latent hostility to the organization ever since it burst into popular imagination and became a favorite hate figure during the first oil shock in 1973. While major consumer countries, to a certain extent, can feel confident that there seems to be a more abundant and variety of energy sources, the same feeling of predictable energy supply security is not prevalent for OPEC producers. For OPEC countries embarking on long-term capital-intensive capacity expansion plans, the issue of *psychological* security is important. The global energy supply system is a vast complex of large, fixed capital assets that take years to plan, approve, and construct, even if governments decide to approve them in face of conflicting socioeconomic priorities. Once in place, these assets are in place for decades and can either add to national wealth or turn out to be costly "white elephant" projects. OPEC today operates in a multipolar energy world, trying to find its way through a maze of complex supplier and consumer relationships with new "petrosuperpowers" like the USA and Russia and consumer powers like China and India, leading to a potential for energy conflict if a stable and mutually beneficial energy market does not evolve to meet conflicting resource objectives.

The analysis of the evolving challenges facing OPEC is important for an assessment of the organization's future role in the world petroleum market since its establishment. There is a finite amount of oil in the world, and while new technology will push back the boundaries of "peak oil," current oil endowments depend on the rates at which oil is being produced and how it is being used for the good of nations. The modern oil era of the twentieth century has brought about many benefits to countries and citizens' way of life and home comforts. High oil prices have brought about shifts in financial assets from consuming to producer nations. At the same time, high prices have promoted energy efficiency measures and the introduction of substitutes, leading to oil supply disruptions and oil price volatility, compounded by other factors that have become important for the future use of oil such as environmental concerns.

OPEC is once again in the limelight and, like all international organizations, is constantly facing some crisis, despite enjoying periods of successes and achievements for its members. However, the current evolving structure of the world's energy sector, with added supplies from shale oil, has introduced a new element compared to OPEC's earlier market dominance of oil supply and prices. The 166th ordinary meeting in Vienna in 2014 was perhaps the most important meeting since OPEC's historical meeting in late 1973 when it decided to take over the pricing of its crude from international oil companies. The significance of the two meetings rests on the fact that their outcomes shifted the paradigms in the market and created new orders. The 1973 meeting transferred the power of pricing crude oil from international oil companies to OPEC. In the 2014 meeting, OPEC decided it will not continue to manage supply in the market to support falling oil prices. By this decision, it signaled to the market that there is no swing producer to the rescue. It was obvious from that decision that OPEC voluntarily transferred the limited influence it has on oil prices to the market and the search started in the market for a new swing producer. Saudi Arabia emerged out of the 166th OPEC meeting as a victor, as it succeeded in forcing a new mandate upon the group that was different from its traditional role. It was unprecedented for OPEC to let the market fully determine prices. It was also unusual for OPEC suddenly not to give support to any floor for prices and let them experience a rapid free fall. Moreover, OPEC decided to transfer the responsibility of setting a price floor to the market. This decision caused a new problem for the market since it was always OPEC's role to determine and defend price floors.

Conventional wisdom has held, until the dramatic oil price collapse of late 2014, that OPEC has the world's consumers in its grasp and that it can manipulate prices by tinkering with its own supplies. The events of late 2014 have proved conventional wisdom to be wrong, for it has shown that indeed "the Emperor has no clothes." OPEC, for the most part, has seen its actions, or more precisely OPEC decisions, lagging behind fundamental changes in oil supply and demand and a new energy paradigm when assessing the rise of shale oil from non-OPEC countries. The events of 2014 can be viewed as a watershed in OPEC's perceived control of world prices which demonstrated that instead of being a masterful and controlling cartel, OPEC was in fact now riding out events, hoping for the best in terms of an eventual oil price recovery. OPEC cannot be properly understood without up-to-date studies of the problems the "cartel" faces and the tensions within the organization's members, many of whom face issues associated with different socioeconomic and political agendas and strategies.

A key problem for OPEC is rooted in an ambition to set both price and production levels at the same time, but as discussed in the volume, this ambition is now more of a hope than a reality and can only be achieved if current marginal non-OPEC producers exit the market. What has evolved over the past decade, with the rise of non-OPEC production, especially nonconventional oil, is that OPEC does not now control the majority of production. As such, the desire to set prices will work only if OPEC sets a price *and* defends it by increasing production if the price is above a "target" and by reducing production if the price is below the target. Our analysis suggests that this is not easy to achieve. The internal allocation of OPEC production and the subsequent adherence to an agreed set of production level or quotas created problems due to nonadherence to what was agreed upon by OPEC members, to the substantial differences of interest concerning acceptable or fair prices, and how great the supply restrictions of each individual OPEC producer should be. In order to be able to influence oil prices in an effective manner, the organization must be able to do two things. First, determine a "fair price" for the group as a whole. Second, determine a "fair production" level for the group as a whole and the allocation of output among members. However, in reality, OPEC today is not the same organization as in its heyday when it was *the* marginal producer of oil and had global capacity and a market clout to influence prices.

This book is about a potential crisis facing OPEC and about policies and strategies it can take to ensure that it has a meaningful future role to play in a shale oil world and other alternative renewable energy supplies from non-OPEC sources. The November 27, 2014, decision by OPEC, despite some objections by smaller producers, to hold current production levels and let market forces determine oil prices, is the beginning of OPEC's realization that its global role has changed. As such, an improved understanding of OPEC's future role and how it faces such non-OPEC energy supplies are covered in this book and would be of great value not only for an academic insight in the field of geopolitics of energy but for all those concerned in different policy-making capacities in the energy sector.

Past history has vividly illustrated that as world markets expand, competitive pressure increases. It was easier to be an "oligopolistic" power in a smaller world market than in a large one characterized by growth in energy demand from emerging BRIC economies, especially that of China. Such rising energy demand has led to non-OPEC marginal production capacity growth which has created a dilemma for OPEC in how to maintain its market share and meet conflicting production quota levels by its members based on different costs, break-even pricing, and socioeconomic objectives while at the same time try to meet another OPEC objective to develop a comprehensive strategy that keeps oil prices constant in real terms at a level low enough to deter non-OPEC production from adding to global reserves. The November 27, 2014, OPEC Vienna meeting might yet go down as one of those epochal strategic moments for the organization. This follows OPEC's decision to maintain its collective production level of 30 mbpd despite falling oil prices and calls from some members to make production cuts, but with OPEC confirming that the organization was now entering a battle for global market share and that oil prices should be left to the market to decide "who blinks first." The organization reaffirmed its commitment to this new market-led policy in its June 2015 meeting.

OPEC is now betting on higher marginal producers, especially in US shale oil, to blink first, on the assumption that a prolonged period of lower oil prices to around $60 pb by 2015/2016 will erode profits for even the cheapest suppliers of crude from US shale, thereby threatening the economic viability of other non-OPEC producers around the world. However, sustained lower oil prices threaten some OPEC members too and put further strain on OPEC's unity. These include Nigeria, Iran, Venezuela, Libya, and Iraq in the short term and others like Algeria, the UAE, Saudi Arabia, and Kuwait in the longer term, given the fact that the latter countries had amassed significant financial reserves during the previous years of high oil prices to see them through future leaner budgetary periods. However, such a policy is illusionary even for the richer group of OPEC countries in the long term. All these countries rely on high oil prices to finance their budgets and generous fuel subsidies to citizens and, for some, such as Saudi Arabia, to continue building and maintaining expensive spare capacity, a policy which might now be questioned in the face of the new reality of maintaining market share in the face of non-OPEC capacity building. The repercussions of this new market-share OPEC strategy, if it is strictly adhered to by all members, are that other key non-OPEC oil producers such as Russia and Mexico might face recessions, as well as pressure on their currencies, forcing them to seek market share through price discounting. This will add even more pressure on fiscal-stressed OPEC members to do likewise to maintain their own market share.

The book analyzes the lessons of past OPEC actions, particularly those of the 1970s, production cuts, sharp rises in oil prices and lessons learned, and whether OPEC has indeed learned these lessons and is now adjusting itself to the fact that the organization is no longer seen as the marginal incremental producer but that this role has now been assumed by non-OPEC producers, most notably the USA. Despite inherent rivalry between OPEC and non-OPEC producers, there are lessons of interdependence between the two. The non-OPEC cooperation of the early establishment period involved resolving differences and wresting national rights from multinational oil companies, but from the mid-1970s, OPEC turned its attention to ever-increasing non-OPEC oil supplies. Contacts began with a host of non-OPEC countries, notably Mexico, Malaysia, Brunei, the UK, Norway, Russia, and Brazil.

While in the short to medium term oil sands and other nonconventional oil surges onto the market diminish OPEC's market share, there is another important longer-term question. Unlike *demand destruction* arising from vagaries in the economic fortunes of nations that depend on domestic fiscal, monetary, or geopolitical factors, such as the sluggish Eurozone economies of the post-2007 financial crisis, *supply destruction* due to a slowdown in exploration, reserve capacity building, and closure of oil fields due to uneconomic costs takes a far longer period of time to readjust.

Countries might have underground proven reserves, but if they are financially stressed and begin to disinvest or postpone new energy investments due to sharply reduced oil prices, then in the long run when demand for oil rises as economies recover, higher global oil demand will be met by *reduced* global supplies, leading to higher oil prices and reversing the "gain and pain" situation of consumers and producer nations to the advantage of the latter.

It is not only accesses to future energy sources that are of concern to consumer nations, but that the economic growth of countries such as China, India, Japan, and the USA also depends on a *sustained* flow of crude oil at *stable* prices. Given the above-stated political sentiments, it is not surprising that the petroleum industry in the twenty-first century started to focus on production of oil and gas from unconventional sources like heavy oils, tar sands, oil shale, and other renewable resources in order to ensure some viable and possible long-term energy independency. For those in the energy sector, technology is key to the continuing success of enhanced oil exploration and production. Further advancements in technology are essential to meet future energy needs at competitive prices. While traditionally more and more oil has been found by drilling deeper, it has been the use of new technology in more challenging geological environments that has turned dormant energy resources into *viable* reserves.

A better insight of the requirements and priorities of the OPEC nations will help us understand the needs of both producers and consumers of petroleum, as OPEC is now seeking a greater assurance *of producer supply security* compared with an earlier era of *consumer demand security*. Such mutual interdependence, based on market forces and the ability to compete in an ever-changing energy nexus of technology, environment concerns, and geopolitics, will ensure that producers and consumers are more closely tied to international trade, monetary policies, effective transfer of technology, and the rational management of nonrenewable resources that meets the massive economic transformation programs that many OPEC nations are facing. Given the global changes in non-OPEC production and capacity expansion, will OPEC have the ability to raise real prices in the future? And how much market power will OPEC have, or is more likely to have, to exert directly or indirectly? Or will the organization transform itself to meet the challenges of a different, market-led energy environment as opposed to sovereign-led energy production management?

Such type of behavior sustains a *rational myth* about OPEC's influence over the world market for oil, and the perceived cartel power of OPEC allows its members to reap political rewards in terms of diplomatic influence and the attention paid to its members, whether they are in the "large" or "small" league. OPEC was unable to coordinate production quotas and increase prices throughout the 1990s, and the situation from 2014 seemed to echo this collective impasse. Earlier periods of collusion were short-lived, based on some production cuts being undertaken, notably by Saudi Arabia. OPEC overwhelmingly controls the proven world's reserves of oil, with nearly 80 % of these reserves concentrated in Saudi Arabia, Iraq, Iran, and Venezuela. Over the years, countries like Saudi Arabia, Venezuela, Kuwait, and the UAE have amassed significant financial reserves, while some had seen their financial fortunes, notably Iraq and Iran, affected by either war or embargoes. Toward the end of 2014, all OPEC countries, including those who had amassed large financial surpluses in the past, faced the same questions: For how long could they cope with weak oil prices and what was their "break-even" oil price level to avoid deepening fiscal stress? Can a "fair" price ever be determined, one that meets the demand and supply security needs of both consuming and oil-producing countries? What are the major confluences of factors that could lead to such price cohesion and acceptance? The book analyzes these issues in detail.

The shale energy revolution is a reality and here to stay. Conventional producers and, more specifically, major OPEC producers have to learn to adapt to this new reality. A key uncertainty regarding non-OPEC supply is the potential diffusion and mass use of shale technology outside the USA, like in Russia. Leaving the issue of the embargo of high-technology end use by the USA on Russian oil companies following the Ukraine-Crimea events of 2014, there is some doubt on whether the US tight oil revolution could be replicated in other countries like Russia. The shale oil development in the USA has come at a great cost by drilling many expensive experimental wells and the accumulation of large amounts of debt, which in periods of low oil prices could and has put such shale operators under strain.

The US "expensive" model might not be easily replicated in other countries. The US shale revolution did not take place in isolation, but a host of accommodating enablers helped to propel this energy sector forward, stamping it as a "unique American model" which might be difficult to replicate in other shale-rich countries like China which reportedly holds some of the world's largest but yet undeveloped shale energy resources. The legal and regulatory framework for the development of unconventional resources in the USA is a mixture of laws, statutes, and regulations at the federal, state, regional, and local levels. There is no disputing that the size of the US oil and gas supply shock has been phenomenal, with tight oil production rising from around 1 million bpd in 2010 to more than 3.5 million bpd in the second half of 2014 and shale gas rising from negligible amounts in 2006 to 11,400 billion cubic feet by 2013.

OPEC faces other challenges. The issue of climate change and environmental obligation of energy consumer and producer nations is an emotive one with significant long-term consequences for both in terms of the quality of life and impact on economic growth. It was only a few decades ago that discussion of the effect of climate change was firmly the domain of so-called "fringe" environmentalist pressure groups like Greenpeace, but by the beginning of the twenty-first century, the debate had entered mainstream economics and politics, and governments throughout the world, to varying degrees of enthusiasm, have subscribed to the notion that not doing anything on global emission was not an option any more. The key is how to obtain not only consensus on emission targets but on actual national and voluntary programs to translate intentions into reality. While there are still some differences of opinion on the scientific evidence of global warming, there seems to be a global consensus to agree on measures to reduce emissions. With this "fact," oil producers, and OPEC specifically, have two choices: either to ignore both scientific evidence and global consensus or to try and seek fair terms for themselves given their limited economic base, bearing in mind that the objective of these oil producers is to establish the basis of a *sustainable* economic growth model in the long run. The US Energy Information Administration (EIA) notes that economic growth is the most significant factor underlying the projections for demand growth in energy-related CO_2 emissions in the immediate future as the world continues to rely on fossil fuels for most of its energy use. This is the argument that OPEC has been making—that oil is a vital component to the world's continuing economic growth and talk on "zero-carbon" emission for this vital energy source is unrealistic although even countries like Saudi Arabia seem now resigned to a possible "fossil-free" world and is taking the lead in OPEC in playing an active role in climate change negotiations to ensure that fossil fuel producers have an equitable hearing in the critical 2015 Paris climate change negotiations due to replace the Kyoto Protocol agreements.

The global environmental bandwagon is moving ahead, maybe at a snail's pace in the view of some, but it is moving along with compromises and long-term goals now set by those that were implacable opponents of the *Kyoto Protocol* in the first place. What has been OPEC's collective position on the matter, and of more importance to the organization, what is the position of OPEC's largest members? It would seem that while OPEC is more than willing as a group to play its responsible share in curbing global emissions, there is a feeling that the organization is being unfairly targeted and stands to lose from more stringent and compulsory emission abatement policies. The issue of taxes on carbon emission is also another area of concern to OPEC members, as this will raise the price of oil to consumers, reduce demand further, and reduce OPEC's market share in the future. Some have pointed out that any form of a "carbon tax" may in fact increase the "rent" that governments in the energy-importing countries have in the oil market and contribute also to a further transfer of wealth from oil producers to consumers. There is no question that the shale revolution has, despite some production fluctuations due to lower oil prices, tipped the balance of power in the oil market by opening up vast new energy resources, not only in the USA but in other frontier countries with future technological advancement and environmental controls in this sector.

The question then is: Is OPEC relevant or, more precisely, is it *as relevant today* as it has ever been since its establishment by the founding fathers? Is it less "powerful" today and where can it still add a meaningful role to ensure that it still holds together? As stated earlier, even if OPEC did not exist, oil-producing nations would still want to discuss oil market issues with one another, like other major commodity producers such as copper or wheat. If OPEC did not exist, it would be necessary to invent another organization much like it. As OPEC tries to navigate through this period of uncertainty and find a role for itself in a shale oil world and tries to cope with a long and difficult period of adaptation for which some of its members are not prepared, questions arise as to who are the most fit to survive?

While all OPEC producers are losing out to the fall in oil prices, some OPEC producers with larger marginal costs of production and higher "break-even" prices are suffering more than those with lower marginal production costs and break-even prices. OPEC in essence is now divided into various camps—those that are "efficient" low-cost producers and can sweat out the increased marginal non-OPEC producers and those "inefficient," high-cost OPEC producers who are struggling to make their voices heard to convince the first OPEC group to cut production and help raise prices. This assumes that *all* OPEC members agree on what they believe OPEC represents as its twin fundamental interests: *fairness and stability*, whereby "fairness" is about the equitable distribution of the global oil surplus in providing a reasonable return to investors and ensuring that neither producers nor consumers claim a disproportionate share of oil rent benefits. "Stability," on the other hand, is something that OPEC aspires to, without the organization being sidelined to a role of a "residual supplier" to ensure market stability, but primarily at the expense of its more efficient, high-production capacity members who end up losing market share.

OPEC has become part of the global international brands. Despite reduced market share and competition from new energy sources, it still facilitates dialogue and helps to educate officials and ministers from the less-advanced member countries on how to deal with international issues. Its senior officials and past representatives are well respected in their field and are not afraid to speak their mind at international forums, often expressing divergent views on what the organization should be doing. This diversity is to be welcomed, but at the same time, there must also be a sense of purpose on what OPEC needs to do to make it more effective and avoid divisions between OPEC members. The book sets out some suggested changes to the OPEC statutes and charter to meet current challenges as well as new objectives to ensure that a re-branded "new" OPEC not only survives but continues to play an important role for many years to come. The news of OPEC's death is truly exaggerated and premature.

Part I
A History of Mistrust and Struggle

Chapter 1
A New Paradigm: Protecting Market Share?

> *We can't solve problems by using the same kind of thinking we used when we created them.*
>
> Albert Einstein

Introduction

Since its inception in 1960, OPEC has made two important decisions. The first decision was in October 1973 when it took over the pricing of its crude from international oil companies. The second decision is the one its ministers reached in their 166th ordinary meeting on November 27, 2014. By refusing to cut output and instead keeping the production ceiling intact at 30 million barrels a day, at a time when oil prices lost close to $40, the organization's ministers signaled that they are seeking a new policy that does not focus on defending oil prices. The decision, in effect, created a new paradigm shift for the oil market as it marked the end of the residual or "swing" producer role for OPEC, a responsibility it assumed since the early 1980s to defend prices. OPEC, or at least its largest members, is now focusing on defending its share in the market. In June 2015, OPEC members, this time without dissent, agreed to continue with the Saudi-led policy of no production cuts and defend market share (Smith et al. 2015).

The new OPEC policy to maintain its market share is a big transition for a group that made prices and maximizing the economic rent from oil sales the center of its policies for four decades. The policy resulted in instability in the market now that technically there is no "swing producer" to keep supply and demand in balance. The effectiveness of the new market-share policy in stabilizing the market was questioned by critics who varied widely in their views on the motives behind the decision. Some believed that the real motives were political, since it is not an economic principle to see a producer accepting deterioration in his revenues. Others saw in the decision an attempt by a "weak" OPEC to safeguard its interest against the threat of shale oil revolution.

The purpose of this chapter is to look at the conditions and the reasons that led OPEC to make such a transition and set the ground for the volume. Understanding Saudi oil policy and the structural developments in the market that occurred on the supply side are key elements in also understanding the radical changes that the group underwent.

OPEC's 166th Meeting: A New Remedy for an Old Illness

After 15 years of depressed oil prices, OPEC was waiting for a new "golden window" to open after the one that opened between 1973 and 1980. The golden window opened and oil went through a commodity super-cycle for more than a decade between 2000 and 2014, with prices increasing by around four times over the period. OPEC countries benefited greatly from the recent super-cycle. The group's revenues from petroleum exports doubled from $611 billion in 2009 to $1.21 trillion, as yearly average of oil prices hovered over $100 for 3 consecutive years ending 2013. This is noted in the price of OPEC Reference Basket (ORB) and the price of Dated Brent in Fig. 1.1 (OPEC ASB 2014). With higher revenues comes higher spending. The weighted average of OPEC members' fiscal breakeven oil price, which is the price they need to balance their national budgets, was around $105 in 2013 (Aissaoui, Sept/Oct. 2013) illustrated in Fig. 1.1.

Fig. 1.1 OPEC ORB vs. Dated Brent Prices (2003–2013).

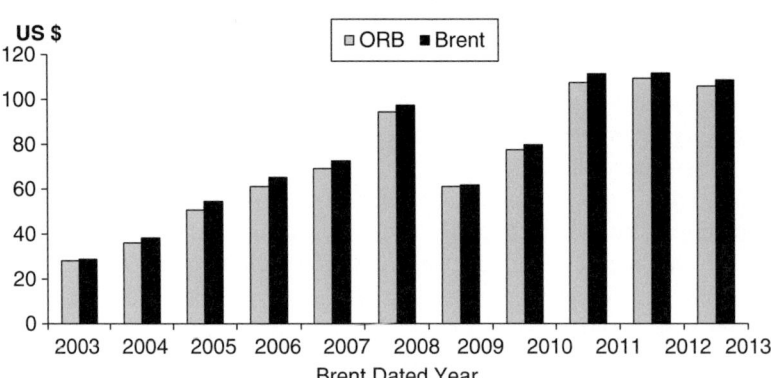

Source: OPEC, ASB (2014)

The extended period of the "super-cycle" drove OPEC countries to set higher ranges for a "fair" oil price that can be acceptable for both consumers and producers. Saudi Arabia, the OPEC de facto leader known for its moderate views on prices, considered oil price of $70–$80 as a fair price (Bloomberg, Feb. 22, 2011) for almost 2 years, before considering $100 as a reasonable price (Bloomberg, March 18, 2013). The $100 became the "ideal" price for a majority of OPEC countries in 2012 and 2013 (Platts, Sept. 18, 2012) now that the fiscal breakeven price was close to the market price as illustrated in Fig. 1.2. Yet the rise in prices over the period

Fig. 1.2 OPEC Fiscal breakeven prices for 2012 and 2013.

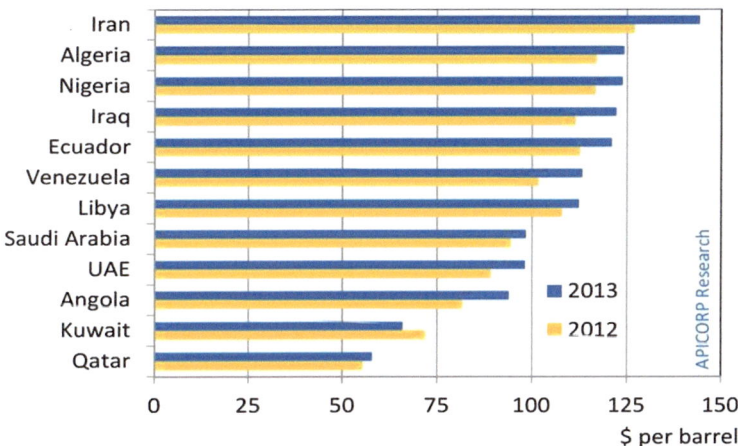

Source: www.apic.comResearch/Commentaries/2013/Commentary_V08_No8-09_2013pdf

2011–2013 was not the product of market fundamentals alone, but it was mainly driven by geopolitical factors and supply disruptions from the Middle East and North Africa (MENA) region. Production came to a halt either fully or partially in countries like Libya, Syria, Sudan and South Sudan, Yemen, Nigeria, Iran, and Iraq. The Energy Information Administration (EIA) estimated that global unplanned oil disruptions grew by 2.8 million barrels per day over the period January 2011 through July 2014 (EIA, Aug. 27, 2014[a]). Figure 1.3 sets out the relative differences for each

Fig. 1.3 OPEC fiscal cost curve for 2013 (Bar width: country's production; bar heights: price estimate ranges)

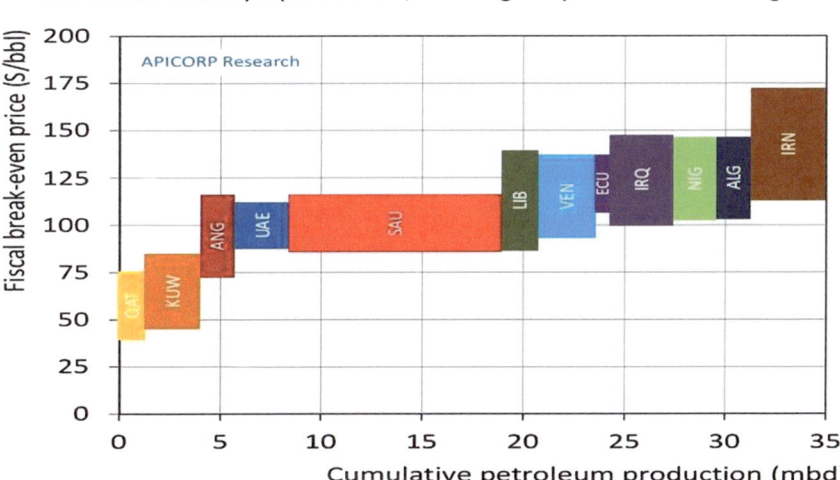

Source: www.apic.comResearch/Commentaries/2013/Commentary_V08_No8-09_2013pdf

of the 12 OPEC members in terms of their individual production levels and estimated fiscal breakeven price bands.

In 2014, as geopolitical tension receded and high prices allowed for increased production globally, the oil super-cycle came to an end with the drastic fall in oil prices. *Brent* went down from around $115 in June to $46.6 in January 2015, as Bloomberg data showed (Bloomberg 2015[m]). At the beginning of the fall, there was a conviction in the market that OPEC will trim its production to defend the $100 oil price it viewed as fair. This conviction lasted until November 27, the day of the 166th ordinary meeting of the organization. On that day *Brent*, the benchmark for pricing half of the world's crude, was trading at around $72. The cut was expected as Standard Chartered noted in a report before the meeting, as the other option for OPEC was to let the market balance itself, a step that would result in an extended period of low oil prices (Standard Chartered, Nov. 25, 2014). Apparently, by agreeing to keep the organization's production target intact at 30 million barrels a day, the OPEC ministers decided to let the market balance itself fully. Oil prices continued their slide following the meeting, hitting 6-year lows as OPEC's decision meant that there is no entity in the market that can act to support any price floor.

Certainly, this is not the first episode for oil prices to record a drastic fall of more than 30 %. The oil market witnessed six significant episodes of price declines since oil contracts were first traded on future markets in 1983. The price fall of 2014 is the third largest drop over the last 30 years after the price drop of 1985–1986, and 2008, as a World Bank report noted (Baffes et al., March 2015). The difference between the three episodes was that OPEC in 2014 for the first time reacted by letting the market fix the imbalances and engage in a price-discovery exercise. The differences/similarities between the three episodes will be discussed in more detail later in the chapter.

As for the 2014 price fall, the oil market was not in balance during the second half of the year due to many factors, among them the surge in production from outside of OPEC driven by an increase in the output of *Light-Tight Oil* (*LTO*), which is also known in industry terms as "shale oil," mostly from North America and other non-OPEC producers as illustrated in Fig. 1.4. US crude oil and other liquids (biofuels, NGLs, and refinery processing gains) grew by 4 million barrels a day between January 2011 and July 2014, of which 3 million barrels was crude oil (EIA, Aug. 27, 2014[a]).

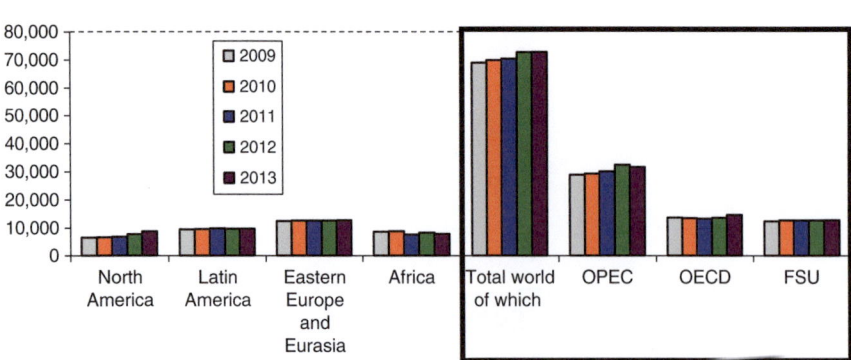

Fig. 1.4 World crude oil productions by region and OPEC (1000 b/d)

Limiting the imbalance in the market at the time of the November 2014 meeting to supply-side factors, however, was a very simplistic approach since the market imbalance was a result of mixture of structural and short-term factors that affected the supply-demand balance. The appreciation of the US dollar has exerted downward pressure on oil prices in the second half of 2014 and complicated the situation for OPEC countries; however, the major contributing factor for the decline was the supply-demand imbalance. From a supply-demand point of view, the imbalance on both sides can be summarized as follows:

On the *demand side*, first, the expected growth in demand early in the year failed to materialize and demand recorded a sluggish growth of below 1 million barrels according to OPEC's own estimates (OPEC MOMR April 2015^c). Second, the price structure for *Brent* was in *contango* forcing crude buyers to store the cargos and thus raise the stock levels, a factor that always adds downward pressure on prices. On the *supply side*, the imbalance occurred because of the structural changes that resulted from the oil market shifting from "scarcity to plenty."

On the *supply side*, first, output from outside the group was booming due to the increase in North American production and, second, output from within the group rose on the return of shut-in Libyan production and improvement in Iraqi output. The complexity of the situation in the market was reflected in the length of the actual meetings of OPEC ministers on November 27. The meeting took longer than usual as it extended for 4–5 h. The group at the end decided to keep the production ceiling intact at 30 million barrels a day, thus applying a new remedy for an old illness, which was the rise in non-OPEC supply with increased prices.

Saudi Arabia emerged out of the 166th OPEC meeting as a victor, as it succeeded in forcing a new mandate upon the group that was different from its traditional role. It was unprecedented for OPEC to let the market fully determine prices and discover a new price floor, yet OPEC yielded to the power of Saudi Arabia at the end and accepted the new order. The reason for the Saudi policy change is discussed later in the chapter.

2014 Price Crisis: Another 1986 or Another 2008?

It is true that history repeats itself, but not necessarily in an identical way as the fall in oil prices in 2014 showed. The 60 % fall between June 2014 and January 2015 resulted from a mixture of (1) weakening demand, (2) increase of supply from outside of OPEC, (3) appreciation of the US dollar, (4) easing of geopolitical concerns, and (4) speculative activities in the oil paper market (Baffes et al., March 2015; OPEC Monthly Bulletin April 2015^c; Sen Sept. 8, 2014).

The speed, magnitude, and longevity of the 2014 price fall has made many in the market to compare it with the two previous price crises of 1986 and 2008. Nonetheless, the 2014 price fall is fundamentally different even though it shares many features with both past price crises. The intriguing question asked this time however is why OPEC did not react to the 2014 fall in the same manner it did in 1986 and 2008, i.e., by adjusting its production to support prices. The answer to this question lies in the market conditions and developments that *prevailed prior to the price fall*. Despite many attempts to describe the 2014 events as a mere market imbalance resulting from oil glut, the recent price fall is by no mean a simple repetition of the past market experiences. In fact, it is more of a mixture of 1986 and 2008 *with new elements added to the mix*.

By looking at the price curve, the resemblance between the two past price falls in 1986 and 2008 and the more recent fall in 2014 is very strong. In all three episodes, prices have fallen down by a large margin (more than 30 %) over a period of more than 6 months (Baffes et al., March 2015). Both price episodes of 1985–1986 and 2014 occurred due to a surplus of crude in the market that followed an extended period of supply growth from outside of OPEC. However, when considering the facts on the ground there are fundamental differences between the 2014 episode and the two previous ones.

A 1986-Like Event?

Starting with distant past developments, some voices in the oil industry such as BP's CEO Robert Dudley (Bloomberg April 21, 2015^e) believe that the market environment in 2014 is similar to that of 1986. It is true that there are similarities between the two events, yet there are profound differences as well.

The first and most important *similarity* is that both price crises occurred after an extended period of high oil prices. The price fall of 1986 was the result of OPEC's attempts to raise and defend high prices between 1973 and 1985. Oil prices increased by four times between 1973 and 1976. Prices during that 12-year period had peaked at $36.8 in 1980, which is equivalent to $104 in 2013 money value. The rise in prices was huge over a short period as it jumped by ten times between 1973 and 1980. Prices were on a continuous decline between 1980 and 1985; however, oil prices in 1985 were still nine times their level in 1973. The situation in 2014 was no different. Brent prices in 2013 averaged $108.6, double their level in 2005, and the benchmark was trading above $100 for 3 consecutive years after 2010 (BP Statistical Review 2014).

The *second similarity* is a result of the first. With high oil prices, supply from new and more difficult resources in non-OPEC producers increased over a long period of time as a result of increased investments. In both 1986 and 2014, the market experienced an oil glut. The challenge for OPEC during both events is the same, with non-OPEC barrels displacing those from OPEC. At the start of the 12-year period that ended in 1985, the share of OPEC in total world's oil production was 65 %. By the end of the period, it fell down to 40 % in 1985 (Chalabi 1989). Turning to the 2014 crisis, the concern of OPEC producers from supply growth from outside of the group was mirrored in a speech by the group's Secretary General *El Badri* in Manama, Bahrain. Since 2008, non-OPEC producers were able to raise supply by almost 6 million barrels a day. In contrast, OPEC's production has been steady at around 30 million barrels a day during that period (El-Badri, March 8, 2015). A World Bank report found that the North Sea and Gulf of Mexico added 6 million barrels a day of crude over 1973–1983, an amount similar to the crude that was added between 2004 and 2014 (Baffes et al., March 2015).

Looking at the *differences* between 1986 and 2014 price crises, one can note that the first and most important difference is the shape and the structure of the demand for oil. Demand was mainly contracting due to the high-oil-price regime that OPEC administered throughout the 12-year period prior to the fall in 1986. The most obvious and steepest contraction appeared after the "second oil shock" in 1979. Oil demand in the OECD fell down from 44 million barrels a day in 1979 to 37.5 million barrels a day by the end of 1985, driving total world demand to fall from 63.9 to 59.2 million barrels a day during the period. On the other hand, world oil demand was growing at an annual compound rate of 1.31 % between 2003 and 2013 from 80.2 million barrels a day to 91.3 million barrels a day. As for the structural change in demand, OECD's oil consumption was no longer the main engine of growth for global demand as it was in the 1970s and 1980s. The rate of growth in non-OECD countries was outpacing that of OECD since 2003. *For the first time in history, oil consumption in non-OECD countries in 2013 surpassed that in the OECD* (BP Statistical Review 2014).

The *second apparent difference* is that supply from outside of OPEC was lagging behind OPEC prior to the 1973 first oil shock and it started to rise gradually. Whereas in the few years prior to the fall in 2014, a large gap between both groups did not exist, it became much easier for non-OPEC producers to catch up with or surpass OPEC on an annual basis. If production from the former Soviet Union bloc is excluded, OPEC's total oil production was 30.99 million barrels a day in 1973, surpassing that from outside the group by little more than half. Thanks to the high-oil-price regime that OPEC defended during the 1973–1985 period, matters have turned the other way around and production from non-OPEC producers reached 23.05 million barrels a day in 1985 compared to 15.45 million barrels a day for OPEC (Chalabi 1989). In today's oil world, OPEC is competing with various sources of supply, and its primacy in the market is becoming something of the past. OPEC's total oil production (crude and other liquids) was estimated in the region of 36.8 million barrels a day in 2013, indicating a decline of 1.8 % from 2012. However, the non-OPEC supply, excluding the former Soviet Union bloc, grew at 2.7 % in 2013 from 2012 to reach 36 million barrels a day (BP Statistical Review 2014).

The two main elements that do not make the 2014 fall a "1986-like" crisis are the growth in demand and the introduction of more difficult oil resources that were once considered to be at the frontier. *In 1986, conventional crude from OPEC was competing with other conventional crude from outside the group*. In 2014, competition became fierce with the advancement in unconventional resources such as shale oil in North America, oil sands in Canada, and sub-salt developments in Brazil as discussed later in the volume. Technology was playing a bigger role in 2014 and areas such as the Arctic became a target for major oil companies with Shell announcing in 2015 that exploration for energy in this frontier region was being considered (Pals 2015). This time around, OPEC is battling both an increase from conventional and unconventional resources.

For OPEC to defend high oil prices in 2014 means that the unconventional revolution will stay for a longer period, exerting more pressure on the world's oil market and prices. The shale oil revolution supported by high oil prices in the period 2010–2013 has allowed the USA to regain some of its lost oil power as its production grew from 7.5 million barrels a day in 2010 to 10 million barrels by end of 2013 (BP, ibid). No country in OPEC was able to add that much of oil over the same period, not even Saudi Arabia which opened the taps to compensate for the loss of production from other OPEC members in the aftermath of the political uprising in the Arab world that started in 2011. Yet apart from the points mentioned above, the 12-year period prior to the 1986 price fall was a period when OPEC used to set crude prices unilaterally. This power does not exist today and prices are now determined by market forces.

A 2008-Like Crisis?

The price collapse of 2008 was an intense episode as oil prices escalated rapidly to $147 in July from the start of the year before falling down sharply to just below $40 by December., and falling to under $40 in August 2015. Unlike the 1986 fall, the price episode of 2008 originated mainly because of the collapse in demand for oil that followed the negative developments in the financial markets. A second reason for the rapid price fall of 2008 was the *financialization* of the oil market and the speculative activities that grew with huge volumes of traded paper oil barrels (Baffes et al., March 2015). Some of the literature of that period linked the price crisis of 2008 to "peak-oil" theory suggesting that the inability of producers to add more supplies resulted in the rise of prices over the first half of the year. Khan argues that peak-oil theory cannot explain the 2008 episode as it can only explain long-term price movements, and it is not suitable for explaining short-term price fluctuations (Khan 2009). Another explanation for the 2008 price crisis suggested that it was a result of the passive role OPEC played in the market in the first half of the year when oil prices started to rise (Fattouh, Oct. 2008). The oil market in the 2000s became a more complex place with the increase in future contracts trading, unlike the 1980s. The enormous financialization of the oil market that has taken place in only 5–6 years before the price collapse of 2008 reflects the role speculative activities played in short-term price movements. The average daily trading volume of oil futures in 2008 was 15 times the daily world production of oil (of around 85 million barrels a day), up from four times in 2002 (Khan 2009).

The resemblance between the 2008 and 2014 price collapse is very clear. First, in both episodes, oil prices fell down sharply in the second half of the year. Second, not only the speed of the price fall was almost the same but also the rise in the level of crude stockpiling supported by a *contango* in oil prices. Yet, there was a key difference between the nature of the two price *contangos*, and this has to do with developments in the financial sector. Barclays Bank explains the difference between the two *contangos*, by stating that "the 2008–2009 *contango* was caused more by cascading effects from the financial system to the oil markets, creating a sharp fall in oil demand, and OPEC supply had adjusted swiftly to balance the market. This time, a stark OPEC policy-driven stance against supply adjustment is set to result in a relatively prolonged supply glut. Also of note is that current global oil demand is not falling this time on an absolute basis, rather it is just growing more slowly" (Mahesh et al., Jan. 26, 2015, p. 2).

The role of financial speculations was a key factor in the oil price fall in 2008, but speculation is a short-term factor. Some officials in OPEC believed that market fundamentals do not warrant the 60 % fall in prices between June 2014 and January 2015 and accused speculators for playing a major role in that fall (El-Badri, March 8, 2015, Al Muhanna, April 9, 2015). There are not enough studies by OPEC or elsewhere to support this claim and apparently this argument did not circulate well in the market.

What really distinguishes the price fall in 2008 from that of 2014 are the developments in the supply part of the equation. The increase in production from non-OPEC producers is instrumental in the development that occurred in the market in 2014. Not only was the size of the increase important, but also the fact that the role of non-OPEC has created a structural change in the market. Due to advances in the production methods of *LTO*, OPEC had to confront a new element that made all past remedies of adjusting its own output to balance the market something of the past. The non-OPEC supply over 2008–2009 was not growing as large as it was in 2014. It only recorded a growth of 600,000 barrels a day in 2008 (OPEC Monthly Report, Jan. 2009). Fuelled by the take-off in shale oil production in North America, non-OPEC supply grew by 2.17 million barrels a day in 2014 from the previous year (OPEC Monthly Report April, 2015[c]).

In 2008, OPEC responded to the price collapse by announcing a cut to its production by 4.2 million barrels a day. This was possible, since demand was contracting that year due to the global economic recession while output from outside the group was not rising significantly. Demand in 2014 was not contracting, but it was not high enough to accommodate the rise in non-OPEC supply. OPEC estimates that oil demand in 2014 was less than 1 million barrels a day (OPEC MTOMR April 2015[c]), while Saudi Oil Minister Al-Naimi stated that the growth in 2014 was around 700,000 barrels a day (Al-Naimi, Dec. 21, 2014).

Another distinctive feature of the supply-side imbalance in 2014 was that the increase in production from outside OPEC was considered more of a structural change and not a short-term phenomenon. The breakthrough in technology that helped in increasing output from North America is a major contributor. OPEC went through a similar situation in the 1980s when high prices unleashed production in the North Sea. Yet, the incremental increase of supply from the North Sea was mainly due to price improvements and not to huge improvements in technology as it was the case with shale oil production in North America. Accordingly, and after taking into consideration both the magnitude of the increase from outside OPEC and the technological advancement that allowed such an increase, it becomes clear that oil prices were impacted heavily by the structural changes that took place on the supply side with oil going from "scarcity" in the 2005–2008 period to "plenty" over the 2010–2014.

In a nutshell, the 2008 price fall was shaped by demand contraction resulting from economic downturn and an overhang in oil inventories that exerted downward pressure on prices. The role of speculation in future markets was yet another contributing factor to the 2008 crisis. Comparing the market developments of 2008 with those of 2014, it appears that both episodes are not identical, despite that inventory levels were high in both years and with *Brent* contracts in *contango*. The 2014 price collapse was a complex situation for the producer group, as both OPEC and non-OPEC were competing for a modest demand growth, with the latter experiencing an expansion in its production. In the light of this condition, curtailing production would have resulted in a loss of OPEC's market share in favor of other producers from outside the group, *a condition that did not exist in 2008*.

A New Paradigm or a New Low-Price Cycle?

The outcome of the 166th 2014 ordinary meeting, although seemingly unanimous, opened the door for a new role for OPEC and a new paradigm for the market now that the oil futures market was becoming a free market for the first time, as noted by a US energy envoy (Bloomberg 2015[i]). However, the decision not to adjust production levels opened the door for criticism, as some observers of the situation believed that OPEC was committing another fallacy that might lead to a long cycle of suppressed oil prices. The justification for such a belief is found in the cycle that the market witnessed between 1986 and 2000, which resulted from OPEC's mismanagement of oil pricing in the first half of the 1980s (World Bank, Jan. 2009; Asharq al-Awsat, Nov. 22, 2014[a]). The collapse of 1985–1986 originated when OPEC changed its policy after Saudi Arabia abandoned its swing producer role to secure a higher market share, by unilaterally raising production and offering competitive pricing for refiners.

The first attempt of Saudi Arabia in 1986 to defend its market share at a time of falling oil prices and rising non-OPEC supply had an adverse effect on prices, and with that in mind, the success of the second attempt in 2014 is in much of doubt. The first attempt was short-lived and it only lasted for the first 8 months of 1986. Chalabi argues that OPEC was not prepared for the new "market-share strategy" and was united to formulate a suitable program that can help the member countries to implement that strategy since there was no defined size for the share it wanted to defend. Chalabi also noticed two problems about the new strategy: (1) it created tension between OPEC and other non-OPEC producers especially the USA, and (2) it allowed competition between OPEC members since there was no defined share for any country. Chalabi argues that the market-share strategy of the 1986 *might* have succeeded in winning back the lost market share, if they had continued with gaining a higher market share, as the numbers showed that OPEC's production rose by 5 million barrels a day in August from January 1986. Output from the USA and the North Sea would have fallen if OPEC had continued with the same strategy for another 2 years (Chalabi 1989). Another factor that can be considered for explaining the reason behind the short life span of the new strategy was the removal of former Saudi Oil Minister *Ahmed Zaki Yamani* from his post.

It is apparent from Chalabi's remarks that OPEC in 2014 was almost facing the same internal challenges to implement the new strategy. The group in their 166th meeting did not define the size of the share each member should secure. As a result of that, OPEC members for the first 6 months following the meeting were operating on the bases of maximizing their share to the most possible extent, as the monthly production numbers of the group issued by the Secretariat in Vienna showed. Saudi Arabia and Iraq, for example, raised production to levels not seen in the past 30 years (OPEC MTOMR, April 2015^c). Another problem with a market-share strategy, which allows the market to balance itself to determine the equilibrium prices, is considered more of a long-term approach to fix market imbalances. A price recovery and market self-balancing may take longer time as noted by *Nasser Al Dossari*, an adviser to the Saudi oil minister and the country's national representative in OPEC's Economic Commission Board (Al Dossari, April 8, 2015).

The IEA argued in its monthly oil report in May that the battle for market share was just starting with all producers in OPEC and outside of the organization either increasing their production or not seeing any impact on their output from the fall in prices by end of April 2015, in which global crude oil supply rose by 3.2 million barrels a day on annual basis (IEA OMR, May 2015^a). However, the IEA acknowledges that shale oil producers started to "blink" in their standoff with OPEC due to months of cost cutting and a 60 % plunge in the US rig count (ibid).

For a group of producers that is known for internal rifts, any decision that OPEC takes regarding long-term strategies is met with doubts. The reason for this is that most of the producers within the organization are known for focusing on short-term gains by maximizing their profits from oil sales. OPEC in 1986 was only able to implement the market-share strategy for 8 months (Chalabi 1989). Assuming, however, that OPEC succeeded this time around in keeping the market-share strategy for a period longer than 8 months, then what would be the features of the new paradigm that resulted from the new strategy?

The first feature as mentioned above is that oil is now trading in a free market, at least until OPEC starts to reverse its new policy. The second feature is that OPEC is now behaving under the influence of Saudi Arabia to be *the* dominant producer. The decision at the 166th meeting gave birth to a new concept regarding Saudi's role within the organization. If Saudi Arabia is no longer the swing producer within OPEC, it is certainly now becoming the *dominant producer* with its efforts to keep its market share from dwindling. Saudi Arabia was also trying to push OPEC to transform into a "dominant producer" in the market, yet its endeavor might not be successful. The new market-share policy was not welcomed by the other producers who favored defending prices. In less than 6 months since the meeting, Iran and Libya called for OPEC to return back to its previous role to defend prices by cutting production (Bloomberg, April 14, 2015^g). Changing the mind-set of other producers within the organization is not an easy task knowing that some like Iraq, Libya, and Iran suffered from low exports due to political or security constraints. Other producers like Venezuela and Nigeria depend heavily on high oil prices to meet their fiscal obligations. Shifting the group from a swing to dominant role to preserve its market share is a cumbersome task in a world where technology has pushed the boundaries of production, and the market shifted from "scarcity" to "plenty" over a few years.

Only Saudi Arabia and its GCC allies were in a strong fiscal position to pursue such a new policy and role and endure its financial consequences, as will be explored in more detail in this volume.

The 166th meeting was a turning point in the history of the organization not only because it transferred a large part of its pricing power to the market, but because it reshaped the way OPEC conducts business in an oversupplied market. The ministers agreed in their meeting that the market was oversupplied as the communiqué showed (OPEC, Nov. 27, 2014), and this condition might persist in 2015 as non-OPEC supply was expected to increase by 1.36 million barrels a day at a time when stocks in OECD countries are already above the 5-year average. In June 2015, all OPEC members agreed to continue with this new market-share defense policy and kept their combined daily production target at 30 million bpd (Smith et al. 2015).

The "Happy Meeting"

The 167th meeting of the OPEC countries in June 2015 was best described as the "happy meeting" by the ministers who emerged from it conveying to the press their satisfaction with its outcome. The meeting ended before the scheduled time and more than one minister stated afterward that there was only one scenario discussed by participants and that scenario was to keep OPEC output intact at 30 million barrels a day. In the author's studied opinion, the word "happy" is used interchangeably with the word "smooth" to describe that OPEC ministers did not have to argue over many issues as usual and that the most of the important issues were postponed until OPEC's next meeting on December 4, 2015. What are these future issues confronting the organization following on from the "happy meeting"?

First, the decision to keep the group's ceiling unchanged is nothing but an attempt by the ministers to buy more time before the perfect storm. The decision in itself meant to outsiders that OPEC has now been temporarily suspended until further notices. With every member country entitled to produce the desired volumes it wants, the ceiling that OPEC imposed on itself is somewhat meaningless. The group's communique issued following the meeting stated that the conference "urged member countries to adhere" to the self-imposed ceiling. The whole adherence concept is just a voluntary act on part of the member countries. The ceiling itself is not a production target, rather a guideline for the group as the Secretary General of the organization told the press following the meeting.

Second, when OPEC was created in the 1960, it was built on the model of *Texas Railroad Commission* to be, inter alia, a price stabilizer that adjusts the production of its members to eliminate any surpluses from the market. However, without production disciplines, the group is turning into an organization similar to the *International Energy Agency* that is research driven and only acts at time of crisis to meet any disruption in the market. Third, there was no mention following the June 2015 meeting of any credible and transparent mechanism to be put in place to monitor and keep production levels at or near the rolled over 30 million bpd and, finally, there was no mention of any individual quotas set for those seeking to reenter the market and claim their "rightful" market share like Libya and Iran. Above all, the OPEC meeting did not seem to take into consideration worrying signs that the demand for world crude was still not on a sustained path mode, backed by genuine economic growth demand and not stockpiling.

Behind all this transformation is OPEC's kingpin Saudi Arabia that seems to be frustrated with the inability of other members to shape up and become efficient producers who can align their fiscal balances to endure lower oil prices. It is after all a very difficult exercise to let prices move up to the levels that will allow shale oil producers, who have been termed "the sprinters," running a 400 m race, to increase production and drive OPEC "the long haul marathon runners" out of the market. Whether the Saudi motives are purely political or entirely economic based, the ceiling rollover means a suspension of OPEC as the world knows it. The suspension will remain until prices stimulate OPEC to resume its market role. Otherwise, OPEC will be transformed into nothing more than a research organization that controls a considerable share in the market individually, *and not collectively*, and the world's oil market will enter into a long cycle of lower investments and subdued oil prices similar to the one that lasted between 1987 until 2002. This is illustrated in Fig. 1.5, indicating that, despite such "happy meetings," there are large discrepancies between current OPEC supply and demand for its crude.

Fig. 1.5 OPEC's large discrepancy between its crude supply and demand.

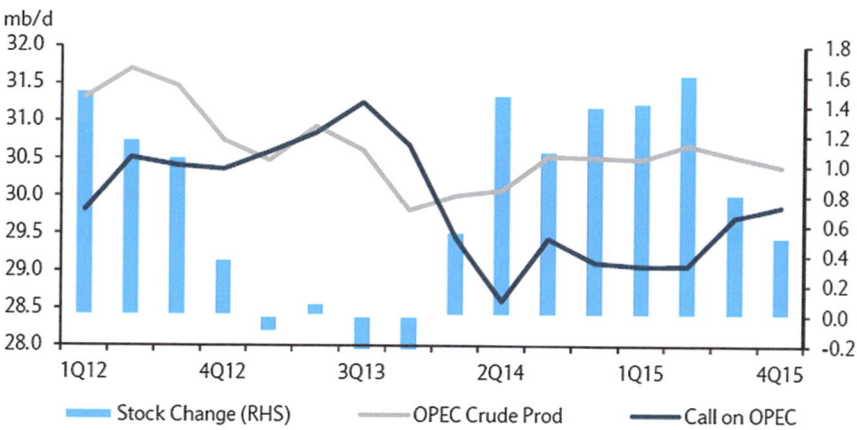

Source: IEA (history), Barclays Research (2015)

At the end, all what it takes for OPEC to lose its influence over the market and prices is for another two "happy meetings" similar to the 167th June 2015 gathering. However, based on OPEC's past experience, the "happy meetings" will come to an end once Iran restores its lost capacity and Iraq reaches its recently stated goal of 6 million barrels a day by 2020 as will be explored in this volume and which will widen the gap between the actual "call on OPEC" and what it is producing. The next two OPEC meetings will be decisive in setting the course of the organization into the foreseeable future and confirm whether it stays relevant.

In a market where non-OPEC production was expected to continue increasing at a pace higher and faster than the rise in demand and OPEC's supply growth, the organization had few alternatives at the time of the meeting. Faced with a glut that was putting downward pressure on prices, some OPEC ministers supported a proposal by Algeria to cut the group's production target by 5 %, while others supported the Saudi-led proposal of defending market share. It was not the first time for OPEC to confront an oil glut, but it was the first time for the group not to participate in restoring the market's balance.

The OPEC market-share strategy initiated a downward spiral in the market in the months that followed the 166th meeting. Investments worth billions of dollars were shelved by international oil majors that might lead to a tightening of the oil market in the near future. The IEA estimated that investments on oil projects will fall by $100 billion in 2015, or 15 %, from 2014 levels (Bloomberg, Jan. 21, 2015). In its annual assessment, the *Oil and Gas Journal* expects spending on US oil and gas projects to drop by 26.8 % this year on annual bases to a total of $261.99 billion, bringing total spending close to its 2008–2009 levels of at $260 billion (Oil & Gas Journal, April 2015).

As a result of spending cuts, around 100,000 jobs in the industry were lost due to low oil prices (Olson, Feb. 12, 2015). The number of oil drilling rigs went down sharply by more than half between October 2014 and May 2015, reaching the lowest level since 2010 (Bloomberg, May 12, 2015[d]). The industry became in complete disarray, and events are still unfolding as a Saudi Aramco executive, *Amin Nasser*, expects the industry to see a delay of around $1 trillion of capital spending over the coming years due to the fall in oil prices (Bloomberg, March 9, 2015[i]). Such disarray is always the case when an old order is giving birth to a new order, a one that was engineered by Saudi Arabia, whether intentionally or unintentionally.

The new paradigm rests on the notion that crude prices no longer have a floor set by producers and thereafter a new floor is subject to the laws of supply and demand. The paper market, however, cannot function without stability in the physical market. Thus the former entered into a search for a swing producer and initiated an exercise of self-discovery for the floor of oil prices. *A new swing was identified, and it was the producers of shale oil*. The paper market's focus shifted to North America to monitor production there, and since January 2015, the future contracts traders reacted heavily to any news on decline in US output or any fall in the number of oil drilling rigs. The role of shale oil producers as a "swing producer" for the international market is discussed later.

No "Swing Producer" to the Rescue

The "no-cut" decision OPEC ministers decided upon on November 27, 2014, made it clear that the producers club, which once controlled nearly half of the world's production, is, first, voluntarily transferring the limited influence it has now on oil prices to the market and, second, abandoning the swing producer role. Saudi Arabia had already abandoned the role years before, but the market was still under the belief that OPEC will continue to be a swing producer or at least its largest member Saudi Arabia will act as one in the 2014 price environment.

The seed for the November 2014's decision is found in the 1973 meeting and the events that followed until 1986. OPEC's decision in late 1973 created an order where the group became responsible of setting the price of its oil instead of the international oil companies that operated within its member countries. Gradually, profit maximization became the group's new objective in an effort by the countries to compensate for the years when the posted oil prices of oil by the majors were kept fixed at certain levels until they were finally lowered, explored more fully in the next chapter of this volume. This new order paved the way for OPEC to become a residual "swing" producer whereby the group must manage the supply in the market to defend a desired price level for its marker crude. However, assuming the role of a swing or residual producer was not an easy task, and it came at the expense of the group's market share during the 1980s (Chalabi 1989).

Setting oil prices in the 1980s was a challenging task for the group as there were *two pricing powers in competition with each other*. On one hand, there was OPEC itself, and there was the "spot" market on the other. The competition from the spot market on OPEC grew strong as it was fuelled by an increase in production from producers outside the group such as Norway and Mexico and from new areas like the North Sea. The share of non-OPEC producers kept growing, thus limiting OPEC's power to balance the market and sustain high oil prices that would allow the group to maximize its profit. For example, in 1980, OPEC's production was 26 million barrels a day while non-OPEC's output was 24.8 million barrels a day. To defend prices in the following years, OPEC had to cut its production year after year and thus it dwindled to 15.8 million barrels a day by 1985. On the other hand, non-OPEC's production increased during the same period until it reached 29.6 million barrels a day in 1985 (BP Statistical Review 2014). Yet, it was Saudi Arabia that took most of the burden to balance the market during that period and the country's production fell down sharply from 10.27 million barrels a day in 1980 to 3.6 million barrels a day in 1985 (ibid). The cuts orchestrated by OPEC did not help to stop prices from falling and spot oil prices averaged $13.1 (*Dubai*) in 1986, down from $35.69 (*Arabian Light*) in 1980 (ibid).

OPEC's role as a swing producer never proved viable at times of falling demand and rising supply from outside of the group. The role was difficult to maintain in the 1980s by the group members, especially Saudi Arabia, as (1) prices kept falling, (2) the organization's market share dwindled, and (3) internal rifts over cuts mounted. The rift widened with two members (Iraq and Iran) went into war with each other. The whole structure did not help as well, with Saudi Arabia taking most of the burden to support prices during the first half of the 1980s. The structure was a "gigantic inverted pyramid," in the words of former Saudi Oil Minister *Ahmad Zaki Yamani*. The problem with that system, as he put it, rested on the ability of *one* large producer to carry the burden. Global energy prices rested on oil prices and oil rested on OPEC and OPEC itself rested on Saudi Arabia. Because of that structure, Saudi Arabia became "*de facto* the world's swing producer. No wonder the whole pyramid soon toppled over!" (Yamani, April 1994, p. 7).

The Saudi experience with OPEC and falling prices in the first half of the 1980s shaped Saudi policy in the years that followed. Abandoning the swing producer role was the one of the first tasks for the new oil minister, *Hisham Nazer*, who was appointed in 1986 by *King Fahd bin Abdulaziz* to replace Yamani. The new policy was communicated clearly and directly to the market by Nazer himself (New York Times, Oct. 18, 1987) that Saudi Arabia will no longer bear alone the burden of trimming its output to balance the market to support prices. The new focus for Saudi Arabia shifted to covering up for any major supply disruption. This policy manifested itself well in 1990–1991 when it increased its output to make up for the loss of Iraqi and Kuwaiti crude. Saudi Arabia demonstrated great ability in ramping up production on a short notice, and it managed to lift its output from 5.7 million barrels a day in July 1990 to 8 million in September and maintain production at above that level for the following year (Parra 2004).

Saudi Arabia stopped playing the role of the swing producer since 1985, and all the cuts to support prices between 1986 and 2014 were made through collective efforts with other producers within or without OPEC. Saudi Arabia's refusal to act alone to support oil prices was reconfirmed by Saudi Oil Minister Al-Naimi in December 2013, when he stated that Saudi Arabia only responds to supply disruptions and meeting demand. He asserted that history showed that whenever there is a call for a cut, the burden was only shouldered by Saudi Arabia and other GCC producers (MEES, Dec. 6, 2013; Alhayat, Dec. 6, 2013). It seemed that the market missed the Saudi signals since 2013 and Al-Naimi's comments left no room for surprises, but market participants assumed that Saudi Arabia will act on the basis that prices will put pressure on OPEC producers knowing that many of them have expanded their fiscal spending programs since 2011. *It was the lack of understanding the signals that Saudi Arabia was sending that contributed to the failure of the market to predict, consume, or accept the outcome of OPEC's 166th meeting.*

What was surprising to many trading houses and analysts in the market of OPEC's decision in November 2014 is the fact that Saudi Arabia was able to persuade other members not to curtail production and the *group as a whole abandoned the swing producer role*. The failure of OPEC to set oil prices in the 1980s, the rise of the spot market, and the fall in OPEC's share of world's total petroleum output had left OPEC with limited ability over setting prices since 1986. Yet OPEC still maintains its profit maximization goal. For the market, however, OPEC was always the "supplier of last resort" and the last line of defense for oil prices. It was still the swing producer, *despite the many signs that the role of OPEC is ever changing and continuously evolving.*

Nevertheless, the market's view of OPEC is the result of long years of price defense attempts and profit maximization. In the year 2000, the price band mechanism was introduced and the group became more focused on attaining a favorable price band of $22–$28 for the OPEC basket of crude. Continuing with that mechanism was difficult, as the market situation became different. OPEC lacked the spare capacity to help it to implement that mechanism (WTRG 2011) and oil prices traded outside that band for over a year, and the group finally decided in its

134th extraordinary meeting in 2005 to suspend the price band mechanism (OPEC Bulletin, Feb. 2005[b]). The abandoning of the price band was another historical example of how OPEC's interaction with the market is dynamic and how its role is always changing based on the internal dynamics of its member countries and external political and economic conditions.

Despite the different shifts in the group's policies and the different mechanisms it applied since its inception, the mind-set of the market did not change greatly on how it views OPEC's role. Until the day of the 166th meeting, some commentators still believed that OPEC will resort to is swing production strategies and trim output to defend prices (Standard Chartered, Nov. 25, 2014). Deutsche Bank said in a note prior to the meeting that an OPEC cut is "inevitable" based on the group's previous attempts to restore prices by cutting output, the persisting supply overhang, and the need of OPEC member countries for higher fiscal breakeven oil prices to support their higher fiscal spending (Deutsche Bank, Nov. 25, 2014[a]).

OPEC's previous attempts to restore prices through output cuts gave strong support to the belief that it will cut output in 2014. The most recent attempt was in 2008 when OPEC decided to take out 4.2 million barrels a day of crude off the market in what was the largest collective cut ever by the group. The market conditions in both 2008 and 2014 were very similar and that was another factor that supported the belief that OPEC will cut in November 2014. Another justification for an OPEC cut was that oil prices at the time of the meeting were already below $80, a level that many countries within OPEC needed to balance their fiscal budgets (Aissaoui 2010, Deutsche Bank, Oct. 16, 2014[b]). Thus, when OPEC met on November 27, 2014, to consider the actions it should take after prices were sliding as a result of an oversupplied market and slow demand, there were still many players in the market who were under the impression that OPEC would not abandon its old role.

OPEC's decision was contrary to the "common wisdom" that was building in the market over the few weeks prior to the 166th meeting. The explanation of the logic behind the decision was not very clear. OPEC at first only stated the preconditions to the decision in its communiqué after the meeting. The communiqué pointed that market conditions were changing because of the increase in production from outside of the group. The communiqué issued by the group's Secretariat said: "The Conference also noted, importantly, that, although world oil demand is forecast to increase during the year 2015, this will, yet again, be offset by the projected increase of 1.36 m b/d in non-OPEC supply. The increases in oil and product stock levels in OECD countries, where days of forward cover are comfortably above the 5-year average, coupled with the on-going rise in non-OECD inventories, are indications of an extremely well-supplied market" (OPEC 2014).

The communiqué also stated that to restore the equilibrium in the world's oil markets, OPEC decided to keep its output ceiling at 30 million barrels a day, a level that has been in place since December 2011. The decision made by OPEC during that meeting not to intervene in the market was called "historic" by Saudi Oil Minister Ali Al-Naimi 3 months later in a speech in Berlin: "I think history will prove that this was the correct path forward" (Al-Naimi, March 2015).

The market needed a shock to wake up to the new reality about OPEC. Immediately after the meeting, the market understood that OPEC no longer wants to be the swing producer and pricing power had now shifted to the trading floors of London and New York. *BNP Paribas's* head of commodity markets Harry Tchilinguirian immediately after the meeting told Bloomberg "OPEC has chosen to abdicate its role as a swing producer, leaving it to the market to decide what the oil price should be" (Bloomberg, Nov. 27, 2014[a]) Michael Wittner, head of oil market research at *Societe Generale*, issued a note following the meeting in which he echoed other voices in the market. The note said: "Today's decision means that Saudi Arabia and OPEC will no longer be the mechanism to balance the market from the supply side. They have relinquished that role. Instead, the market itself—prices, in other words—will be the mechanism to rebalance the market" (Wittner, Nov. 27, 2014, p. 2). The shock from OPEC's decision was clearer in the words of *Bank of America Merrill Lynch* in a note under the title "The End of OPEC," in which the bank considered the group's failure to reach an agreement on a cut as a sign it is "effectively dissolved" (Bank of America, Nov. 27, 2014[b]).

Confronted with a new role, the market's initial response was mainly defined by panic. *Brent* lost around $5.17 a barrel after the OPEC meeting and was trading at around $72. The Canadian dollar fell and the Russian *ruble* tumbled, while shares of energy companies were the biggest losers that day on global stock market (Bloomberg Data, Nov. 2014[e]). In the first 3 weeks that followed OPEC's 166th meeting, the market still had hopes that a fall in prices will force OPEC to reassume its role as a swing producer. The market at first tested the $60 level, expecting to see the group changing course if prices reached that level. The media was full of speculative stories citing banks and analysts that suggested OPEC would be calling for extraordinary meetings to discuss a possible cut. At the end, the market lost hope with oil ministers from Kuwait, UAE, and OPEC's kingpin Saudi Arabia all stating late in December 2014 that they will not cut production no matter how low prices reach. The three ministers made it clear that OPEC will now try to protect its market share in face of the rise in production from outside the group. OPEC's Secretary General said in a speech on December 14, 2014, that the group was not sure at the meeting that prices will rise even if they considered a cut of around 1–1.5 million barrels a day (Bloomberg, Dec. 15, 2014[b]).

With OPEC's main producers signaling they would not cut down production, the market went into another wave of panic that sent prices lower and markets were now testing levels in the $50's range. The market was on a new path to search for a floor price and for a new swing producer. Bank of America Merrill Lynch issued a report on December 4, 2014, under the title *Saudi puts no more* saying: "Saudi Arabia at the helm, giving up on its mission to 'ensure the stabilization of oil markets' last week, the cartel has entered a new era. What does that mean in practice? Imagine the Fed effectively giving up on its mandate to keep US inflation in a steady band around 2 % and arguing that consumer prices across the economy will 'balance' themselves overtime!" (Bank of America Merrill Lynch 2014[a], p. 3). This comparison puts Saudi Arabia's role in perspective.

A New "Oil Order": The Search for a New Swing Producer

The decision not to cut production at the OPEC November 2014 meeting was painful for both OPEC and the market. *Brent* oil prices lost more than 50 % of their value by December 2014 after peaking at $115.71 in June of the same year. By the end of January 2015, oil prices were trading at a level lower than all what OPEC countries needed to balance their fiscal budgets, forcing some members like Iraq and Kuwait to review downward the breakeven prices for their next budgets. With oil prices losing 50 % of their value and no swing producer in the market to set a price floor, and with the non-OPEC, mainly shale oil companies looked upon by the market as the new swing producers, it was indeed a new era for the market.

The change initiated by OPEC was fundamental and structural to the extent that had Citibank form a new theory for the market where it replaced the *call on OPEC* with the term the *call on shale* (Citi, Dec. 18, 2014). The call on OPEC is basically the amount of crude required to be supplied by the group to balance the market. Citibank now argues that there is a new paradigm under which it is the shale oil producers who will be balancing the market. However, the *Citi* report acknowledges that there are differences between the dynamics of OPEC and shale producers who would use to balance the market with the capital market being at the center of a shale oil market-balancing mechanism.

First, shale producers can adjust relatively rapidly in response to prices; and second, access to credit plays a great role on the amount oil shale producers will bring to the market as many of them have been cash flow negative on aggregate (ibid). Citibank's assessment, seemingly unique, provides a significant focus on shale oil producers and little attention to the *call on non-OPEC* in general. The UAE and Saudi oil ministers made it clear in their statements that high-cost oil producers are responsible for the fall in prices and the glut in the market and that they should be the ones to balance the market and not the low-cost and more efficient producers.

Goldman Sachs argued that shale oil producers have become the marginal producers; thus, they have more influence on prices than OPEC. The bank said in a report issued 1 month before the meeting: "a tight global oil market had until now required strong OPEC production and US shale production growth. While getting to a point where the market shifted back into surplus was only a matter of time, as US shale oil production grows by Libya's capacity every year, we now have higher confidence that a structural transition has been reached and that US production growth needs to slow" (Goldman Sachs, Oct. 26, 2014, p. 1).

Goldman Sachs argues that core OPEC members lost their pricing power. Pricing dynamics of oil have shifted to US shale oil producers as shale production was exceeding OPEC spare capacity. Another shift in pricing dynamics in the oil market was the move from the "dominant firm's production decision and towards the marginal cost of US shale oil production" (ibid, p. 1).

The other main feature of the new paradigm is that *shale oil companies are also among low-cost producers* and that they can compete directly with OPEC on a cost basis; they can drive the prices needed to balance the market lower than the prices that OPEC countries need to balance their fiscal budgets. Venezuela considers it be in OPEC's interest to stabilize oil prices at $100 per barrel in the medium term (Bloomberg, May 15, 2015[c]). If correct, then this is also a radical new reality for OPEC to meet a future battle for market share *based on cost efficiency*. Despite OPEC reiterating its stand of no production cuts during the June 2015 meeting, which continued its previous controversial November 2014 decision, some analysts argue that no matter what OPEC decides about its own production target, the outcome is for more oil to be produced by *all*, irrespective of OPEC quotas. This would principally come from Iraq and Iran which will add pressure on the organization which pumped 31.58 million bpd in May 2015, well above its 30 million bpd production target. It is not only OPEC members who will pump as much as they could if they have the spare capacity, but also non-OPEC conventional and nonconventional producers locked in a battle for market share with OPEC as both sides hope they prove to be the most efficient source of production and remain resilient (Rowling and Arnsdrof 2015).

Being a swing producer is one of the most difficult roles to be assumed by any producer, argues Kemp (2015). The role is normally associated with power and control over the market, but in practice, "the swing producer often becomes the passive absorber of shifts in market supply and demand" (Kemp, May 14, 2015). The role of the swing producer was assumed by a group of American independent producers for almost four decades ending in the early 1970s under the umbrella of the *Texas Railroad Commission*. This role was transferred to Saudi Arabia in the 1980s before Saudi Arabia transferred it to the shale oil producers in November 2014, but the latter may find "it as uncomfortable as the Texas independents and the Saudis did previously" (ibid).

In conclusion, the oil market has witnessed a major transformation and the birth of a new oil order in the aftermath of the 166th OPEC meeting. At the core of this new order are the US shale oil producers who are now forced to be the new swing producers. However, the structure of the shale oil industry is totally different than that of the *Texas Railroad Commission* or OPEC, where coordination for planned cuts is more possible and effective to a large degree. The shale oil producers are fragmented, and they are highly indebted and sensitive to price fluctuations and to developments in interest rates and the financial sector, as will be explored in more detail later in this volume. The new oil order is also one where the free market is the entity responsible for balancing the market and shale oil companies are the new swing producer. For a free market to balance the market is a long-term process, thus the effectiveness of this new order is still in doubt. The main question that needs to be answered is how did Saudi Arabia manage to create all these shifts within OPEC?

How Did Saudi Arabia Transform OPEC?

Persuading OPEC members as a whole to forgo what left of their power to influence the price of crude oil and accept a new order where prices are fully determined by the market at levels below the level OPEC countries needed to break even their fiscal budgets is indeed a Herculean task, yet it was easier to a greater degree for Saudi

Arabia to persuade its GCC allies (Kuwait, UAE, and Qatar) within OPEC to follow suit and defend their market share. The four countries formed a united front in the organization's 166th meeting as they shared many things in common that would help them endure any negative impact of that decision for a relatively long period of time in comparison to other producers.

Firstly, the four countries enjoy strong fiscal positions and all of them have put aside billions of US dollars in the form of foreign reserves. Saudi Arabia alone accumulated around $734 billion until December 2014 (Sfakianakis, Dec. 2014). Secondly, the four countries enjoy low cost of production per barrel; therefore, they can sustain production and make profit at any cost above the cost of production which is estimated to be in single digits. The estimated cost of producing a barrel of oil in Saudi Arabia, for example, is around $4–$5, according to the Saudi oil minister (MEES, Dec. 2014). Thirdly, the four GCC states historically had relatively low fiscal breakeven oil prices compared to other OPEC nations (Aissaoui 2010, Deutsche Bank 2014[b]). Qatar's fiscal breakeven oil price per barrel for its 2015 budget is estimated by *Ashmore* to be around $65, while that for Kuwait to be around $55. Saudi Arabia has the highest budget breakeven price among the resilient Gulf economies at $101. UAE is second to Saudi Arabia at $82. Despite the high breakeven price, Saudi Arabia has the fiscal buffers and official reserves to sustain a deficit for 2015 even if its fiscal breakeven is in the low $90s (Sfakianakis, Dec. 2014).

With the four GCC members agreeing on protecting their market share, it was not an arduous task for the Saudi oil minister to obtain the consent of the other OPEC producers at the November meeting. Iraq's Oil Minister *Adel Abdulmahdi* explained that it was difficult during the meeting to go against the will of the four GCC producers as they combined half of OPEC production, and without getting them to agree to a cut, the whole idea of curtailing the group collective production target would be impossible.

What has changed profoundly in recent years is not OPEC's relevance to the market. *The real change is that OPEC is now under a strong influence from Saudi Arabia with many OPEC decisions being made in Riyadh.* At times of high oil prices, Saudi Arabia considers policies that are suitable for it. An example is when it started to increase production unilaterally in the second half of 2011 after it failed to reach a collective decision in the June meeting that year to raise the group's production ceiling. At times of low oil prices, Saudi Arabia looks for collective cuts as it did during the *Oran* meeting in 2008 and is still doing after the November 2014 meeting. The question that needs to be answered is what are the reasons that made Saudi Arabia abandon the old policy of the group and instead accept a new policy that goes against the general wisdom of the 12-member organization?

Why Did Saudi Arabia Relinquish Its OPEC Role?

There are five main drivers behind Saudi Arabia's push for OPEC's historic decision not to intervene in the market to support prices. First, the Saudi reluctance to solely bear the burden of balancing the oil market by giving up its market share; second, the shrinking OPEC market share in front of the rise of supply from high-cost producers outside the organization; third, the future outlook for demand; fourth, past experiences of Saudi Arabia and OPEC with failure of non-OPEC producers to fulfill their pledges to cut production at times of market turmoil and collapse of prices; and fifth, the developments within OPEC and the potential threat it sees from growing supply from other producers such as Iraq and Iran.

First driver: The Saudi position is best described in the words of Oil Minister Al-Naimi in his speech on March 4, 2015, in Berlin. In his speech the minister explicitly said that Saudi Arabia has finally learned the lessons from the 1980s and it will not repeat the mistake of letting go its market share to defend prices. The minister stated: "over the past eight months, though, with the market in surplus, it is Saudi Arabia that is called upon to make swift and dramatic cuts in production. This policy was tried in the 1980s and it was not a success. We will not make the same mistake again" (Al-Naimi, March 4, 2015). Saudi Arabia finally came to the realization that it cannot act as the de facto world's swing producer. The Saudi position is reinforced by developments in the market. The world is awash in crude oil and technology is helping OPEC and non-OPEC to unlock more resources at a time when demand is starting to slow down or flatten in some of the major consuming nations. The Saudi policy makers are also aware of the fading power of OPEC and Saudi Arabia over prices in recent years. The last time Saudi Arabia and OPEC had a real power over prices was in 2008. The cut at the *Oran* meeting testified that OPEC unilaterally was able to lift oil prices, yet that required a concerted effort and a cut that is double the size of the growth in non-OPEC supply in 2014.

The conditions in 2008 were all in support of OPEC to exert control over prices. The market was tight, peak-oil theory was the norm in the industry, demand was still growing at healthy and high rates from emerging markets despite a short interruption in 2009, and OPEC countries, and Saudi Arabia in particular, *were the marginal producers*. The power that Saudi Arabia enjoyed that year was a result of its expansionary oil program, while other major producers did not announce any significant additional capacity increment. The Kingdom assured the world in the Jeddah Energy Meeting in June 2008 that it will raise its capacity to 12.5 million barrels a day within a year's time from 10 million barrels a day from new increments such as *Khurais*, and it will add an additional 2.5 million barrels a day on top of that to reach 15 million barrels a day if there is a need for that (Al-Naimi, June 2008).

Since 2008, many things have changed in the global market. Peak-oil theory did not seem valid as will be explored later in the volume, as supply kept expanding from different parts in the world. Global supply grew largely from the country where the theory originated, the USA, as technology allowed for hydraulic fracturing of tight oil formations to produce light tight oil. Additionally, on the demand side, economic growth in China was no longer in the double digits, and in the EU, oil consumption was declining and so was that of the OECD as a whole. The EU, hit by economic woes that affected growth in the Euro zone, consumed 1.9 % less oil in 2013 compared to a year earlier, while the OECD's oil consumption fell by 0.4 % over the same period (BP Statistical Review 2014).

Saudi policy makers acknowledged these new realities and turned them into an action plan for state-owned *Saudi Aramco*. The company will no longer consider expanding its maximum production capacity beyond 12.5 million barrels a day. Aramco's then CEO *Khalid Al Falih* (who was promoted to the post of Minister of Health in a major government reshuffle in April 2015, while also becoming Chairman of Saudi Aramco) explained that it is not feasible for the company to proceed with plans to raise its oil output capacity to 15 million barrels a day at a time when expansion plans in other producing countries such as Iraq and Brazil should be enough to satisfy world markets (WSJ, Oct. 2011).

From a policy point of view, Minister Al-Naimi had stated at an event in Washington DC in 2013 that his country will be "lucky to go past" 9 million barrels a day by 2020, as new petroleum production from other countries comes into the global market. The "call on Saudi," a term used to show the amount of oil required by Saudi Arabia to balance the market and meet global demand, will not be more than 11–11.5 million barrels a day by the year 2030 or even 2040 (CSIS, April 2013). There is a policy limitation, however, under this scenario that limits Saudi Arabia to expand its market share. As a policy, Saudi Arabia needs to keep between 1.5 and 2 million barrels a day of its capacity idle at all times to serve as a spare whenever there is a major supply disruption or a sudden surge in global demand that cannot be met from other producers (Al Moneef, April 2011). Instead of being a swing producer, Saudi Arabia decided to *become a producer of last resort* and, to achieve this, it needs to keep its spare capacity ready at all times. Thus under this scenario, it is difficult for Saudi Arabia to pursue a larger market share or forgo its exiting share for the purpose of securing future income. The other limitation that Saudi Arabia faces is that its internal consumption of crude oil is on the rise and supplying local market with crude is a priority for the country instead of expanding its share abroad. With this in mind, the future income of Saudi Arabia will depend on developments in global oil prices and not in selling more crude (ibid).

For a country, like Saudi Arabia, that does not want to expand its production capacity, it is difficult to "juggle more than one ball" at the same time. Thus a compromise must be reached between meeting local and global demand obligations in the long run at a time when its share in the global market is expected to shrink due to competition and other market forces. Minister Al-Naimi made it clear in his Berlin speech that Saudi Arabia's position within OPEC to protect its market share is part of its obligation to meet customers' demand (Al-Naimi, March 2015). Another major factor that can be added to the discussion is the effect of the slowdown in Chinese imports of crude oil from Saudi Arabia in 2014, discussed in more detail later in the volume. This factor is probably a very strong element behind the decision of Saudi policy makers not to let go of their market share in Asia that was home to more than 60 % of Saudi Aramco's oil exports in 2014 (Saudi Aramco, May 2015). Saudi oil shipments to China fell on annual basis in 2014 by 7.9 % from 2013, while it rose from Russia, Iraq, Kuwait, Iran, Angola, UAE, South Sudan, Colombia, and Oman. Oil imports from Russia and Colombia surged by 36 % and 156 %, respectively (China Customs General Administration, Jan. 2015).

To get a clearer picture why Saudi Arabia is desperate to retain its market share in Asia, one has to look also at the country's share among other competitors in the Chinese market. Saudi shipments to China, the world's second largest buyer of crude, fell to 15 % of the nation's oil imports in 2014 from 21 % in 2011, according to China customs data. Saudi Arabia's annual exports to China have fluctuated between 4.5 million metric tons a month and 4.7 million, even as total China crude imports have risen steadily from 22 million metric tons a month to 30.3 million by the end of 2014 (ibid).

The fall in Chinese imports along with other factors such as a 17 % drop in crude demand from the USA over the same period and an abundance of West African crude makes it little surprising that Saudi Arabia is now defending its market share as a national policy. These developments also explain to a great extent why Saudi Arabia and other Gulf states have lowered their official selling prices to Asia to record lows since October 2014 (Bloomberg Briefs, Jan. 17, 2015[h]). Defending market share is a revisited line of thinking for Saudi policy makers. Another possible factor behind this development within the Saudi oil decision-making machine are the changes within the Saudi Arabian OPEC team. Prior to 2012, the Saudi team was mainly made up of academics who were recruited as advisers at the Saudi Ministry of Petroleum. The OPEC team, however, became more market oriented in recent years with the recruitment of many of its key personnel from the state-owned Saudi Aramco. As of November 2014, the team consisted of five key personnel. The oil minister is the first key person, followed by Deputy Oil Minister *Prince Abdulaziz bin Salman*. The Prince holds a B.S. in industrial management and an MBA from King Fahd University for Petroleum and Minerals at Dhahran, Saudi Arabia. He was a member of the team since 1987 and always served as the second senior figure in Saudi OPEC team under two ministers. The other two key members are the OPEC Governor and the National Representative to OPEC's Economic Commission Board. Last but not least is the Adviser to the Minister, *Ibrahim Al Muhanna*, who formerly taught international relations at King Saud University in Riyadh and is mainly responsible for handling the ministry's relations with the media but does not have any background in energy economics or crude sales and marketing.

The team underwent a major change since 2012 with the appointment of *Nasser Al Dossari* as a national representative, replacing *Ahmed Al Ghamdi*, and the appointment of *Mohammed Al Madi* as Governor in place of *Yasser Mufti* who took over the post briefly from *Majid Al Moneef*. Mufti, who joined Aramco in 1995, was also the OPEC national representative between 2003 and 2009. Most of the former Saudi team members did not have any practical experience with an oil company prior to joining the ministry. With the exception of Al Moneef, none have any significant scholarly contribution in the field of energy economics. In 2012, Mufti and Al Dossari were seconded from Saudi Aramco to the ministry. By the end of 2013, Al Madi joined in the team as a Governor after spending around 7 years in China and Korea where he was responsible for marketing Saudi crude and managing Aramco's businesses in the two countries.

This shift in the background of the team members is significant for Saudi policy-making process since the team now includes more people with direct interaction with the oil market. All three, Al Madi, Mufti, and Dossari, have technical backgrounds that range from petroleum to chemical engineering. Among the three, Al Madi's background is the most important as he spent a long time with Aramco selling, marketing, and pricing crude oil before moving to two of Aramco's key markets. He also has a PhD degree in petroleum engineering from Beijing University and fluent in the Chinese language, a fact that demonstrates his understanding of the Chinese system and its culture to a large degree. Al Madi's involvement with the Asian market, mainly China where he helped in increasing Saudi crude sales between 2007 and 2012 (Asharq al-Awsat, Nov. 12, 2014[b]), is essential for the team to better understand the dynamics of the Asian market.

The Saudi OPEC team lacked over the past two decades members with deep understanding of market dynamics and the technicality of oil production. The brief involvement of Mufti in the late 2000s with the team as national representative bridged some gaps in the team. It was not until 2012 that the two major Saudi representatives in OPEC were young technocrats from the industry with hands-on experience in oil marketing and corporate planning. With the appointment of Al Madi as Governor, the market expertise of the OPEC team was enhanced and policy formulation at the ministry became more complex than before. The new situation was reflected in the communication of the ministry with the market through the speeches of the minister himself. Concepts such as market competition between high-cost producers and low-cost producers now dominated the official language of Saudi officials. It is on the back of these new developments in Saudi oil policy and in demand, and the past experience with OPEC and the market in the 1980s, that Saudi Arabia finally decided to focus on maintaining its market share and was against any cuts to support or lift prices when the group met in November 2014.

Second driver: the shrinking OPEC's market share in front of the rise of supply from high-cost producers outside the organization is another reason for the Saudi stance in November 2014. In 2008, OPEC met in *Oran*, Algeria, and decided to cut production by 4.2 million barrels a day to lift slumping oil prices at that time. The cut was possible due to the limited growth in supply from non-OPEC. Many commentators and OPEC ministers described that meeting as one of the best the group held in years, as they were all united and did not argue about the need to cut. In the words of OPEC's Secretary General Abdallah El-Badri, "all the ministers were on the heart of one man" at the *Oran* meeting (El-Badri, April 9, 2014). The market conditions in 2014 were different than in 2008, with the advent of non-OPEC high-cost producers. Saudi Arabia found it difficult on its part to engage in a self-balancing act of the market.

The differences in the situation between the recent market rout and earlier ones such as those in the 1980s or 2008 are best described in the words of Al-Naimi himself in his Berlin speech: "today, it is not the role of Saudi Arabia, or certain other OPEC nations, to subsidize higher cost producers by ceding market share. And the facts on the ground are very different anyway. Non-OPEC supplies are much larger than they were in the 1980s and a much more multi-national approach is required" (Al-Naimi, March 4, 2015). In 2008, OPEC was still the marginal producer (i.e., the group that has the last supplied barrels entering the market), and even though the market was tight, demand was still growing in emerging markets.

The Secretary General of OPEC, El Badri, mirrored the same concerns of Saudi Minister Al-Naimi in his own speech delivered to the Society of Petroleum Engineers in Manama in March 2015. El Badri explained that oil prices were hit by an unexpected slowdown in oil demand growth at a time when global supply was on the rise from outside of the group and OPEC supply was kept nearly unchanged. The result of this situation was an oversupply of 2 million barrels a day in 2014 (El Badri, March 2015).

Thus, at OPEC's 2014 meeting, the minority GCC states within the group came to the realization that the organization was being cornered by the non-OPEC producer. Since 2008, a combination of higher and steady oil prices in addition to improvement in oil recovery technology had led to a huge increase in non-OPEC production.

Third driver: The *Oran* meeting in December 2008 provides a historical explanation of the Saudi refusal in November 2014 for OPEC to act without the support of non-OPEC. In the *Oran* meeting, non-OPEC producers, Russia and Azerbaijan, pledged to cut their production in support of OPEC, but that pledge never materialized in 2009. On the contrary, Russia was surpassing Saudi Arabia in oil exports during 2009 for the first time since the Soviet Union's collapse as Russia exploited OPEC production cuts to gain more market share (Bloomberg, Sept. 8, 2009). Exports of crude oil and refined products from Russia rose to 7.4 million barrels a day in the second quarter, from 7.25 million in the first quarter, while Saudi shipments fell to about 7 million barrels a day, from 7.39 million, during the same period (ibid).

OPEC's experience with Russian pledges goes back to many years before *Oran*. In 2001, Russia and other non-OPEC pledged to cut exports by 150,000 barrels a day effective January 1, 2002. OPEC was committed to cut 1.5 million barrels a day for the first 6 months of 2002. This cut follows a cut of 3.5 million barrels a day in 2002, which would bring the total cut to 5 million barrels a day in 2002. A cut from non-OPEC equals to 10 % of the 5 million barrels that was needed in order for OPEC to move forward with reducing production (OPEC, March 15, 2002[a], March 1, 2002[b]). The Russian pledge in 2002 did not materialize, and Russia raised its production for 6 consecutive months during that year (Mahdi, March 2015[b]). The organization's cooperation with Russian and other non-OPEC producers on collective cuts to support prices was not always a successful endeavor as Al Madi noted. OPEC had 19 attempts to cut production to support prices, with six of the attempts made in coordination with non-OPEC. Not all the attempts yielded cuts and not all the cuts resulted in an increase in oil prices, as Al Madi argued. Only eight attempts out of the 19 attempts led in the end to a cut that resulted in an increase in oil prices (ibid).

Fourth driver: Demand started to slowdown from major consumers of oil especially China that no longer sees economic growth in double digits. After years of rapid growth, the Chinese engine started to cool down and the country was battling to meet its annual growth target of around 7 %. India, another major oil consumer in the group of countries known as the *BRIC* emerging markets, also experienced a period of slowdown in its annual GDP growth similar to China. The OECD countries, which make up the bulk of global oil consumers, are not in better economic conditions than China and India, and the group as a whole is, as of 2015, still sending mixed growth signals.

According to World Bank data, over the past 10 years, Chinese GDP growth peaked at 14.2 % in 2007 before it fell to 9.6 % in 2008. The economy then witnessed another year of double digit growth in 2010 at 10.4 %. Since then, the growth rate has fallen and it recorded 7.7 % in 2012 and 2013. The Indian economy is no different than the Chinese when it comes to its rate of growth. Indian GDP grew by 5 % in 2013, half of its growth rate in 2010 (World Bank 2014). In the OECD, the picture is different than in the emerging markets as the economic bloc's annual GDP growth rate fluctuated between the positive and negative over the past 5 years. The GDP growth rate for the OECD as a whole recorded 0.3 % in 2008 before going into the negative the following year at (−3.5) percent. The growth rate went back to positive again in 2010 at 3 % and has been below 2 % since then. Within the OECD, European members are still not showing signs of continuous and steady economic growth. Between 2008 and 2013, OECD Europe recorded 2 years of negative GDP growth in 2009 at (−4.4) percent and again in 2012 at (−0.3) percent. The group recovered in 2013 with a modest growth rate of 0.3 % (OECD 2014).

The state of health of the world economy is an important factor to determine the future outlook of the global oil industry. OPEC's El Badri stated in his speech in Manama that the growth of the world economy, which continues to offer mixed signals as of 2015, is the biggest challenge that his group faces with Europe not improving greatly and China seeing a slowdown in GDP its growth rate (El Badri, March 2015). Indeed, the world's economy may have showed signs of recovery over the period 2010–2014, yet it did not go back to consume oil at the same pace prior to the financial crisis and the global downturn of 2008/2009. The slowdown in the world's economic growth will probably linger and extend into the second half of the current decade. The whole situation will be visible on the world's oil consumption over the period 2014–2020 as indicated by the IEA in its *Medium Term Oil Outlook Report* (*MTOMR*). Over the stated period, the IEA expects global demand growth to average 1.2 % per year, a level below the 1.9 % that the world witnessed in the period 2001–2007 before the start of the global economic meltdown. Despite the lower growth rate, the world's oil consumption is set to grow reaching around 99.1 million barrels a day by the year 2020 from 92.4 million in 2014, according to IEA's calculations. This represents aggregate demand growth of 6.6 million barrels a day for the 6-year period ending 2020. The world demand for oil is set to outpace available supply capacity that will grow by 5.2 million barrels a day over the same period (IEA, MTOMR 2015[b]). As for Chinese demand growth, it is expected to slow over the period 2014–2020 to less than 300,000 barrels a day annually, down from average growth of 440,000 barrels a day in 2009–2014. The IEA attributes the slowdown to Beijing's decision to reorient the economy away from manufacturing/exports. Demand growth from other non-OECD Asian countries will consistently overtake China for the first time since the mid-1990s (ibid).

Another issue related to the state of global oil demand is the attempt of many developing nations to fill their *Strategic Oil Reserves*. In a complex oil market where major buyers now are starting to build strategic reserves to join those who already have one, it is important to note that oil consumption or imports is not the best indicator for the future trend of oil demand. GDP growth is still a more comprehensive indicator to reflect structural changes on the demand side, as oil imports by major customers can be misleading and can hide weaknesses in fundamentals. With oil prices falling to 6-year lows over the period 2014/2015, many countries were encouraged to fill up their strategic reserve tanks. This trend was very clear in China, where oil imports continues to increase even as economic growth is slowing down. China has cut its growth target rate for its economy in 2015 to 7 %, the lowest in two decades (Economist, March 11, 2015[a]), yet the country will still increase its crude oil imports during the year (IEA OMR, March 2015[c]). The reason for the rise in crude imports was to fill up storage tanks with cheap oil. Thanks to Chinese efforts to keep their strategic reserves full, crude oil demand in 2014 was revived in the last quarter after being lower than expected throughout the year. Oil demand growth in 2014 was below the original projection of OPEC early in the year, as it turned out to grow by "just below" 1 million barrels a day (El Badri, March 2015).

In 2015, crude purchases for the purpose of filling strategic reserves in China for the large part, and India to a lesser extent, would become a main driver for global oil demand revival. The IEA expects China, which has been storing crude for strategic purposes for many years, and India, which will start its strategic reserve program in 2015 to increase oil imports in 2015. Other countries that do not have strategic reserves such as Vietnam have also used lower oil prices to boost commercial stockpiles held at refineries (IEA OMR, March 2015[c]). In the oil industry, there are two general types of crude oil storage, *strategic* and *commercial*. The *commercial*, or discretionary storage, refers to oil stored as part of normal day-to-day operations in refineries or oil terminals or that stored by professional or semi-professional oil consumers such as power plants, or even that stored by end-users. *Strategic* oil reserves, on the other hand, are those kept for the purpose of meeting supply disruptions (Downey 2009). Few governments store crude themselves while most governments mandate oil producers and refineries to carry larger storage. The USA and China are among the few who store crude themselves. The USA started its *Strategic Petroleum Reserve Program* (*SPR*) in 1977 in response to the Arab oil embargo of 1973–1974, while China started its program in 2006 and has a capacity of 102 million barrels. The SPR had a capacity of 727 million barrels as of 2009, and President Bush announced in his State of the Union address that the government had plans to raise that to 1.5 billion barrels by 2027, highlighted later in the volume on the efforts of the USA to ensure a semblance of energy "independence."

The IEA said China was "expected to again stockpile crude in 2015" as it completes new tanking capacity (IEA OMR, March 2015[c]). Beijing in November 2014 for the first time revealed details of its crude storage program, saying it held about 91 million barrels in four different locations. The USA holds 696 million barrels of oil in its emergency reserve, the largest in the world (Bloomberg, March 13, 2015[f]). The IEA said the Indian government has approved a $338 million budget to cover the filling of its first emergency tanks, and at *Brent* oil prices of around $60, that would amount to 6.5–7 million barrels of crude (IEA OMR, March 2015[c]). The Indian *Strategic Petroleum Reserves Ltd.*, the company in charge of the stockpile, has already built a tank farm in eastern India capable of holding 10 million barrels of crude. Two more facilities in western India, adding a combined 28 million barrels of capacity, are expected to be finished by the end of 2015 (Bloomberg, March 13, 2015[f]).

China and India have said in the past they need to build strategic reserves to offset the risk of a disruption in supplies, mostly from the MENA. This pattern of stockpiling crude will add to global oil consumption growth, the IEA said, offsetting "current weak fundamentals" of supply and demand and potentially boosting prices (IEA OMR, March 2015^c). However, it seems that some OPEC members, like Saudi Arabia, are aware of that disguised demand trend, and they did not take it into consideration when advocating the defending of Saudi market share in OPEC's November meeting. The numbers above justify that. An oil market sell-off since June 2014, resulting in dramatically lower spot crude and product prices and lower future prices, is expected to have a mixed impact on economic growth, but overall to provide only a modest net boost to global oil demand.

On balance, expectations of world oil product demand growth are more subdued than prior to the recent oil price drop. A combination of cyclical and structural factors stand behind this softer demand outlook, including, but not limited to, significantly reduced expectations of global economic growth for the early part of the forecast period. Toward the latter part of the forecast, a structurally-driven reduction in the oil intensity of the global economy, supported in part by fuel switching out of oil and increased energy efficiency, somewhat blunts the demand impact of forecast economic growth (IEA MTOMR 2015^b).

Fifth driver: OPEC is a heterogeneous organization, and within it, there is always a rift on leadership. The producer group is known for discords on key issues such as setting production targets, agreeing on and abiding by assigned quotas, and agreeing on a reasonable and fair price for oil, issues explored in more detail later in the volume.

Saudi Arabia was the hegemonic member within the organization since inception due to its massive oil reserves and production capabilities. This hegemonic ability helped the Kingdom to steer the organization through turbulent waters on many occasions and to bridge the gap between its members. However, the Saudi hegemonic position within OPEC is under obvious threat now that Iraq is trying to expand its production capacity to a level between 8 and 10 million barrels a day, although the country may only seem to be able to reach 8 million barrels a day by 2025 under the best circumstances. The threat to Saudi Arabia's position in OPEC is twofold. First, if Iraq was able to eat into some of Saudi market share in Asia and, second, if Iraq's new production capacity can help the country challenge Saudi leadership in OPEC. Another country that has long-term plans to expand its production capacity is Iran. The nation has a very ambitious plan to expand capacity to 5.7 million barrels a day by 2019 (Iran Daily, Jan. 13, 2015) despite the technical difficulties associated with its mature oil fields. There are many contradictory figures on Iran's oil production target plans. One recent target is 3.96 million barrels a day by March 2016 up from the current capacity of 3.8 m b/d in March 2015. The target was announced by *Roknoddin Javadi*, deputy oil minister and managing director of *National Iranian Oil Co.* (Bloomberg, May 7, 2015^j). Analysts such *Sadad Al Husseini* argue that Iran's capacity can only raise to 3.5–3.6 million barrels a day by end of 2017, provided that the country can overcome all the technical challenges and international sanction on its sector are lifted (Bloomberg, March 31, 2015^a).

In all cases, if Iran's capacity was lifted to somewhere around 4 million barrels a day and Iraq was able to achieve the 8 million barrels, then the combined production capacity of the two countries can equal that of Saudi Arabia (excluding the Saudi share of capacity from the Neutral Partitioned Zone with Kuwait). The combined capacity of the two countries can put them in a better position to challenge Saudi Arabia hegemony in OPEC. Oil historian Daniel Yergin argues that in defending their market share in November 2014 meeting, Saudi Arabia and its GCC allies were looking at competition not only from American shale oil and other non-OPEC producers, but most immediately, they were looking at Iraq and Iran. The GCC bloc did not want to give up markets to Iraq, a country they see as an Iranian satellite, and they did not want to make way for lost Iranian crude to come back to the market after 3 years of sanctions (Yergin 2015). The June 2015 OPEC meeting that agreed to maintain crude oil production at 30 million bpd in effect meant a license for both OPEC and non-OPEC producers to increase output as they see fit. In May 2015, Russia extracted 10.7 million bpd compared with Saudi Arabia's 10.2 million bpd and was the first time Russia took the global lead since 2010, and the US oil industry showed that it was not intimidated by OPEC's "market-share" strategy, and in May 2015, it produced 9.4 million bpd, the most since 1972 (Bershidsky 2015). Saudi Arabia is aware of all these issues and looming competition and acknowledged the reality when its Oil Minister Al-Naimi stated that "oil production is a sovereign right" when asked about the 30 million bpd production target (Chmaytelli 2015). These looming issues to OPEC's long-term survival as a viable organization are explored in the chapters that follow.

Chapter 2
Oil and Geopolitics in the 1970s and 1980s: OPEC's Boom and Bust Eras

> *It is not the strongest of the species that survives, nor the most intelligent that survives. It is the one that is the most adaptable to change.*
>
> Charles Darwin.

Introduction

The purpose of looking back into the history of the Organization of Petroleum Exporting Countries (OPEC) is for understanding the mechanism that lead to its failure in managing the world's petroleum market since the year 1973 when it took over the responsibility of setting oil prices from the group of international oil companies (IOC) known as the *Seven Sisters*. As discussed in the opening chapter, understanding the history of OPEC is important to understand its current situation and the future direction of its policies under the current market realities. The years between 1973 and 1986 witnessed a series of trials and errors for OPEC with more failures than success in managing the organization's affairs and the world petroleum market. OPEC took over a world petroleum order that was in a long state of stability in the period prior to 1973 under the former structure of the *Seven Sisters* (*Esso, Mobil, Gulf, Shell, BP, Texaco, and Socal*). Since OPEC members started nationalizing their oil industry, it was never capable of putting a strategy that can prevent crisis and ensure the stability of the oil markets. Additionally, it was never possible for OPEC to preside over the world petroleum order in the same way as the IOCs who were engaged deeply in a system of cooperation and sharing of producing assets mainly in the upstream.

The Rise and Fall of the Major Oil Companies

Beginning in the mid-1860s, under the single-minded ambition of one man, *John D. Rockefeller*, Standard Oil was transformed into a massive company the *Standard Oil Trust* that controlled up to 90 % of the US oil industry and dominated the global market until, in 1911, the US Supreme Court ordered the company to be broken up into 34 separate companies (Yergin 2011, pp. 88, 89). And yet, as will be discussed in this volume, the single-minded vision of Rockefeller seems to live on with the mega mergers of the IOCs during the last two decades of the twentieth century and in 2015 following the sharp fall in oil prices from 2014.

From that early period when commercial production of oil started in the USA in 1959 in *Titusville*, Pennsylvania, oil demand grew rapidly in the following decades starting a new industrial revolution and, by the turn of the twentieth century, world consumption was 25 times the level of that of 1870. Standard Oil was based on a strong position in the downstream segment, not on access and control over resources. Rockefeller saw that the best way to control the industry was through the ownership of the refining industry, since the exploration for crude oil carries a great risk than the supply of the product and that if Standard Oil could control refineries, it could automatically control the demand for crude oil and crude oil products, since producers could not go to consumers without going through the refiners and the suppliers of products to consumers. For other competitors the control of reserves (upstream) was more important, as did financial strength to develop reserves in underdeveloped areas, and this led to a focus on international oil reserves, not only among oil companies but also among industrial nations (Claes 2001, p. 52). In the Middle East, the Germans were the first to show their interest in oil concessions and in 1911 the *Turkish Petroleum Company* was established with the purpose of exploring for oil in Iraq. After the First World War, this national resource rivalry intensified and access to foreign oil supplies led to the creation of large oil companies, especially after the division of the *Turkish Petroleum Company* in 1928 which negotiated the *San Remo* Anglo-French oil agreement to take over the assets of the Turkish Company which later included US interests. This is illustrated in Table 2.1.

Table 2.1 The division of the Turkish Petroleum Company in 1928

Counter	Subsidiary	Share
Anglo-Persian Oil Co., in which the British government held 51 %	D'Arcy Exploration Co. Ltd.	23.75
Royal Dutch/Shell (Royal Dutch: 60 %; Shell: 40 %)	Anglo-Saxon Petroleum Com. Ltd.	23.75
Compagnie Francaise des Pètroles (CFP), in which the French government held 35 %		23.75
Standard Oil Co. of New Jersey, 25 %; Standard Oil Co. of New York, 25 %; Gulf Oil Corp., 16.66 %; Atlantic Refining Co., 16.66 %; Pan American Petroleum and Transport Co. (subsidiary of Standard Oil of Indiana), 16.66 %	Near Eastern Development Corp.	23.75
C. S. Gulbenkian	Participation and Investments Com.	5.00

Source: Anderson (1988). Aramco, the United States, and Saudi Arabia: A study of the dynamics of Foreign Oil Policy, 1933-1950, p. 115

By 1928, more than 50 % of oil production outside the USA was controlled by Exxon, Shell, and British Petroleum, who met secretly and worked out market-sharing deals, the so-called "As-is-agreement which was an agreement to keep the respective percentage of market shares of sales in various markets. Later most of the other US companies joined the agreement" (Claes 2001, p. 53). After the end of the Second World War, the international oil market was denominated by seven companies, popularly known as the *Seven Sisters*, and these companies determined the order of the oil market, highlighted in the preceding chapter. The Sisters organized their operations in the Middle East through a consortium in which all the major companies were engaged in *at least two countries*. In this way, the Sisters stood seemingly stronger against possible regulation or nationalization moves by the producing countries as none of them would be totally dependent on the will of one Middle East government only. This ownership structure is illustrated in Table 2.2.

Table 2.2 Ownership shares in the Middle East: production distributed to companies (%)

Company	Iran	Iraq	Saudi Arabia	Kuwait
Exxon	7	11.875	30	
Texaco	7		30	
SoCal	7		30	
Mobil	7	11.875	10	
Gulf	7			50
BP	40	23.75		50
Shell	14	23.75		
CFP	6	23.75		
Iricon	5			
Gulbenkian		5		

Source: Claes (2001, p. 56)

Unlike other Middle East countries, the Saudi oil concessions were all American. The Standard Oil Company of California obtained a concession in Saudi Arabia in 1936, Texaco joined in the same year, and in 1947 Standard Oil of New Jersey and Mobil also joined. These four companies created the *Arabian American Oil Company (Aramco)*. It was in Saudi Arabia in December 1950 that the principle of 50–50 profit sharing was implemented in the Middle East between the oil companies and their host countries. These so-called *oil concessions* were the governing contractual relationship between the companies and countries and usually these concessions covered vast areas and gave the companies the right to explore, produce, develop, refine, and export all or any quantities they might find. The oil companies had the sole authority to decide upon the volume of production, the price, and the destination of oil exports. Most of the laws, taxes, and regulations of the country were not applicable to the companies (Ghanem 1986, p. 11). What seems more astonishing in today's world of jealously guarded "sovereignty" issues is that, except for Venezuela, sovereignty of the producing countries was overridden when the governments agreed that they could not change any clause in the law or the concession without any prior consent of the company concerned. Arbitration in case of dispute was to be presented to international bodies and local courts were disqualified from handling any dispute between the country and the company. In oil-producing countries, government income from their oil was limited to the 50 % of the company profits calculated on the basis of the price *determined* by the companies after deducting all costs and amortization. The end result was that, even as late as 1970, the oil producer's dependency on the oil majors was almost complete, ranging from 88 to 100 % as illustrated in Table 2.3.

Table 2.3 Producing countries' dependency on the majors, 1970 (%)

	Exxon	Mobil	SoCal	Texaco	BP	Gulf	Shell	CFP	Majors	Others
Iran	6.46	6.46	6.46	6.46	37.01	6.46	12.95	5.53	87.77	12.23
Iraq	12	11.9	0.00	0.00	23.73	0.00	23.73	23.73	95.00	5.00
Qatar	6.61	6.61	0.00	0.00	12.93	0.00	58.33	12.93	97.41	2.59
Abu Dhabi	7.89	7.89	0.00	0.00	38.00	0.00	15.94	26.89	96.62	3.38
Kuwait	0.00	0	0.00	0.00	50.00	50.00	0.00	0.00	100.00	0.00
Saudi Arabia	30.01	9.97	30.01	30.01	0.00	0.00	0.00	0.00	100.00	0.00
Libya	22.72	4.19	6.02	6.02	5.16	0.00	3.96	0.00	48.07	51.93
Algeria	0.00	0.00	0.00	0.00	0.00	0.00	9.08	26.6	35.68	64.32
Nigeria	0.00	0.00	0.00	0.00	31.34	37.31	31.34	0.00	100.00	0.00
Indonesia	2.95	2.95	33.61	33.61	0.00	0.00	16.27	0.00	89.39	10.61
Venezuela	41.65	3.17	1.58	4.84	0.00	10.84	25.92	0.00	88.00	12.00

Source: Claes, the Politics of Oil Producer Cooperation (2011, p. 62)

Given such heaven-made concession agreements and rising global demand for oil, competitive pressures against the majors increased as it was easier to be a monopolist in a small market than in a large one. The integrated and jealously guarded structure that the oil majors put together might have created barriers to entry at the company level, but when global demand and markets grew, these barriers became impossible to maintain. Daily crude oil production rose for virtually all the countries that later on were to become members of OPEC as illustrated in Table 2.4.

The change in favor of OPEC's interests was enhanced by the continuous increase in oil demand during the 1960s, illustrated in the above table, and in addition, there were no substitutes for oil in the short term and there had been no incentives to conserve energy due to the low price structure imposed by the oil majors.

The application of the profit-sharing principle gave a special importance to the issue of oil pricing as producer's income was directly affected by any oil price fluctuations in the price of oil and volume of production. When oil companies emerged as global players, they *posted* their oil prices which came to mean the prices published by the companies at which they were willing to sell their crude or refined products in cargo lots to consumers worldwide. However, the posted prices were not necessarily the real prices which oil companies received from their oil sales, known as *realized* prices which were the prices at which the transactions of the crude oil take place between the companies and buyers and between the independent players. Major oil companies usually transfer the crude and refined products to their affiliates on the basis of the posted price (Ghanem 1986, p. 12).

The OPEC Era: Gaining Market Power

It was the issue of posted prices and competitive discounting that created a dilemma for the companies' relations with the host countries, as it was the posted price that the basis for the producer countries' taxes and royalties. Table 2.5 sets out the evolution of crude oil posted prices and the introduction in 1974 of producers' official selling prices.

Both in 1959 and 1960 the IOCs *cut* posted prices which angered the oil-producing governments and triggered the establishment of OPEC in September 1960. It was Libya that took the first step to challenge the power of the international oil majors when in 1970 Libya told the oil companies operating in the country that the government wanted negotiations about a price rise. This was followed by a joint statement by Algeria and Libya demanding an immediate increase in the price of oil and Iraq proclaimed its support for the Libyan action. The Libyans were eventually successful because their strategy was to negotiate with the companies *individually* and not as a bloc and also to play the *independent oil companies* such as *Occidental* against the multinationals. This succeeded, as Libya was in a somewhat different situation than other OPEC producers as it was less dependent on the oil majors as illustrated in Table 2.3, where independent oil companies accounted for nearly 52 % of Libyan production. After Libya's success, Iran and Venezuela increased their

Table 2.4 Daily crude oil production of member countries 1960–1970 ('000 bpd)

	1960	1961	1962	1963	1964	1965	1966	1967	1968	1969	1970
Indonesia	409.6	424.3	453.4	444.0	456.6	480.6	464.6	505.4	600.7	724.3	853.0
Iran	1067.0	1201.0	1334.0	1491.0	1710.7	1908.0	2131.0	2603.0	2839.0	3375.8	3829.0
Iraq	972.0	1007.0	1009.0	1161.0	1255.0	1312.0	1392.0	1228.0	1503.0	1521.0	1548
Kuwait	1692.0	1735.0	1958.0	2096.0	2301.0	2360.0	2484.0	2499.0	2613.0	2773.0	2989.0
Libya	–	18.0	182.3	442.0	862.4	1218.8	1501.0	1740.5	2602.0	3109.0	3318.0
Qatar	174.0	177.0	186.0	191.0	215.0	232.0	291.0	323.0	339.0	355.0	362.0
Saudi Arabia	1313.0	1480	1642.0	1786.0	1896.0	2205.0	2601.0	2805.0	3042.0	3216.0	3799.0
UAE	–	–	14.2	48.0	186.8	282.0	360.0	382.0	496.0	427.8	779.0
Venezuela	2846.0	2919.0	3199.9	3247.0	3392.0	3473.9	3371.0	3542.0	3604.0	3594.0	3708.0

Table 2.5 Evolution of crude oil posted prices and official selling prices for Arabian Light oil (1948–1982)

Year and month	Posted price	Official selling price
1948 April	2.18	–
1949 April	1.84	–
1953 Feb.	1.93	–
1957 June	2.08	–
1959 Feb.	1.90	–
1960 Aug.	1.80	–
1971 June	2.29	–
1972 Jan.	2.48	–
1973 June	2.90	–
Dec.	5.04	–
1974 Nov.	11.25	10.46
1975 Oct.	12.38	11.51
1977 July	13.67	12.70
1979 Nov.	25.81	24.00
1980 Nov.	34.41	32.00
1981 Jan.	34.41	32.00
1982 Jan.	34.00	34.00

Source: Al Nasrawi (1985, pp. 30–31)
US$ per barrel

share of profits, and a game of oil price rise "leapfrogging" began, ushering a new era of "offensive" OPEC power compared with an earlier defensive power, but without any *structural* change in the international oil market up to the early 1970s. However, this was more a matter of change in the *balance of power*, with the IOCs not uniting to stop the oil producers from taking over price control. The focus then turned to what objectives the OPEC countries would pursue, given their achieved market power (Claes 2001, p. 64).

The major obstacle for OPEC as a group was its focus on raising oil prices to capture higher economic rent, whereas the IOCs were focusing on keeping prices stable and sometimes subduing it. The result of OPEC's rent-seeking policies since its inception have resulted in a price war that had led to a long period of low oil prices which ended in the early 2000s when new market players joined the table. The rent-seeking policies that started in the 1973 additionally gave rise to production in the North Sea. To protect the economic rent, OPEC played the difficult role of the world's swing producer. It was hard for OPEC to play that role, and the whole structure must sooner than later come to collapse as explored earlier in the volume. OPEC's history proved that the organization was never successful in creating price stability and cannot control both prices and market share in face of raising supply from non-OPEC producers at a time when demand falters. To achieve one objective, it must always forgo the other. As history showed, defending oil prices, especially at higher levels, was always the more difficult and the least advantageous approach for the organization's survival.

The Pre-embargo Oil Order

The period prior to 1973 and the first Arab oil embargo was a very stable period in the history of the oil market as illustrated earlier, thanks to the role played by the IOCs and the structure of cooperation the seven major companies, the *Seven Sisters*, put together. The companies, while competing against each other in the downstream, were able to cooperate both vertically and horizontally in the upstream. A major feature of the system was that these companies were locked in partnership relations to jointly develop the oil resources in the Middle East. They were able to plan the amount of crude necessary for the market as they controlled the downstream side as well. Yet the most important aspect of that structure was that the companies did not use or take advantage of the system *to target higher oil prices* unlike OPEC members later on. One good example of how the system worked was in 1955 when the world's market was very well supplied and the companies had to limit the growth in Iranian oil production, despite having an agreement to double production from 302,000 bpd in 1955 to 608,000 bpd in 1957 (Parra 2004).

OPEC countries were never able to reach that level of cooperation among member countries since the organization's inception. The *Seven Sisters* were better in keeping the market well supplied, and they were able to compensate for any supply loss or disruption from a member country of the group without engaging in a market-share struggle. OPEC countries, on the other hand, worked against each other due to political differences, and they lacked the vertical and horizontal integration that the *Seven Sisters* had. The political competition between Saudi Arabia and Iran over leadership in the region during the 2000s and the Iran-Iraq war of the 1980s did not allow the organization members to best coordinate on prices and production levels. The hostility of some members toward Western nations was behind their insistence on price hikes even at times when market conditions could not support such moves.

Oil prices were stable in the post-Second World War era and that stability ended in 1973 with the Arab oil embargo and the fierce resource nationalization period that followed. From 1948 through the end of the 1960s, crude oil prices ranged between $2.50 and $3.00 that is equivalent to $17 and $19 during most of the period when viewed in 2010 dollars. From 1958 to 1970, prices were stable near the $3.00 per barrel, but in real terms, the price of crude oil declined from $19 to $14 per barrel (WTRG Economics 2011). The *Seven Sister's* system of posted oil prices and the structure they put in place had helped the world economy to enjoy a period of healthy growth rates in the years between 1950 and 1973. The world's oil consumption was growing at around 7 % per annum for the 20 years until 1970 (Parra 2004). The stability in prices had allowed the world's oil consumption to grow over the period 1965–1973 by around 80 % (BP Statistical Review 2014).

The posted price system of the *Seven Sisters* had to come to collapse in 1973 as illustrated earlier in Table 2.5, because there were many political and social developments within OPEC countries that forced governments to search for higher prices for its scarce and most valuable commodity. Moreover, the companies were resisting increasing the posted price of oil for OPEC countries thus depriving them of additional income. The issue of the posted price was at the heart of the struggle between OPEC governments and companies particularly in the light of a widening margin between posted prices and spot market prices. There was also the issue of the fall in the value of the dollar and how it increased the cost of importing goods for OPEC countries. The companies did not take into account of the need to adjust posted prices to a level that would help OPEC member countries to meet the rising cost of their import bills.

It was in the light of these backgrounds that in 1973 OPEC governments decided to take over the role *of setting posted prices from the companies*. OPEC's Gulf states—Saudi Arabia, Iraq, Iran, Kuwait, Qatar, and the United Arab Emirates—issued a communiqué on 16 October to announce the unilateral increase of posted prices of *Arabian Light* (f.o.b Ras Tanura) by 70 % from $3.011 per barrel to $5.119 in an effort to match posted prices with market prices that were at $3.65 a barrel at that time. OPEC, therefore, put an end to the post-Second World War system and marked the rise of a new system where OPEC countries were now the *price setters* (Terzian 1985). The companies at first protested against the new system on the basis that raising the posted price will lead to a rise in market prices. The whole system, it was argued, would be unworkable because it would lead to further increases in posted prices, and so on ad infinitum (Parra 2004).

In a few years, oil prices were subjected to price hikes by OPEC members that were propelled by many political upheavals in the Arabian Gulf area and military interventions in the Middle East. Yearly average oil prices went up from $3.29 in 1973 to $11.58 in 1974. Prices kept increasing year after year until reaching $14.02 in 1978 before reaching to a new historical record of $36.8 in 1980, equivalent in value to $104 in 2013 dollars. The effect on OPEC pricing power can then be noticed in the period 1973–1980, compared with the earlier period when oil prices were stable at $1.8 between 1961 and 1970 (BP Statistical Review 2014).

OPEC took over a very stable oil market from the *Seven Sisters* and seemingly failed to build on that stability. The result was a period of high prices that lead to demand destruction *and ultimately unleashed the non-OPEC supply forces*. The group tried not to be an "official cartel," given the negative and political connotation, but it slipped into that role "almost unintentionally" (Parra 2004), although as discussed later in the volume, OPEC never really fulfilled the characteristic of a cartel. The group did say in its meeting at end of 1973 that it believed that the price of oil should be "market oriented" like any traded good. However, it was difficult, as some argued, that an organization like OPEC with oil revenues of its members being protected from market fluctuations since its inception can really have a feel of how to administer prices (Parra 2004). OPEC made an endeavor to be the world's "price setter," but it was not successful at the end in that attempt as its members lacked the required fiscal and production discipline to allow them to achieve any common *desired price level*. The group was not as successful as the *Seven Sisters* in coordinating needed supplies to keep prices stable and the market in balance.

The Post-embargo Period: The Rise and Fall of OPEC

The year 1973 marked a historical success for OPEC as the group finally rose to power after it took over the pricing of its crude from the *Seven Sisters*. The Arab producers within OPEC were also able to impose an embargo on their oil shipments to some Western nations in response to the Arab-Israeli conflict, signaling another sign of the change in the group's status, although, as explored later in the volume, setting in motion a train of political calls for *energy supply independence* by key consumer nations, principally the USA. OPEC was finally able in 1973 to raise prices to unseen levels and the result was the *first oil price shock*. Following the first shock, OPEC had become a powerful body, and it started slowly to transform into what many believed it to be a cartel. However, that success did not last for very long as OPEC soon started to discover the consequences of the high-price policy it was pursuing. The changing world of oil and politics is set out in Fig. 2.1.

Fig. 2.1 US world events and oil prices

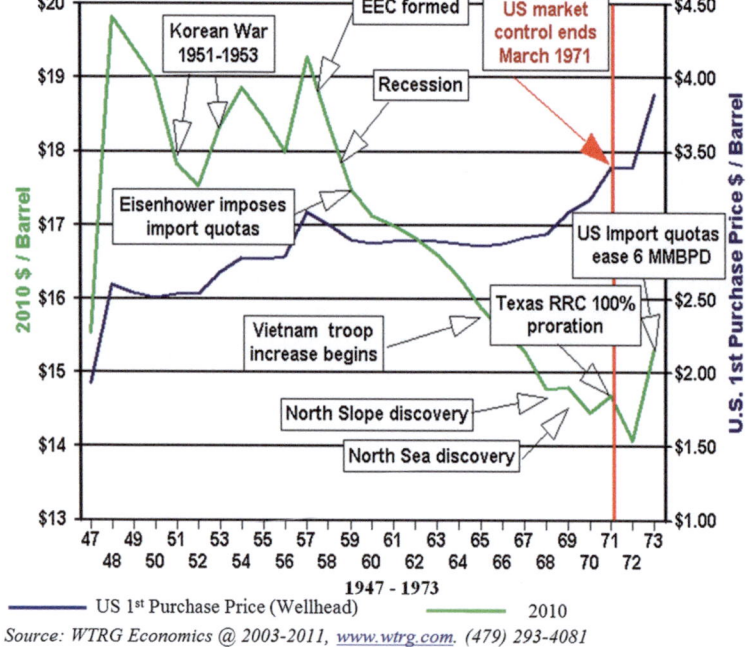

Source: WTRG Economics @ 2003-2011, www.wtrg.com. (479) 293-4081

The high-price policy had led to a slowdown in demand for oil since 1973, and many Western nations started to implement energy conversation measures and initiated the search for alternative energy sources (Terzian 1985). Between 1965 and 1973, the world's oil consumption grew by 80 % from 30.8 million bpd to 55.6 million bpd. The period of high growth had ended with OPEC taking over and consumption grew only by 10 % over the period 1973–1980 (BP

Statistical Review 2014). Slowdown in demand was not the only problems associated with OPEC's new era. With spot oil prices for OPEC's marker crude *Arabian Light*, rising from $2.8/bbl in 1973 to $10.4/bbl in 1974 and then to $29.75/bbl in 1979, the doors were wide open for producers outside OPEC to increase their production. In 1973, OPEC's output was 29.9 million bpd and by 1980 its production hit 30 million bpd. On the other hand, production outside OPEC grew from 19.9 million bpd in 1973 to 24.2 million bpd in 1979 (BP Statistical Review 2014).

The major problem that occurred in the period following the Arab oil embargo and the first oil price shock in 1973 was that OPEC countries did not have a clear view on which policies to pursue other than defending a high oil price. The organization's members could not focus on building the right strategy as they were enjoying a massive transfer of wealth from consuming countries. The group's aim to set prices between the late 1970s and 1986 was in the end, an unsuccessful endeavor. The reasons for this were, first, the continuous divide among members on the *right price* level for its marker crude and, second, the inability of the group to implement a system of quota to defend the desired oil prices as discussed in more detail later in the volume, leading to suspicion and recrimination among the group. It was not that natural for a group of countries who were only receivers of taxes and rents from companies to fill in the role of "price setters" easily and quickly. Although OPEC's creators were inspired by the *Texas Railroad Commission* (*TRC*), the organization itself never reached the degree of cooperation on production quotas that the TRC had enjoyed. The major issue for OPEC in the post-Arab oil embargo period was that its group members were always divided on the price level that was to be defended. OPEC held eight full Conferences between 1974 and 1978 and all of them centered around attempts to increase prices with a divide on by how much should the group increase them every time, without apparent unanimity.

In December 1973, Saudi Arabia opposed the *Shah of Iran's* move to increase oil prices. Again in the January 1974 meeting in *Quito*, Saudi Arabia advocated for a cut of $2 per barrel, whereas the majority wanted a $1.5 rise, and the discord left the group with a compromise decision to freeze prices. The year 1975 saw another divide at the group's meeting in Vienna, as Saudi Oil Minister *Ahmed Zaki Yamani* argued for a 5 % rise in prices in face of Iran's and other members demand for a 15 % rise. What resulted out of that meeting was a 10 % compromise increase that lasted for 9 months. The divide continued in 1976 and it became acute as Saudi Arabia refused to accept the 15 % increase agreed by the group's members at their meeting in Doha, finally settling for a 5 % rise (Yamani 1994). The inability of OPEC to agree on prices at that meeting gave rise to a period of *a two-tier price system*.

The emergence of the short-lived two-tier pricing system was enough evidence for how OPEC was deeply divided over pricing its crude oil. Saudi Arabia and the United Arab Emirates, on one hand, agreed to a 5 % rise in the marker crude price while the other 11 members of the organization agreed to a 15 % rise to be made in two steps. Saudi Arabia was trying to be the moderate voice in OPEC as the country took into account the negative effects of high prices on other economies at a time of soft manufacturing activities in the USA and high oil stocks at developed economies.

Saudi Arabia believed that by raising prices by 5 %, this would adjust the marker prices to the spot prices which were above the former (Galpern and Keefer 2013). The two prices were only unified in July the following year and were frozen until the end of 1978 (Parra 2004). Yet there was another development within OPEC in the year 1978 and that was that the group started to price crude not according to market conditions and the balance of supply and demand in the market, *but to the budgetary needs of its members*. This was manifested in the statement following its December 1978 meeting in Abu Dhabi in which it stated: "in order to assist the world economy to grow further, and also in order to support the current efforts towards strengthening the US dollar and arresting inflationary trends, the Conference has decided to correct only partially the price of oil by an amount of 10 % on average over the year 1979."

The year 1979 brought with it regional political instability in the Middle East that contributed to the *second oil prices shock*. The ousting of the Shah of Iran and the ascending of *Ayatollah Khomeini* had resulted in supply disruptions from Iran that left the whole market in panic and helped in increasing the price of crude in spot market from $12.8 in September 1978 to $21.8 in February 1979. In response to the increase in spot prices by producers in the North Sea and industrial nations, OPEC found it an opportunity to raise its crude and some members started increasing prices above the level agreed on in the December 1978 meeting in Abu Dhabi. With spot prices taking off, OPEC found it easy to raise prices and the organization seemed to believe that it could this from now on. *However, the price increases in the 1970s contained in them the seeds of weakening the group's grip on the world's market in the 1980s.*

What changed between the first and the second oil price shocks is that in 1973 OPEC imposed a price increase on the market, *while in 1979 the market imposed a price increase on OPEC* (Parra 2004). It became clear in the second oil shock that OPEC's pricing power was weakening as spot market became significant thanks to the increase in the volumes traded on the spot oil market that came mainly from the North Sea. The pricing power of the North Sea and the spot market were the main reasons for the breakup of OPEC's unity on pricing in the early 1980s when Nigeria was the first OPEC member to decide to match its prices with North Sea production and depart from OPEC's agreements. OPEC had inherited a stable system with many advantages, but it failed to make use of it to its own advantage. Once again, lack of unanimity and individual actions of OPEC members when it suited their national, as opposed to the group's collective interests, was the reason for such disarrays and which, as the opening chapter highlighted, continues to this day.

However, one great advantage for OPEC, *and probably its continued key strength*, although this is now being challenged by shale oil producers, was the fact that the group was the world's marginal producer of crude as most of the largest supply increments of the day came from its fields. OPEC's production went up by 115 % from 13.9 million bpd in 1965 to 29.6 million bpd in 1973. During the same period, non-OPEC output grew only by 52.6 % and OECD countries by 34 % (BP Statistical Review 2014). OPEC, however, had lost that advantage with the rise of production from the North Sea and other producers outside of OPEC in 1973. Another advantage was that demand for OPEC oil from industrial nations was increasing due to the stable prices of crude up until 1973. OPEC, however, could not put the right policies and system during the 5 years ending in 1978 to better run the world's oil order that they took over from the companies (Parra 2004). The two oil price shocks of 1973 and 1979 brought with them enough structural changes to weaken the group.

The post-Arab oil embargo period carried the seeds for the group's price war in the 1980s and the demise of its ability to control prices during that decade. One factor that weakened OPEC heavily was its inability to take into account the rise of the spot market and the power that this entailed. OPEC always seemed to feel insulated from major developments in the world's petroleum market. Another factor that weakened the group was the political divide among its members. Some member countries like Iran, Libya, and Algeria had painful colonial experiences and that led to enmity in their policies toward Western consuming nations, reflected by asking for higher prices for their crude. OPEC as a whole was not able then to separate politics from economics despite the continuous efforts of Saudi Arabia and other members to moderate the impacts of the group's policies on consumers. As the volume will discuss, this situation has not changed much from those earlier decades, with politics still very much entwined with economics in OPEC's *DNA*.

The Price War of the 1980s

After it won control over the world's oil market in the 1970s and became responsible for pricing its own crude, OPEC gradually started to lose that control starting from 1980. The mismanagement of the market by OPEC members during the period 1980–1986 had led to a price war within the group at first and then against producers outside it. A major attribute of OPEC's behavior in this period is that it failed to put in place the right structure to cope with the structural changes in the market that had resulted from the two price shocks of 1973 and 1979. The first structural change OPEC had to deal with in the years following the second oil price shock was the *decline in demand* from Western nations. This decline was *structural* as it was the product of new energy efficiency and conservation policies. In 1979, the OECD countries were consuming around 44 million bpd. This number went down to 36.6 million bpd in 1984 (BP Statistical Review 2014). The other structural change was that high oil prices had led to the rise of other fuels sources such as *coal and nuclear power*. During the period 1979–1984, around 3 million bpd of oil had been displaced by other sources of energy (Chalabi 1989). Still, the major structural change that shackled the foundation of OPEC in the first half of the 1980s was the rise in production from non-OPEC producers. The rapid rise of output outside the group was never possible without the high oil prices that OPEC sought in the 1970s. Between 1980 and 1985, production outside OPEC grew from 24.8 million bpd to 29.5 million bpd; whereas in OPEC, output fell from 26 million bpd to 15.8 million bpd during the period.

The retreat in OPEC's share in the world oil market was not only the result of higher prices, but also the result of the structure under which OPEC was operating to defend high oil prices. The group was assuming the role of the *world's swing producer* (or residual supplier of energy) as was coined by Chalabi (1989). Ever since OPEC took over the pricing of its oil in 1973 from the *Seven Sisters*, "its pricing policy has been based on fixing the price at a certain level below which no member country would sell its oil, while letting the volume of its sales be determined by the market." This policy and structure at the end forced consumers "to resort first to the relatively cheaper energy supplies, oil and non-oil, from outside OPEC before taking OPEC oil, which is rigidly priced at a comparatively higher level" (Chalabi 1989).

OPEC felt the consequences of its high-price policy early in the decade. The decline in demand and oversupply in the market made it difficult for the group to obtain any desired price level. *To obtain higher prices, OPEC had no option but to become a supplier of last resort.* In March 1982, OPEC met in Vienna and decided to impose a *production ceiling* for the first in order to defend a $34 price for *Arabian Light* that the group agreed on in 1981. The ceiling was set at 17.5 million bpd with Saudi Arabia producing no more than 7 million bpd (New York Times 1982). The problem that OPEC started to confront since the early 1980s was that North Sea producers were cutting the prices of their high-quality crude oil. This imposed tremendous pressure on light oil producers in OPEC such as Nigeria, Libya, and Algeria who had to compete head to head in the market with barrels of the same quality oil from Norway and the United Kingdom.

The New York Times wrote a story on March 16, 1982, titled "OPEC: *TRYING TO BE A CARTEL*" in which it stated the problems that the group was facing at the time. The article said "in essence, OPEC, which is producing less oil than in any year since 1969, is failing in its mission of setting and defending oil prices. The main reason is a sharp drop in petroleum demand that has caused a persistent worldwide oversupply." The article mentioned that the break in OPEC prices in 1982 would be the first since it took over in 1973. It even quoted analysts such Daniel Yergin saying that OPEC was facing its first crisis since 1973 (New York Times 1982). It was obvious from the beginning that defending a higher price for OPEC's marker, *Arabian Light*, will impose a greater responsibility on Saudi Arabia. The whole structure on which OPEC rested in the 1980s was built on the group's ability to play the role of the swing producer. However, due to the lack of unity and cooperation among the group members during these years, it was Saudi Arabia who had to play the role of the swing producer by cutting its production to balance the market. Due to this structure, Saudi Arabia saw its oil production dwindling from 10.27 million bpd in 1980 to 3.6 million bpd in 1985 (BP Statistical Review 2014). The price war of 1985–1986, like all crises, had roots in earlier years.

But it was not until Nigeria started to break up from OPEC's pricing system in the first half of the 1980s that matters escalated triggering the price war within the group and is illustrated in Fig. 2.2. Nigeria was the first country in OPEC to break ranks on prices, and in February 1983, it lowered its oil prices by $5.5/bbl to match cuts in North Sea crude. Nigeria was under severe economic pressure as its production had fallen to 500,000 bpd in 1983 from 2 million bpd. Instead of a cut in prices, OPEC wanted Nigeria to raise its prices by $1.5/bbl over the $34 marker price to reflect the higher quality of its oil (New York Times 1983[a]). Saudi Arabia had always tried to avoid a price war, but it was under pressure from Aramco to lower official selling prices of *Arabian Light* below the $34 marker as oil prices in the spot market were much lower (Parra 2004). To maintain OPEC's unity, Saudi Arabia had to sacrifice its own production. This was clear in the statement of Saudi Oil Minister Ahmad Zaki Yamani after a meeting in December 1983 when he said "unfortunately, this they wanted us to do," referring to pressure on his country to become the group's swing producer.

Fig. 2.2 Middle East, OPEC, and oil prices

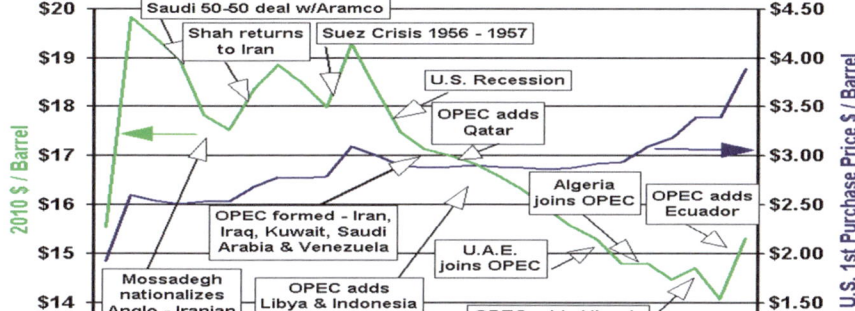

Source: *WTRG Economics @ 2003-2011, www.wtrg.com. (479) 293-4081*

Before going into a price war, OPEC had many attempts to confront the situation in the market. The group made an agreement in London on March 15, 1983, named therefore as *London Agreement*. According to the communiqué, OPEC agreed, first, to lower Arab Light marker prices to $29 from $34, second, to maintain existing differentials among OPEC members at the same level agreed in 63rd extraordinary meeting in Vienna in March 1982; third, to establishing a production ceiling for the group of 17.5 million bpd; fourth, Saudi Arabia will have no quota allocated to it and it will act as the group's swing producer; and fifth, members shall avoid giving discounts in any form (Reuters 1983). It was a significant list of things to achieve in a disparate group with multiple conflicting objectives. The *London Agreement* was significant however, as it was the first price cut accord for the group and established Saudi Arabia's role as a swing producer. The agreement only worked until 1985 when Saudi Arabia decided to abandon it and sold its crude under new arrangements known as *netback* agreements. The *London Agreement* collapse was inevitable as fixing prices or trying to obtain a certain price at a time of lower demand and oversupply is burdensome for any producer. It was "mission impossible" from its inception.

In 1984, prices fell down and OPEC held an emergency meeting in October in which it agreed to lower the ceiling to 16 million bpd effective November 1. The fall in prices, however, was steeper than OPEC's ability to cope with it. The official selling price in early 1985 was $28, still $1–$2 above market prices. In May 1985, Saudi Arabia felt more pain and it was producing around 2.5 million bpd with 1.4 million of that as exports, and by August of 1985, the Kingdom was producing 2.2 million bpd (Parra 2004). The significant reduction in Saudi production compared with other OPEC producers is illustrated in Table 2.6.

Table 2.6 1983 oil production compared to the quota assigned by OPEC (1000 bpd)

Country	Quota before November 1984	Quota November 1984	Production during May 1985	Average production for 1984
Algeria	725	663	600	645
Ecuador	200	183	280	256
Gabon	150	137	150	157
Indonesia	1300	1189	1200	1482
Iran	2400	2300	1800	2106
Iraq	1200	1200	1200	1194
Kuwait	1050	900	900	1154
Libya	1100	990	1100	1075
Nigeria	1300	1300	1450	1387
Qatar	300	280	300	400
Saudi Arabia	*(5000)*	*(4353)*	2500	*4651*
United Arab Emirates	1100	950	1050	1138
Venezuela	1675	1555	1.000	1820
Total	17,500	16,000	14,130	17,365
Neutral Zone			400	
Total			*14,530*	

Source: OPEC, Eleventh Report of the Secretary General, OPEC, Vienna (1985)

Facing this difficult reality, Saudi Arabia started in late 1985 to sell crude under the *netback* contracts and by February 1986 it was selling 3 million bpd of crude under *netbacks* and by July the country's output was hitting 5.5 million bpd (Parra 2004). The flooding of the market with Saudi oil elevated OPEC's total output in 1986, and as a result of the additional high supply, prices went down further in the spot market. What is interesting is that while OPEC was targeting high prices that year, Saudi Arabia was targeting its market share, which came back, full circle again, as the Kingdom's policy after the 2014 oil price collapse, as explored in the introductory chapter.

The group during 1986 started to be realistic about the desired oil prices and members seemed content with a price between $17 and $19 instead of $28. Yet, prices kept sliding and *Brent* oil in the spot market was selling at below $10/bbl while Dubai oil was selling at $7/bbl (New York Times 1986[a]). OPEC met again in August 1986 to find a solution for the sharp fall in prices and there was only one path in front of the ministers and it was to cut output. The task was not easy as Saudi Arabia was determined to get back its market share and Iraq and Iran, both in a war, were trying to secure more income to finance their military engagement. Iran was accusing the Saudis at that time of using the low-price policy to cripple its 6-year-old war against Iraq (accusations echoed during the 2014 oil price collapse period and in 2015 when Iran started to negotiate an end to Western imposed sanctions on the country because of its nuclear program, discussed later in the volume). At the end, ministers agreed to make a cut effective September 1. The ministers voted to reinstate a system of quotas that would effectively limit OPEC's output to about 16.8 million bpd from 20.5 million barrels that members were producing at that time. The ministers also said other non-OPEC producers had promised to contribute to a total of 400,000–500,000 bpd of cuts to help OPEC to support prices. Britain and Norway, however, appeared unlikely to help OPEC by reducing production of their North Sea oil (New York Times 1986[b]).

The OPEC accord was viewed as an end to the Saudi-led price war. After the meeting was concluded *Belkacem Nabi*, Algeria's oil minister told the press: "once a war has been stopped, it is difficult to start it up again." The Iranian minister, *Gholam Reza Aghazadeh,* was quoted as saying: "they have lost the war," referring to the Saudi-inspired price war (New York Times 1986[e]).

For years OPEC has lacked the unity to help it control prices and the 1986 accord had restored some power to the group on pricing. The accord had created an alarm in the USA and the *Reagan Administration* criticized the OPEC agreement the following day on the basis that it will allow the organization to gain control again over oil pricing and that it was an interference in the free market and a possible threat to the US national interests. Energy Secretary at that time John S. Herrington was quoted as saying: "the re-establishment of the dominance of OPEC as it existed in the 1970s is unhealthy for Americans. Not only that, it is unacceptable." The White House and State Department were in favor of letting the market determine the prices. Charles Redman, a State Department spokesman said: "we do not think governments should interfere with markets by attempting to set oil prices" (New York Times 1986[c]).

The group met again on October 6, 1986, to agree on a more permanent system of quota to support prices once and for all. The meeting turned out to be the longest ever in OPEC's history *as it dragged on for 22 days.* OPEC agreed to raise its official production ceiling to 14.961 million bpd on November 1 and to 15.039 million barrels after December 1. This compares to 14.8 million bpd reached in the August accord. The ceiling applied to only 12 of the group's 13 members, *excluding Iraq*, which was estimated to be producing at around 2 million bpd at that time, and Iraq's exclusion from future quota agreements became a long-term issue for OPEC to handle, as explored later in the volume. Under the new quota agreement, Saudi Arabia kept its quota unchanged from August at 4.353 million bpd. Kuwait had an increase of 21,000 barrels to 921,000 bpd, while Iran had a 17,000 barrel rise to 2.317 million bpd. The group also agreed to return back to the fixed-price system, and it was now targeting a price between $17 and $19 per barrel. This price was higher by about $4 from spot prices during that time (New York Times 1986[f]).

The meeting was followed by the dramatic news of the removal by *King Fahd bin Abdulaziz* of Saudi Oil Minister Zaki Yamani from his post after 24 years in that position and with *Hisham Nazer* becoming the third oil minister in the history of the country. With a new minister in place, Saudi Arabia was more focused on defending an oil price of $18/bbl and was trying to convince other members in OPEC to adhere to the production quotas to support that price. When the group met in December 1986, it was a miracle for its members to agree on cutting their output from more than 17 million barrels to 15.8 million bpd, hoping to force up prices from the $14 to $16 level to $18 at a time when there was around more than 300 million barrels of surplus oil on the market (New York Times 1986^f).

A New Era After the Price War

It was very clear from the beginning after the removal of Yamani that Saudi oil policy had entered a new era. The change in policy was reflected in a speech delivered to students in Dhahran by *King Fahd bin Abdulaziz* after the group's meeting in December 1986 when he said: "I believe that the OPEC states this time became fully convinced that reduction in output is the only way to absorb the surplus…This does not mean that $18 is a final price. It is not the maximum but the price must not be less than $18" (New York Times 1986^f). With Yamani out of office, the price war was over and prices started to rise and the group became "married," in the words of Nigerian Oil Minister *Rilwanu Lukman*, to a new floor of $18 (ibid). Yamani had certainly dealt with the OPEC crisis in a different fashion from his successor and that was noted by the New York Times in an article on December 21, 1986, under the title "THE WORLD: OPEC Reaches An Agreement." The article stated that Yamani had supported "free-market pricing as a way to force higher-cost producers out of the market and stop OPEC members from cheating on their quotas" (New York Times 1986^e), an uncanny echo of the policy that seemed to be at the center of Saudi oil policy that was adopted by Saudi Oil Minister Ali Al Naimi following the 2014 oil price crash, as explored in the first chapter. The end of the price war, however, resulted in oil prices going into a prolonged cycle of stable but low oil prices below $20 that remained for many years, until another structural change made itself clear in the year 2000 with the rise of the emerging economies in Asia.

The first half of the 1980s was a period of crisis and there are many lessons for OPEC to learn from that turbulent period in its history and the history of the oil market. OPEC made many mistakes among them *its policy to defend a fixed price and an output ceiling at the same time*. It could not achieve both at the same time and they had to choose between the two objectives (Al Moneef 2012). Yet, the main feature of the entire period was the struggle between OPEC and non-OPEC producers to secure a share in global market. OPEC would not have been able to stabilize the market during that turbulent period without it playing the swing producer. To play that role in a shrinking oil market, OPEC had to make great sacrifices. The non-OPEC producers benefited substantially from OPEC's role in defending prices in the 1980s by undercutting OPEC's prices to maximize their sales (Chalabi 1989).

Chalabi had aptly described the major flaws with the structure under which both OPEC and non-OPEC were operating in 1980s and how producers outside behaved in response to OPEC being the world's swing produce, when he stated "in fact, those countries believe that OPEC has no other choice but to continue with its policy of defending the price through cutting back production, a policy which secures for the non-OPEC exporters high revenues in the short run and provides, at the same time, a guarantee for their future investments to find new oil. Therefore, unless those countries face real difficulties in their pursuit of maximizing both volume and price, and unless OPEC changes its stance vis-à-vis the role of residual supplier, it is unlikely that they will be ready to cooperate with OPEC in stabilizing the market on a fair basis" (Chalabi 1989).

The history of OPEC is repleted with cases in which OPEC actions were inspired by decisions already taken by individual countries, or as analyzed for the earlier decades and most recently after 2014, by instances in which individual members opted to disregard decisions made by OPEC. These divergences between individual members' interest and that of the group as a whole should not be interpreted to mean that there are no common grounds or interests among member countries as the November 2014 OPEC decision not to cut production highlighted. However, as the history of OPEC has also shown, this can only happen to the extent that the economic and political goal of a member country conforms to those of the group. As repeatedly highlighted in this volume, this convergence of interests is very rare. The moment the perception exists that an individual country's interests and goals are not served by an OPEC decision, the interest of that country can be expected to supersede those of OPEC's common objectives.

The Rebirth of the IOC's and Producer Cooperation

The IOC's are now making a comeback as "super-majors" with mega mergers between once rivals to ensure that they survive in a globalized world where scale and financial power as well as diversification of business line ensure survival of the fittest. The role of the IOC's has evolved over the past decades, whereby they were truly masters of everything they surveyed in the energy sector in the 1950s and 1960s, as illustrated in Fig. 2.3.

Fig. 2.3 Oil producers and the majors: changes in control over the vertical production chain

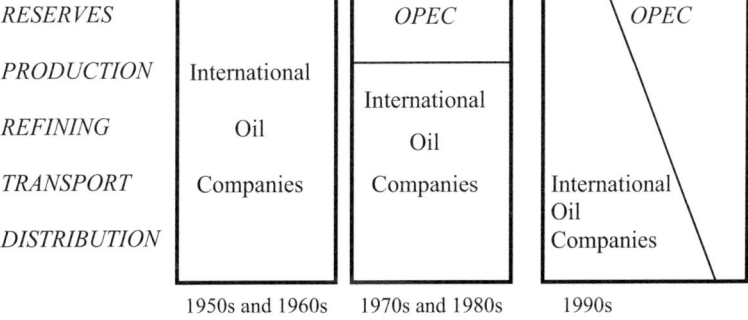

In the earlier decades, the majors dominated the whole production chain, but as resource nationalism emerged during the 1970s and 1980s with OPEC's establishment and the desire of the organization to have a greater say in both production and prices, the major oil companies started to focus on refining, transport, and distribution although some IOC's continued to own reserves, especially in non-OPEC countries. By the 1990s, many OPEC countries had gained enough experience or acquired specialized energy and distribution companies to be involved across all the vertical production chain, often in partnerships with companies that they had nationalized in their home country. The era of the 1980s and 1990s saw some major downstream acquisitions by key OPEC and non-OPEC oil producers, illustrated in Table 2.7.

Table 2.7 Principal foreign downstream acquisitions of petroleum-exporting countries, 1983–1990

Purchaser	Acquired capacity (b/d)	Country	Seller/partner	Activity	Share (%)	Date
S. Arabia	600,000	USA	Texaco	Ref./dist.	50	1988
Kuwait	75,000	Benelux	Gulf Oil	Ref./dist.	100	1983
	70,000	Den./Swe.	Gulf Oil/BP	Ref. /dist.	100	1983, 1987
	35,000	Italy	Gulf Oil	Distribution	100	1984
	70,000	UK	Hays/Ultramar Nafta	Distribution	100	1986, 1987
	100,000	Italy	Mobil	Ref./dist.	100	1990
Venezuela	145,000	Germany	Ruhr/Oel/Veba	Ref./dist.	50	1983, 1986
	305,000	USA	Citgo/Southland	Ref./dist.	100	1986, 1989
	50,000	Sweden	Nynas	Refining	50	1986
	135,000	USA	Champlin/U. Pacific	Ref./dist.	50	1987
	147,000	USA	Unocal	Ref./dist.	50	1989
	44,000	USA	Sea View	Refining	50	1990
Libya	110,000	Italy	Tamoil/Amoco/Finterm.	Ref./dist.	70	1983, 1987
Norway	30,000	Sweden	Exxon	Ref./chem.	100	1985
	45,000	Denmark	Exxon	Ref./dist.	100	1986

Source: Finon (1991, p. 265)

Oil producers and oil majors companies saw this new relationship as a "win-win" for both parties, with producers acquiring refining and distribution expertise and access to new markets, and with the oil companies obtaining access to oil production in a direct supply relationship that was more assured and not subject to contentious disagreements on production volumes, and supply scarcities which was the companies nightmare of the 1970s era triggered by the Arab oil embargo and the fear it left behind, however unfounded that this would be repeated. To ensure such cooperation is now established on a sound footing, according to analysts, producing countries need to "establish new rules under which foreign companies will be allowed direct access to oil reserves under conditions that will guarantee the government's political control" (Luciani 1995, p. 49). Some OPEC members have started to revise their laws and regulations and become more open to production sharing agreements with foreign firms as in the case of Algeria, Iran, Iraq, Libya, and Qatar opening up to foreign equity sharing (Ismail 1995, p. 18). This is a partial reversal of the earlier nationalization era, but as major OPEC oil producers gain more confidence in managing a fully integrated energy production chain, their national oil companies will also gradually become integrated IOCs, with downstream operations increasingly in line with their upstream assets, and reserves located not only in their country of origin but abroad (Finon 1991). In effect, the national oil companies will be the reverse side of the same coin as a multinational oil company. At the same time, the interests of the producing countries and consumer countries come closer as producers with large downstream assets in refineries such as Kuwait, in Table 2.7, could become less affected by lower oil prices on their crude sales since low input prices in refineries increases the profitability of owning such assets. Internationally diversified upstream and downstream oil producers can then benefit in situations of both high and low prices. In 2015, Kuwait made a bid for a portfolio of North Sea assets being sold by EON SE valued at around $ 2 bn, with these assets located in both the UK and Norwegian waters (Nair, 2015). This brings about an almost 360 degree turn in the relationship between OPEC producers and countries like the UK and Norway that had been instrumental in eroding OPEC's power in the earlier eras. The recent statements of Saudi Aramco to become a global integrated oil company reflects this new strategy (Saudi Aramco 2014), and it comes as no surprise that the company is said to be planning to spend as much as $80bn on international investments and acquisitions over the period 2015–2020 (Martin and Blanchi 2015).

Part II
OPEC and the New Reality

Chapter 3
A False Dawn: Myths and Realities of OPEC's Power

Better bend than break.

Scottish Proverb.

OPEC's New Reality

Conventional wisdom has held, until the dramatic oil price collapse of late 2014, that OPEC has the world's consumers in its grasp and that it can manipulate prices by tinkering with its own supplies. The events of late 2014 has proved conventional wisdom to be wrong, for it has shown that indeed "the Emperor has no clothes." As discussed in an earlier chapter, OPEC for the most part has seen its actions or more precisely OPEC decisions, lagging behind fundamental changes in oil supply and demand and a new energy paradigm when assessing the rise of shale oil from non-OPEC countries. The events of 2014 can be viewed as a watershed in OPEC's perceived control of world prices which demonstrated that instead of being a masterful and controlling cartel, OPEC was in fact now riding out events, hoping for the best in terms of eventual oil price recovery.

The OPEC meeting of November 27, 2014, was indeed a remarkable one in many ways and a wake-up call for the organization. On the surface it seems to have brought about a new strategic thinking to OPEC's decision-making process, by re-directing OPEC away from short-term oil pricing toward greater commercial transparency and market-led global oil pricing. This new policy has been in effect forced on the organization's varied members by OPEC's "Big Three"—Saudi Arabia, Kuwait, and the UAE. These countries have the spare capacity and financial reserves to put forward a bold new approach, much to the reluctance and public displeasure of some other financially stressed OPEC members.

What brought about the sudden reality check for the "Big Three?" It had become inevitable to all OPEC members, whether in the "big" or "small" league, that a combination of a weak global economy and softer oil demand, compounded by an abundant supply from non-OPEC countries, had created a new reality for the organization. OPEC had faced other major defining moments in its history, such as between 1980 and 1985 when Saudi Arabia cut back on its production from 10.5 million bpd to as low as 2.2 million bpd by year end 1985, only to see its allocation eaten away by others. This time round in 2014, it was Saudi Arabia that decided not to budge and kept to its 9.6 million bpd average production, and later increasing it to over 10 million bpd, with veteran Saudi Oil Minister Ali Al Naimi vowing that OPEC would not cut production "even if oil prices fall to $20 a barrel" (Arab News, Dec. 24, 2014[c]). How much discipline OPEC has to muster to back these words will be assessed later on when OPEC's quota sharing system is analyzed. By June 2015, when OPEC met again to review their albeit grudging acceptance of the Saudi diktat, the mood had changed to one characterized as being of a "happy meeting" compared to the gloomy November 2014 meeting. The 30 million bpd production target for OPEC was rolled over again, but this time some element of realism had crept in, aptly put by OPEC Secretary-General El-Badri when he was quoted as saying that "OPEC must now accept a new reality of oil prices being below $100 a barrel" and "that is a fact" and that the OPEC output ceiling was an "indicator" not a quota (Rudnitsky and Chmaytelli 2015).

Monopoly Profits and Market Power

A wide range of academic literature has addressed the issue on whether OPEC seeks to maximize profit as a price-making or price-taking cartel. Results tend to vary, depending on the state of global energy demand and supply, as well as whether individual OPEC members are low-cost or high-cost producers (Dahl and Yucel 1991; Smith 2005) or whether OPEC members target certain revenue flows in accordance with investment needs (Griffin 1985). Other studies assess the impact of dominant or price-leader models to describe OPEC's actions and the oil market (Bochem 2004) and assume the supply side of the oil market to be characterized by a dominant producer, like Saudi Arabia, Kuwait, and UAE, who then impose a selling price on a competitive group of OPEC followers who are price takers (Prokop 1999). More recent studies (Hansen and Lindhollt 2008) seem to confirm that OPEC as a whole does not fulfill the condition for being a profit-maximizing dominant producer, but has affected market price. According to Hansen and Lindholt (ibid), there may be various reasons why OPEC can be characterized as a "weak cartel." This is due to the fact that the group consists of countries with different ambitions and intentions and that OPEC may have been uncertain about the true nature of elasticities facing them, particularly demand elasticity which measures the degree of responsiveness of demand to a change in price. Earlier studies of the 1970s and 1980s period have generally found that higher global demand and non-OPEC supply price elasticities indicated that OPEC might have been afraid of losing market share if they were to curtail production. Another important factor, to be discussed below, is that OPEC was finding it increasingly difficult in reaching production consensus among members.

According to some (Mabro 1986), the genesis of previous OPEC price crisis is that fiscal and price measures are compounded when decisions have to suit a group of governments, with OPEC not explicitly concerned with the distribution of gains among members but rather on the premise to "accept that significant gains, unequally distributed, are better than no gains" (Mabro 1986, p. 115). The above scenario works well when prices are rising, but does not hold when prices are under pressure and puts in doubt the layman's assumed branding of OPEC as a cartel.

If OPEC should behave according to conventional cartel theory, it should set a fixed price based on the group's perceived demand for its commodity and assumed production outside the cartel. The cartel then is assumed to increase production if the price rises above the group's agreed level and decrease production if the price falls below the agreed level. The above implies that the cartel's role in the market is primarily based on two factors: *production quotas and collusive price setting*. Others, such as Noreng, emphasize the political aspect and describe OPEC as an "international interest group or trade union of raw-material producers" and a forum for political discussions and a platform for common demands (Noreng 1978). All OPEC members share the successes of the organization and, by default, also share the failures of its policies as will be discussed in quota compliance. According to some (Noreng, ibid 1978; Claes 2001, 2011), the oil exports of the OPEC countries are the result of policies and budgetary decisions formulated in the context of national and international politics by a limited number of governments which determine these countries' long-term political interests related to oil and foreign policy concerns. The use of oil as a political tool to drive *down* oil prices to hurt other countries perceived to have different political objectives to those of leading OPEC members, such as Saudi Arabia's, was categorically denied by Saudi Oil Minister Al Naimi (Arab News 2014[b]) who dismissed claims of a Saudi "plot" to push down prices to hurt countries like Russia and OPEC member Iran, with the oil minister insisting that the Kingdom's oil policy was "based on pure economic principles."

What then constitutes economic principles for OPEC and are they applicable for *all* members of the organization? Most OPEC analysts seem to agree that the key economic factors behind oil exports are income requirements, size of oil reserves, population, and market conditions (Noreng 1978, 1982; Claes 2001, 2011). However, as discussed below, OPEC members differ on the weights to be applied for these economic variables, given their relative advantage in one economic factor or another, thus complicating quota production agreements. According to OPEC observers, the governments of the organization have always placed economic and political national considerations above the common economic interests of the group and that, in the final analysis, it is these national interests that render "fruitless any attempt to categorize an organization like OPEC" (Al Nasrawi 1985). A major flaw in attempting to label OPEC as a cartel stems from trying to impose the behavior of a firm, which has profit maximization as its most important goal, on a group of political entities, each with numerous political, social, and economic objectives. More recent empirical studies on OPEC to test for cartel behavior (Moguera et al. 2011) sheds some light on the literature's lack of consensus regarding OPEC's cartel stability and cooperative and noncooperative collusion. The words of Al Nasrawi seem apt when he succinctly states that "OPEC is simply OPEC, and to attempt to force it into any of the frameworks of conventional economic analysis is a futile exercise" (Al Nasrawi 1985, p. 89).

OPEC-Non-OPEC Cooperative Limits

OPEC, ever since its early founding years, has professed cooperation between its members and non-OPEC producers. Article 2 of the 1960 OPEC statute lists, among other objectives, the following:

> Due regard shall be given at all times to the interests of the producing nations and to the necessity of securing a steady income to the producing countries; an efficient, economic and regular supply of petroleum to consuming nations; and a fair return on their capital to those investing in the petroleum industry. (Al Fathi 1990, p. 1)

The non-OPEC cooperation of the early establishment period involved resolving differences and wresting national rights from multinational oil companies, as explained in Chap. 2, but from the mid-1970s, OPEC turned its attention to ever-increasing non-OPEC oil supplies. Contacts began with a host of non-OPEC countries, notably Mexico, Malaysia, Brunei, the UK, Norway, Russia, and Brazil. However, as Table 3.1 illustrates, there are now many non-OPEC producers, whether these are classified as category *I Majors* with production capacity over 2 million bpd, or category *III Minors*, with production capacity of 0.1–0.3 million bpd.

Table 3.1 Categories of non-OPEC producers

Production (mbd)	Category I Above 2	Category II 0.3–1	Category III 0.1–0.3
Share of			
Non-OPEC			
Reserves	About 70 %	Nearly 25 %	About 5 %
Production	About 60 %	Above 20 %	Above 5 %
Countries	Canada, UK, Norway, USA, Mexico, China, Russia, Brazil	India, Oman, Egypt, Argentine, Angola, Malaysia, Syria, Colombia, Australia, Vietnam, Kazakhstan, Yemen, Gabon	Congo, Denmark, Vietnam, Peru, Italy, Azerbaijan, Brunei, Romania, Trinidad and Tobago, Sudan(s), Turkmenistan, Cameroon, Chad

Source: Claes (2001, p. 282, BP 2014, p. 8)

What Table 3.1 also highlights is the heterogeneity of the non-OPEC countries in relation to their political and economic systems, making collective action among them even more problematic compared with OPEC's own diverse political and economic grouping. The only feasible action that OPEC can take is to "target" the most prominent non-OPEC producers to cooperate on a bilateral basis, by inviting them as observers to OPEC meetings and hoping that common interests on pricing and, more importantly, on production levels prevail. Sometimes it takes extraordinary events to get OPEC and non-OPEC to try and cooperate as the price collapse of 1986 provided a shock to all producers, especially those high-cost non-OPEC producers like the United Kingdom and Norway who were producing at nearly full capacity and started to cut back on price, while others like Mexico were more willing to have a dialogue with OPEC (Claes 2001). The first of what constituted a price war between OPEC and some non-OPEC members, led by Norway, foreshadowed later events during 2014 when oil prices again tumbled, as illustrated in Fig. 3.1.

Fig. 3.1 Crude oil prices 1970–2014 (US$ per barrel)

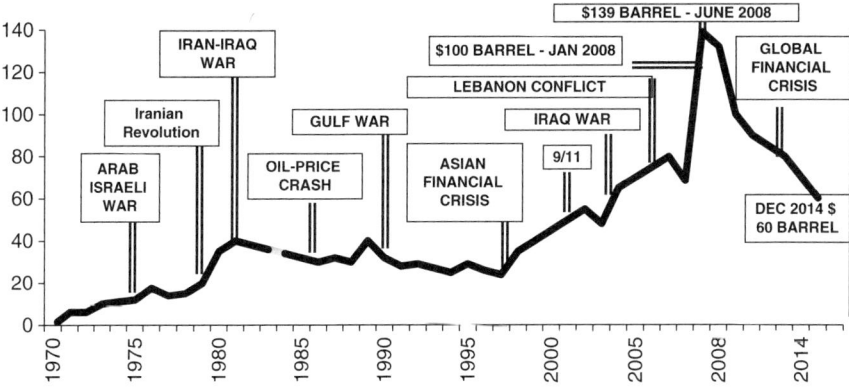

Source: BP (2014), Bloomberg

During the period of the first major oil price crash of 1985/1986, there was some, albeit reluctant cooperation between OPEC and non-OPEC members to reach production cuts of around 183,000 bpd for non-OPEC and 700,000 bpd for OPEC members, signaling broader producer cooperation in the market (Claes 2001). This spirit of mutual cooperation was not much in evidence during the sharp fall in prices during 2014 which saw oil losing nearly 50 % of its trading value to around $60 pb from $110 to $115 levels at the middle of the year. If there was one country that should have signaled a cooperative pact with OPEC, it was Russia, as no country was suffering more from plunging oil prices than that country, with estimates of losses of around $100 billion a year and with President Putin quoted as saying further "catastrophic" slumps in the economy was possible (Bloomberg Businessweek 2014). And yet Russia said "no deal" to OPEC in cutting oil production, despite talks with OPEC representatives Venezuela and Saudi Arabia ahead of the organization's November 27, 2014, Vienna meeting. According to analysts, there were several reasons for the Russian *nyet*, namely, that the country could not turn off production in the Siberian oilfields where it was more difficult during the winter and that, unlike other OPEC exporters, the Russian economy had been somewhat cushioned from the oil price decline because the value of the Russian currency, the *ruble*, had also fallen sharply. According to Bloomberg Businessweek analysts, even though Russian dollar value of exports had fallen, their value in *rubles* has remained relatively steady so that the Russian government could still collect enough taxes to cover domestic budgetary obligations (Bloomberg Businessweek 2014). The primary reason though is that both OPEC and non-OPEC members did not want to lose market share if they unilaterally cut back production. The ghosts of 1985 still lingered in OPEC's memory and, as Fig. 3.2 illustrates, OPEC members had every reason to worry, as they have seen their share of the world's energy production fall to around 36 million bpd or 41 % (including crude oil, natural gas liquids, condensates) out of total world production of around 87 million bpd (BP 2014, p. 8). In 1976, OPEC's share stood at around 52 % or 35 million bpd out of a global production of around 65 million bpd.

Fig. 3.2 World liquid production (*total liquid production includes crude oil, natural gas liquids, condensates, and other liquids*) 1946–2013 and OPEC share (Millions of barrels per day)

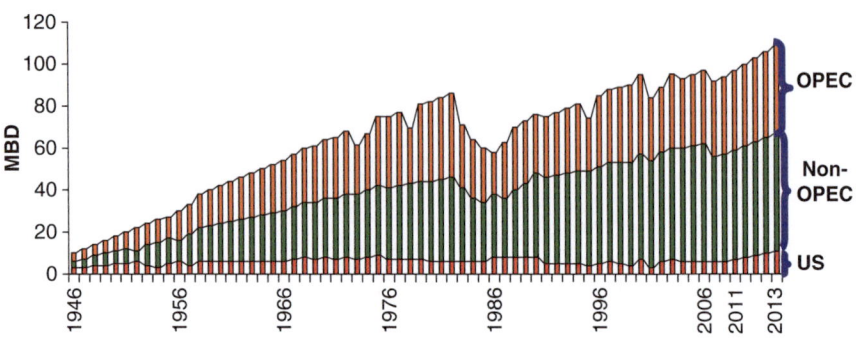

Source: Yergin (2011, p. 240), BP (2014, p. 8)

From the above figure, the future for OPEC retaining its market share seems to be a daunting task. Once again, as in the mid-1980s and late 1990s, the organization is hoping that the sharp drop in oil prices will either force some tacit OPEC-non-OPEC producer agreement on production cuts, *with pain to be shared by all*, and not just by dominant producers like Saudi Arabia, or that the sharp fall in prices sees high-cost producers exit the market. This is a high-stake risk strategy that seems to be based on the following factors. *First*, that the incremental production of world liquid production coming from US shale oil and gas production will fall back from 2013 levels of over 10 million bpd, compared with 7.3 million bpd in 2003 as world prices fall (BP 2014); *second*, that OPEC maintains unanimity and does not break rank with opportunistic production increases; *third*, that both OPEC and non-OPEC members do not engage in a round of deep price discounting to maintain their market share in lieu of production cuts; and *finally* that no substantial non-OPEC production comes on stream in the medium to long term. The above issues are addressed later in this and other chapters, but to add to OPEC uncertainties, the organization could also very well be facing a future energy superpower in Brazil, to add to the recent emergence of Russia as a key OPEC competitor.

Brazil: One More OPEC Headache

Brazil, Latin America's largest economy and seventh in the world, discovered its deep-water oil fields in 2006. According to latest data, Brazil is today the seventh largest oil consumer and tenth largest oil producer in the world, with 15.6 billion barrels of proven oil reserves—the second largest after OPEC member Venezuela's 298 billion barrels (BP 2014). In terms of production, Brazil produced an average 2.1 million bpd in 2013, compared with 1.5 million bpd in 2003, but analysts predict that Brazil's energy landscape could change dramatically over the next two decades to add to OPEC's diminishing market-share problems. In March 2013, Brazil launched a 10-year energy plan that aims to expand oil production to over 5 billion bpd and oil exports of 2.2 million bpd by 2021. Some believe that Brazil could become a net oil exporter and a top ten producer from 2015 if the country overcomes technical hurdles to develop its deep-water offshore reserves which are buried under a layer of sodium chloride (hence the fields known as *pre-salt*) up to 2 km thick, which some analysts say could hold up to 50 billion barrels of high-quality crude (Tamimi 2013).

The International Energy Agency (IEA) is bullish about Brazil's long-term contribution to global oil production, as illustrated in Fig. 3.3, with deep-water production increase taking the lion's share of Brazil's forecasted production expansion.

Fig. 3.3 (**a**) Contributions to global oil production growth: 2013–2035. (**b**) Brazil oil production: 2012–2035

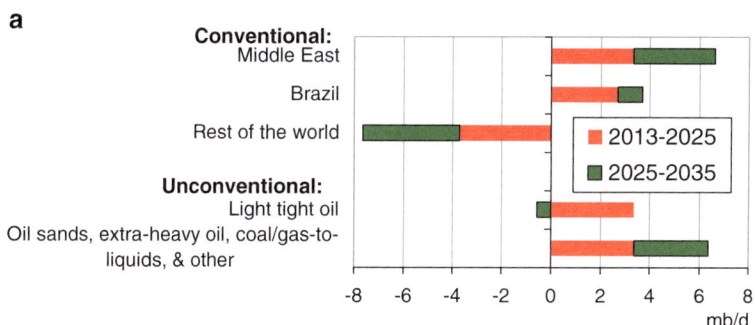

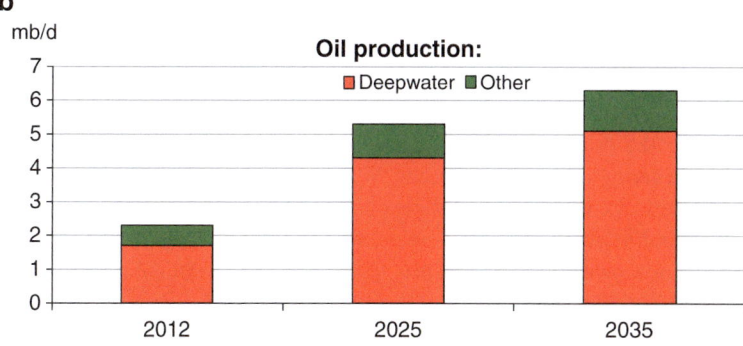

Source: IEA (2013)

From Fig. 3.3a, we note that the Middle East remains critical to the longer-term outlook to 2035, but that both the USA, in the production of light "tight oil," and Brazil in deep-water oil are important contributors to global oil production until the mid-2020s. However, in its *2014 World Energy Outlook* published in November 2014 during the sharp decline in world oil prices, the IEA adds a cautionary statement that lower prices were beginning to curtail upstream spending plans, with implications for future supply and that "overtime, squeezed cash flow could constrain the capacity of North America and Brazil to act as engines of global supply growth" (IEA 2014). According to industry experts on deep-water production, *pre-salt* is very difficult and expensive, and the technology is not as well developed as tight shale oil as in the USA and that *Petrobras*, the Brazilian oil company, has not done shale exploration and has amassed over $114 billion in debt thus making *Petrobras* the most "heavily indebted" national oil company in the world (Business Monitor International 2013). According to the IEA, the above factors could accelerate reliance on low-cost producers in the Middle East, primarily Saudi Arabia, and bring to reality one of the key assumptions on why OPEC, and Saudi Arabia in particular, did not opt for a production cut in November 2014. However, this still does not address another key assumption raised earlier—namely, how will OPEC ensure compliance on current quotas among its members and price discipline? These are addressed next.

Quotas: Did They Ever Work?

A major feature of OPEC is the uneven production and distribution of oil reserves among its members. As such, OPEC can be divided into two groups using the above two classifications to help better understand the relative power and market influence some members have compared to others. Table 3.2 sets out oil production (which includes all types of crude oil but excludes natural gas) and proven reserves of the OPEC countries for 2013.

Table 3.2 also sets out the hypothetical ratio of reserves to production (R/P), which estimates the total number of years of production left on the basis that current production levels are maintained. The table groups OPEC members into two. It is noticeable that out of the seven countries in *Group II*, the dominant players, with the exception of Venezuela, the remaining six are located in the Middle East, with Saudi Arabia being the dominant OPEC member in terms of production, but second to Venezuela in terms of proven reserves. What is also interesting from Table 3.1 is that while OPEC's share of global production stood at around 42 % in 2013, in terms of global reserves the organization accounts for around 72 %, notwithstanding possible new production and reserve finds like those in Brazil mentioned earlier. Despite the 2014 price falls, OPEC's proven reserves still makes the organization a powerful force in future energy markets and geopolitical calculations.

Grouping the OPEC members into two broad categories creates different economic interests for these countries. While there are other factors to be considered, splitting OPEC countries by size of their proven reserves highlights some common economic interests and follows on from earlier work done in this field (Noreng 1978).

Group I: These are countries with smaller reserves and production profiles of around 30–40 years, but generally have ambitious plans for economic development, large income requirements and an interest in maximizing oil income in *the short run*.

Table 3.2 OPEC countries oil production, proved oil reserves, and ratio of reserves to production—2013

Country	Production[a] (million tons)	Proved reserves (1000 million barrels)	R/P ratio[b]
Group I			
Ecuador	28.2	8.2	42.6
Algeria	68.9	12.2	21.2
Angola	87.4	12.7	19.3
Nigeria	111.3	37.1	43.8
Qatar	84.2	25.1	34.1
Total Group I	**380.0**	**95.3**	
Group II			
Libya	46.5	48.5	More than 100
Venezuela	135.1	298.3	More than 100
Iran	166.1	157.0	More than 100
Iraq	153.2	150.0	More than 100
Kuwait	151.3	101.5	89.0
Saudi Arabia	542.3	265.9	63.2
UAE	165.7	97.8	73.5
Total Group II	**1,360.2**	**1,524.0**	
TOTAL WORLD	**4,130.2**	**1,687.9**	**53.3**
Of which OPEC (%)	*1,740.2 (42.1%)*	*1,214.3 (71.9%)*	*90.3*

Source: BP (2014, pp. 6, 10)
[a]Includes crude oil, tight oil, oil sands, and natural gas liquids and excludes natural gas
[b]**R/P Ratio**: If the reserves remaining at the end of any year are divided by the production in that year, the result is the length in time that those remaining reserves would last if production were to continue at that rate

Group II: These are OPEC members with large reserves and production profiles of more than 50 years, often more than 100 years, an interest in maintaining a market for oil over *the long term*—(this will be explored later in the volume concerning environmental cost of fossil fuel emissions and disagreements between major oil producers like Saudi Arabia and others), and a desire to maintain their long-term political influence derived from oil. Allied to the above, seems to be a common characteristic in their seeming inability to develop alternative sources of income and diversify their economies away from oil.

Irrespective whether they are in Group I or II, all OPEC countries are faced by the same question: *to produce or not to produce given their capacity constraints*? The answer lies in managing three different but inter-linked perspectives—economic, political, and global energy balances in terms of quota agreements. From a political viewpoint, the problem is often to secure the amount and the kind of economic growth that is required to maintain social and political stability without dislocations caused by either rapidly rising or falling incomes, creating unequal distribution of income. This applies to both OPEC and non-OPEC oil producers, like Mexico, where research indicated that unequal distribution of income meant maximizing oil production, which in turn lead to rising social expectations (Noreng 1982, p. 199).

OPEC's Quota Setting Formalities

Before assessing the history and use of quotas and their effectiveness, it is important to highlight the mechanism by which OPEC determines such policies. OPEC has three administrative and supervisory organs: the *Conference*, the *Board of Governors*, and the *Secretariat*. The Conferences are OPEC's supreme authority and consist of delegations representing the member countries, each of which should be represented at all Conferences. A quorum of three-quarters of member countries is necessary to hold a Conference, and each country has one vote and all decisions of the Conference *require unanimous agreement of the full members* as per Article 11 of the statutes (Claes 2001, p. 143). According to OPEC, the Conference meets twice a year, but extraordinary meetings of the Conference may be convened at the request of a member country or by the Secretary-General after consultation with the President and simple majority approval. OPEC has also a Board of Governors nominated by member countries and, according to Article 25 of OPEC, the Board of Governors direct the management of OPEC affairs and in practice runs the organization between the conferences (OPEC 1990, p. 37). By implication, this leaves little autonomy to the OPEC Secretariat, represented by OPEC's Secretary-General, leading to the observation that the Secretariat's "effectiveness has depended on the quality of its staff, rather than any particular structure of the organization" (Skeet 1988, p. 237). The main organs of OPEC's organizational structure are set out in Fig. 3.4.

Fig. 3.4 Organizational structure: the main organs

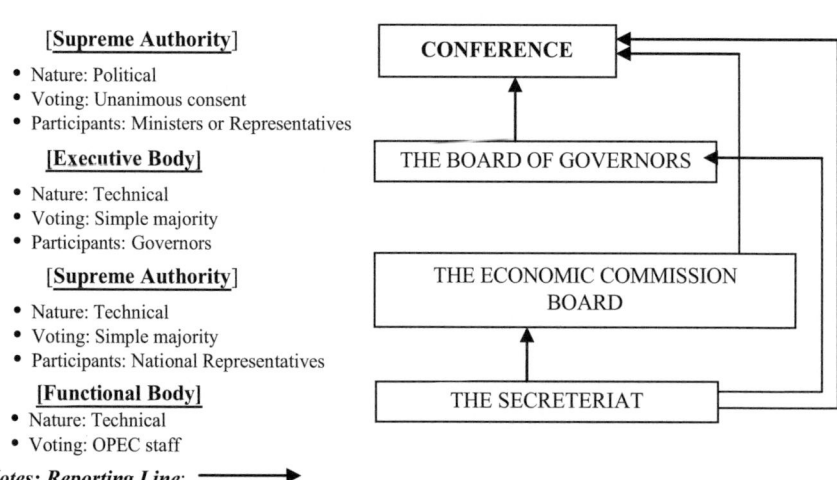

Source: Al Saif, p. 92

From Fig. 3.4, we note that OPEC has established an Advisory body called the *Economic Commission Board*. It has an important role to play as it was set up to assist OPEC decision-making in promoting stability in world oil prices by collecting economic data and reports on world oil market conditions for the Secretary-General.

The above OPEC organizational and functional hierarchical structure is underpinned by a complex voting system referred to earlier. Within OPEC, voting takes place at three levels of authority:

1. All decisions taken at the **Conference** level, *other than on procedural matters*, require the unanimous consent of all full members.
2. All decisions taken at the **Board of Governors** level require a *simple majority* vote of attending Governors (*at least two-thirds of all the members*).
3. All decisions taken at the **Economic Commission Board** level require a simple majority vote of attending national representatives (with quorum of at least two-third of all members).

Each representative of a member country, whether at the Conference, Board of Governors, or Economic Commission Board, *has only one vote, equally weighted, regardless of whether the member is a "founder" or "full" member*.

Given the above functional limitations, according to OPEC observers, the direction of OPEC is determined by what each member country does or proposes to do, leading to instances in which individual members opt to disregard decisions made by OPEC (Al Nasrawi 1985; Al Saif 1996; Chalabi 2010). Under these institutional parameters, are there any mutual beneficial functions that accrue to all OPEC members? Claes (2001, pp. 145–148) sets out six beneficial aspects for OPEC membership:

First is that OPEC provides an arena for exchange of information and that in the absence of an arena for communication, cooperation would be impossible.

Second, the provision of information about member's positions and the oil market in general is in order to reduce uncertainty in the relationship between member states.

Third, creating decision-making rules that affect the subsequent outcome of OPEC bargaining, on the basis that such decisions are the results of shared interests of the member countries.

Fourth, monitoring agreements and verifying accusations of breaking such agreements. This assumes that the more verification there is on a transparent basis, the more likely are states committed to cooperative agreements.

Fifth, framing decisions and establishing shared perceptions among OPEC members.

Sixth, establishing a collective identify among OPEC members in that they belong to a unique "club" that bestows prestige and privilege compared to nonmembers and also identifying themselves as the shield or spearhead of the developing countries of the world.

On reflection, it would be hard to disagree with most of the above beneficial aspects of OPEC membership. As analyzed later, implementing principles is often very difficult, with the possible exception of the last item, namely, accrual of some political benefit to OPEC as a whole and to some individual members' domestic and global political agendas. Venezuela is a case in point, whereby the late President *Hugo Chavez* sought to take credit for the country's economic fortunes so long as OPEC was viewed as a powerful cartel, giving him a significant political asset in Venezuelan domestic politics (Wilpert 2007, pp. 93–94). Such type of behavior sustains a *rational myth* about OPEC's influence over the world market for oil, whereby the perceived power of OPEC allows its members to reap political rewards in terms of diplomatic influence and the attention paid to members, whether they are in the "large" or "small" league (Colgan 2012, p. 12). Empirical analysis carried out indicates that OPEC membership was strongly and positively correlated with levels of diplomatic recognition than comparable non-OPEC members, with, on average, OPEC membership correlated with an increase in diplomatic representation from *nine* additional states compared to an equivalent country that is not an OPEC member. Small OPEC members like Ecuador, Angola, and Gabon are examples, although Ecuador joined OPEC in 1973, suspended its membership in 1992, and rejoined in 2007, while Gabon left OPEC in 1992 (Colgan 2012). In periods of high oil prices, being a member of OPEC was almost akin to belonging to the exclusive nuclear weapons countries' club.

Setting Quotas: Theory and Realities

On the basis that OPEC behaves like a cartel, it must carry out the following:

1. Determine a price for the group as a whole
2. Determine a production level for the group as a whole
3. Allocate output among members
4. Detect and punish cheaters
   Cremer and Isfahani (1991, p. 30)

The first two aspects relate to the cartel's external market relations, while the last two relate to internal bargaining and compliance problems. If OPEC should behave in line with the above, then the required action it has taken is simple: *it should set a fixed price based on world demand conditions and assumed non-OPEC production level and then increase production if the price rises above agreed levels and decrease production if price falls below the desired level*. Price setting and production quotas are the two instruments for OPEC. This begs two fundamental questions: (a) how does OPEC actually set the desired quota, and (b) does the set quota and member allocation system actually matter in current conditions of diminishing OPEC world energy market share?

Figure 3.5 illustrates a schematic overview of the OPEC quota negotiation process.

From the above diagram, the process of quota setting seems straightforward enough, with the OPEC Secretariat preparing an economic forecast of world energy

Fig. 3.5 OPEC quota negotiation process

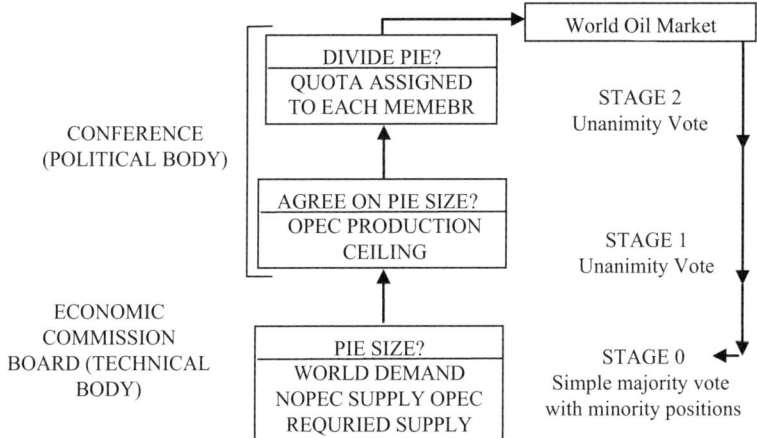

Source: Al Saif (1996)

demand, non-OPEC supply, and OPEC members' supply, which national representatives discuss at the Economic Commission Board meeting. If this expected OPEC production passes by simple majority, then the Secretariat sends it to the oil ministers to start the negotiation process to determine each member's quota for the next one- or two-quarters (stage 2), with a *unanimous* vote required to approve the ministers' decisions. A key element for the quota allocation process, in order for it to be successful, is *accurate forecasting* of global and each member's energy supply and demand conditions. OPEC forecasting is carried out by both the OPEC Secretariat and by individual OPEC members. This raises a host of issues as to the reliability and accuracy of the data used to make quota allocations. Incorrect estimations could lead to over or under quota allocation, assuming that individual OPEC members are *willing* to share their own private energy situation with other members, thus diminishing one of the beneficial functions of information sharing by OPEC members highlighted earlier.

Review of OPEC literature does not provide an answer on the existence of specific allocation formulas which can shed light on what OPEC actually uses as specific factors and their weights to reach quota allocation decisions (Chalabi 1989; Gately 1979; Mabro 1989). What transpires is that members initially agree on specific economic and social "parameters," but not their weights. These initially involve reserves, historical production growth rates, country size, population, and the proportion of oil income in government revenues, with member countries putting heaviest emphasis on the factors which produces the highest quota for it. Over time, the above factors were further refined, but again without specifying weights, especially for population, given that some OPEC countries have larger populations but smaller production or reserves than others and therefore wished to emphasize the population factor over other factors.

By 1990, two distinct "economic" and "social" parameter groupings had been proposed (Al Saif 1996). In the *economic* parameters were (a) proven reserves, (b) oil production capacity, (c) historical shares of total OPEC production, (d) historical rate of output growth, (e) production cost, (f) oil quality, (g) new discoveries, (h) expenditure on oil exploration, (i) domestic oil consumption, (j) external debt, and (k) realized price level.

The *social parameters* included (a) population, (b) area, (c) government development on certain projects, (d) proportion of income in government revenues, (e) social needs, and (f) so-called special circumstances, such as wars, civil unrest, natural disasters, etc.

Given the complexity of accurately measuring, let alone quantifying some of the "social parameters," most OPEC members disagreed on social parameters, compared to economic parameters. However, as noted earlier, most data was provided by sovereign members with no *independent audits, especially on proven reserves*. OPEC realized that the credibility of the whole quota system was at stake. In June 1987, an onsite production monitoring system was proposed but not fully implemented, and in the same year, the organization agreed on hiring an external auditor to report on member's actual reduction. This too floundered.

With all the above parameters to consider, OPEC quota allocation decisions faced several options to reach "fair" allocation. This involved either using some combination of economic parameters or a combination of economic and social parameters, or applying equal/unequal weights to selected members, or applying the allocation to only one part of the quota "pie" and allocating another portion to members facing special circumstances like war or embargo and excluding them from quota allocation, as happened with Iraq. The possible combination permutations seem endless, raising a key question: did the quota system matter and was it effective at all, even when implemented by OPEC? Table 3.3 summarizes OPEC's attempts at both production quota and price-setting regimes.

Table 3.3 OPEC quota and price-setting regimes 1965–2014

Period	Price-setting	Production quotas
1964–1966	No	Yes
1966–1973	No	No
1973–1979	Yes, fixed	No
1979–1981	No	No
1981–1982	Yes, fixed	No
1982–1985	Yes, fixed	Yes
1985–1986	No	Yes
1986–1990	Target/weak form	Yes
1990–1992	Target/weak form	Abandoned/Gulf War
1992–1997	Target/weak form	Yes
1997–2001	Target/weak form	Yes
2001–2003	Target/weak form	No
2003–2004	Target/weak form	Yes
2005–2014	Target/weak form	No

Source: Claes (2001, p. 247), Authors own research

The above table illustrates the inconsistent use of quotas, which in fact when agreed, were seldom set for more than one- or two-quarters, but have been a part of OPEC's policy option since their introduction in 1982. However, it is interesting to note that the first OPEC production quotas were put in place long before that in 1965. The reason is that because ownership and control of oil production in OPEC was in the hands of the major oil companies in 1965, the member countries could not increase their production according to the terms of the 1965 quota agreements. First Libya then Saudi Arabia withdrew from the agreement arguing that they felt the quotas allocated to them was too low and that the powerful international oil companies operating in their countries would be against these quotas and that there were many other oil producers such as Nigeria, Algeria, and Abu Dhabi who were not OPEC members yet in 1965 (Al Saif 1996, p. 30).

Price Signaling, Compliance, and Cheating

Analysts of crude oil markets have argued that OPEC has used its announcement of production changes and quotas to signal future direction in price, with such announcements allowing for oil prices to move toward OPEC's *desired price target* in anticipation of production changes (Al Hajji 2014a; Moguera et al. 2011; Cuervo 2008).

Returning to the question on whether the quota production system is now effective, the answer seems to be negative, given that key OPEC members, specifically Saudi Arabia, are no longer willing to play the "swing producer" role and have their market share eroded. Along with Kuwait and the UAE, these are the "real" players of any future OPEC quota agreement because they have the excess production capacity. From 1985/1986 Saudi Arabia has signaled its unwillingness to compensate for others' overproduction and cheating at its own expense. A strong cartel would have little cheating, as research on quota compliance or noncompliance has shown. In OPEC, cheating seems endemic as over the period 1982–2009, the organization as a whole "overproduced a staggering 96 % of the time and all but two members over-produced in more than 80 % of the time" (Colgan 2012, p. 8). This is illustrated in Table 3.4.

Table 3.4 OPEC quotas and production 1982–2009

OPEC member	% months production exceeds quota
Algeria	100
Iran	72
Iraq[a]	82
Kuwait	90
Libya	83
Nigeria	88
Qatar	90
Saudi Arabia	82
UAE	96
Venezuela	71
OPEC 9 (excl. Iraq)	96

Source: Colgan (2012, p. 9)
[a]Up to March 1998 only. Iraq was not assigned an OPEC quota after March 1998

It would seem that quota compliance can only come due to external, unplanned events as "even on the relatively rare occasion when member countries are not overproducing, the root cause is often voluntary production constraints such as strikes or accidents, rather than a conscious decision by the governments to obey OPEC quota" (Colgan 2012, pp. 8–9).

The November 2014 OPEC Conference meeting aptly demonstrates the current futility of either imposing or enforcing any meaningful quota rules. Despite pressure by other OPEC members on Saudi Arabia to cut back on its production to help boost falling prices, Iraq announced soon after the meeting that it planned to boost its crude production to 4 million bpd in 2015. This announcement was followed by an agreement between the Iraqi Federal Government and Kurdish authorities to raise Iraq's crude oil exports by 250,000 bpd, undermining compliance with OPEC's decision to maintain its production quota at 30 million barrels a day (Bloomberg, Dec. 3, 2014[c]).

This lack of compliance and futility of trying to get OPEC members to implement on what they agreed to at the highest political levels was highlighted in a remarkable anecdotal style by the Iraqi former OPEC Deputy Secretary-General *Fadhil Chalabi* in 1987, when he went on a tour of all OPEC member states with OPEC's Secretary-General to plead the case for compliance directly with OPEC Heads of State. His conclusion in 1987 might as well have been written in 2014 when he stated that "… regrettably for all the expenses involved in the 1987 'quota mission', the OPEC delegation returned absolutely empty-handed … That it would all end up as a 'mission impossible' was a foregone conclusion before we even embarked on the great expedition. Here was an example of OPEC's extravagance" (Chalabi 2010, p. 239).

Oil Market Crude Pricing

The dramatic volatility in oil prices seem to often catch laymen and professional analysts by surprise in trying to understand why changes in such a vital commodity can be large and abrupt. Following on is determining accurate price forecasts for future years, as getting oil price forecasts right is essential to the health, not only of national economies, but to individual consumers who try to adjust their consumption and saving habits accordingly. As explained later, oil markets are not only influenced by supply, demand, and inventory considerations but also by the actions of financial commodity traders. Table 3.5 sets out current oil trading mechanisms and the terms of oil trading to balance world demand and supply and achieve "optimum" pricing for both consumers and oil producers.

Producer countries' preference is for more stable and long-term oil contracts to allow for more predictable national economic planning to take place (Noreng 1978; Chalabi 1980). However, the continuous mismatch between global energy demand and supply conditions, as was apparent from late 2014, has created a fertile ground for the development of a "spot" market, characterized by short-term contracts.

Oil Market Crude Pricing

Table 3.5 Anatomy of oil trading

Type	Volume	Duration	Price	Physical condition
Term				
Standard term	Multiple cargoes	Months or years	Government-determined: Official Sales Price (OSP)	Wet
Equity	"	Years	Special arrangements	"
Market-related	"	Months	Adjustable	"
Spot				
Single spot	Single cargo	One-time basis	Set by prevailing market conditions, below or above OSP	"
Daisy chain	"	Multiple transactions of same cargo	"	"
Tertiary	Multiple cargoes	Weeks or months	Discounted below OSP ($1–2/barrel)	"
Netbacks	Multiple cargoes	"	Spot product price-related	"
Forwards	Single-cargo	One-time basis	Set by expectations of future market conditions	"
Futures	Less relevant	Less relevant	"	Paper
Derivatives Swaps	"	"	Set to reduce price	"
Hedging	"	"	"	"

Source: Claes (2001, p. 79)

Under increasingly competitive market pressure, oil producers' end up adopting pricing formulas linked to day-to-day fluctuations in oil prices and discounting below government determined *Official Sales Prices* (*OSP*). This occurred following the November 27, 2014, OPEC Conference meeting which failed to agree on production cuts. Saudi Arabia, and Iraq following, cut their OSP to Asian buyers to try to maintain market share. Under such conditions, individual producers, whether in OPEC or outside OPEC, have no guarantee concerning the long-term loyalty of their customers who now happily find themselves as "price makers" instead of "price takers" in periods of supply shortfall, similar to the conditions following the 1973 Arab oil embargo. The current situation faced by OPEC confirms the feeling expressed in earlier eras, especially during periods of sharp price falls, that OPEC's major actions on prices were no more than a series of reactions to *external events* which took place and had the effect of reshaping the oil market before OPEC as a group could decide to undertake any real change (Chalabi 1980). According to this belief, OPEC's actions and policies did not reflect a strategic thinking toward setting long-term common objectives for member countries. Trying to implement a quota system to achieve such a common objective and the dismal results achieved as analyzed earlier attests to this lack of strategic thinking.

The question therefore boils down to the following: will OPEC continue to play an effective role in the world's price regime, or will the market completely take over, as the most recent statements coming out of leading OPEC countries like Saudi Arabia and the UAE seem to indicate? (Arab New 2014[a]). Even if OPEC's major players are saying this to explain weaker oil prices, yet every OPEC member, whether those in the "top" tier or the "second" tier, understands very clearly what is at stake for them. Their ability to administer oil prices in the past, *albeit* for certain periods, yielded all OPEC members considerable financial benefits and political rewards in terms of diplomatic recognition well above their standing. A surrender of this oil price administering role to "anonymous market forces" might deprive some low reserve endowed OPEC members of international "free-rider" gains in the future. This was illustrated earlier, with a seeming "revolving door" OPEC membership policy adopted by some oil producers, depending on their perception on whether OPEC commanded high or low prestige on the world's stage.

Not All Oil Is the Same

All crude oils are not the same, with some being light fluids versus near solids at room temperature, varying in color from brown to black and containing varying quantities of impurities. Because of this differentiation, the commercial products that each type of oil yields can be different and can command different market prices. Crude oil also contains varying amounts of non-hydrocarbon impurities such as sulfur, nitrogen, and oxygen, and different quantities of these not only affect the quality and price of oil, but can be hazardous. The quality of oil also affects the type of refinery technology and investment needed for different types of crude, with the more modern refineries able to blend and refine different grades of oil compared to older, single blend refineries. In general, the oil industry labels oil with less than 0.5 % sulfur *sweet,* while oil with more than 0.5 % sulfur as *sour*. Given this differentiation, it is not correct to classify all OPEC members as producing a homogeneous product, as some countries predominantly produce "sweet" crude/such as Nigeria, Libya, Iraq, and Algeria, while others generally produce "sour" like Saudi Arabia, Venezuela, Ecuador, and Kuwait, although some produce quantities of both types of crude, like Saudi Arabia with its more recent "sweet" finds in the *Shaibah Rub Al Khali* fields. According to latest estimates, over 160 different crude oils are traded globally, and these are generally compared in terms of their quality and price by using three common "benchmarks" from Texas, the North Sea, and OPEC itself (Gorelick 2010). These benchmarks are graded as follows:

- *West Texas Intermediate* (*WTI*) is sweet light crude and has 0.24 % sulfur and is refined to produce a large proportion of gasoline.
- *Brent Blend* reflects 15 oil fields in the North Sea and is also a sweet light crude, but with 0.37 % sulfur.

- The *OPEC Dubai basket* is an average of OPEC nations oil (except Angola) and ranges from heavier sour oil originating in the Gulf like Abu Dhabi/Dubai, Saudi Arabia, to lighter quality Algeria oil. It is high in sulfur at 1.8 %.

In general, despite an average price premium for *WTI* over the OPEC basket of around 10 % over the last decade, and *Brent* at between 5 and 10 % premium over OPEC, the prices of the oils of various qualities track together the ups and downs of the global market (Fattouh 2006; Gorelick 2010).

Futures, Hedgers, and Speculators

The futures market is meant to provide those trading in the physical commodity market with the opportunity to offset business risks and reduce price uncertainty and the possibility of arbitrage. Some even regard the futures market as an alternative form of contracting in the oil market (Sykuta 1994). Others see the role of the futures market in reducing buyer's costs by assuring access to a commodity, and likewise producers' assured access to market outlets (Claes 2001; Roeber 1994). As Table 3.5 illustrated, futures are deemed *paper* transactions as opposed to *wet* commodity physical transactions. None of the OPEC countries are actually involved in futures trading, thus depriving them of information and possibilities to understand the formation of oil pricing (Claes 2001, p. 81). This makes it more complicated for OPEC to influence oil price movements.

Despite not participating in the futures market, OPEC members use futures prices to derive the price of their *reference* crude and try to signal their preference for certain price ranges as discussed earlier, whether this signal is in a "strong" or "weak" form. Instead of using spot or dated *Brent*, some major OPEC countries such as Saudi Arabia, Kuwait, and Iran rely on the *IPE Brent weighted average* (*BWAVE*) as the basis for pricing crude exports to Europe, with the *BWAVE* representing the weighted average of all futures exchange (IPE) during a trading day (Fattouh 2006, p. 68). However, unlike long-term contracts and spot transactions, the main purpose of the futures market is not to provide a mechanism for actual delivery, but rather a mechanism that allows market participants to spread the risk of price volatility. The volumes traded under these contracts can be huge, far exceeding the actual volume of oil production with the Senate Committee on Homeland Security noting that of 5 billion barrels being traded over a 7-year period on the futures market, only 31,000 barrels were actually delivered (Fattouh 2006, p. 69). Who are the players then in the futures market and what power do they really hold on oil prices?

Table 3.6 shows the ownership structure of outstanding *NYMEX* crude oil futures contracts by sector for 10 months during 2000.

Table 3.6 Ownership of outstanding NYMEX crude oil futures contracts by market parties (January–October 2000)

Oil traders	40 %
Financial institutions	13 %
Integrated oil companies	13 %
Refiners	9 %
Investors	7 %
Investment funds	5 %
Floor traders	5 %
Marketers	5 %
Producers	1 %
End users	1 %

Source: Fattouh (2006, p. 70) from the Senate Committee on Homeland Security & *Government Affairs* downloaded from http://www.senate.gov/~gov_affairs/psisec3pricingofcude.pdf

The above table illustrates a wide diversity of players in the futures market, including financial institutions, investment funds, and oil traders, with end users representing only 1 %. While some of these contracts are legitimate hedging against future price movements, there is a feeling among some senior OPEC officials that such activity is merely speculation initiated by market "manipulators" (Yergin 2011). Following the price crash of late 2014, Saudi Oil Minister *Ali Al Naimi* stated that he was not surprised by the extent of the price drops as "we knew the price would go down because there are investors and speculators whose job it is to push it up or down to make money" (Arab News, Dec. 24, 2014[c]).

It is not only OPEC ministers who feel that investors have played a role in global oil price volatility but those who closely follow events in the oil market. Currie et al. differentiated between two primary types of investors, "speculators" or "active" investors including hedge funds and swap dealers, and "index" or "passive" investors comprised of pension funds, endowments, and other real money investors (Currie et al. 2010). According to this analysis, index or passive investors have little, if any, impact on prices because such investors pursue commodity allocations for strategic diversification and are paid to take risk off the balance sheet of commodity producers. Speculators, however, bring fundamental views and information to the market, thus impacting physical supply management and facilitating price movements, with prices rising when speculators buy and falling when speculators sell. Minister Naimi's feelings about speculators are somewhat confirmed when new data suggests that speculators increased the price of oil by $9.50 per barrel on average during the 2008 price rise, to reach nearly $140 per barrel (Currie et al. 2010, p. 7).

Equilibrium Price of Oil: In Search of the Holy Grail

The massive swings in oil prices over the past few years, notably 2008/2009 and 2013/2014, raise questions on whether it is possible to cut through the volatility and identify an equilibrium price of oil to ensure some more predictability to the oil

Equilibrium Price of Oil: In Search of the Holy Grail

market. Commodity prices adjust to expectations, but expectation models often fail the empirical test. One still needs conceptual frameworks to better understand complex pricing behavior which, in the energy sector, can influence investment decisions on whether to add future production capacity or not (Aissaoui 2014[a, b]). Such research involves understanding the composition of a commodity's forward price curve, which tries to capture the series of sequential prices for future delivery. Figure 3.6 illustrates the different components of a commodity price curve to explain whether prices are entering "soft" or "firm" future price trends.

Fig. 3.6 Decomposition of a commodity forward curve

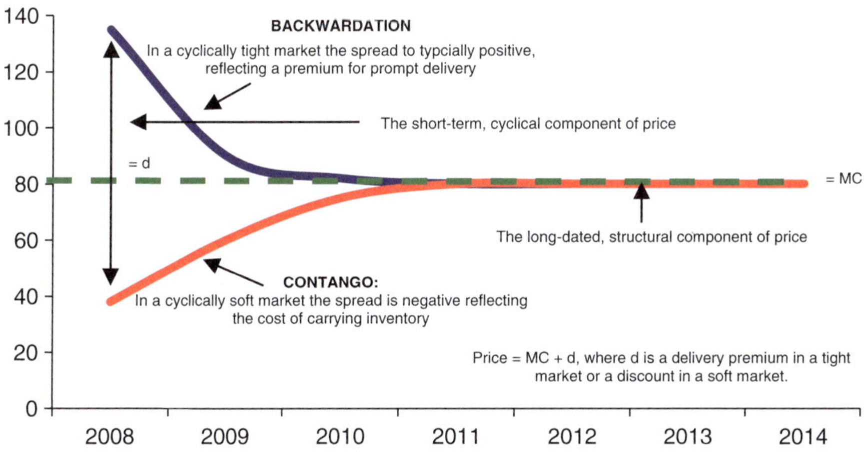

Source: Currie et al. (2010, p. 10)

In their work on commodity pricing, Currie et al. stated that, on balance, the key to commodity price movements is marginal cost and inventories. As such, the forward curve, as in Fig. 3.6, is discomposed into a short-term cyclical component and a long-term structural component with the former driven by fluctuations in inventory levels, while the latter is determined by the cost of bringing the last or marginal— needed unit of the commodity output to the market. The dynamics of oil prices are then observed in the shape of the forward curve in the above figure, with *backwardation* due to tight market conditions and a price premium and *contango* occurring in a "soft" market characterized by discounts to offset the cost of carrying inventories. In the case of *backwardation* the cost of adding the one more unit—the marginal cost of production—is driven by geology, technology, and politics and is usually of a long-term nature, as supply is generally slower to adjust than demand, given the capital and time-intensive nature of production investments. This contrasts with the shorter cyclical component of price movements adjusting to changes in demand.

For OPEC, the preferred position is for low inventories and steep *backwardation* allowing for premium prices to be charged, but it now finds itself on *contango* due to high inventories from non-OPEC marginal producers. However, war premiums and risk premiums of all kinds can reverse this situation to the benefit of OPEC as illustrated by the small rally in oil prices on news of the escalating conflict between different Libyan groups that damaged the country's oil storage tanks capable of holding four times Libya's daily production. Brent futures rose as much as 1.6 % on this news (Bloomberg, Dec. 29, 2014[d]). However, such modest price reliefs cannot hide the situation that OPEC is now facing a long-term *contango* market. Only a confluence of uncertainty surrounding oil technology, political, and economic factors that might push up "marginal cost" and cut back on existing or new production capacity by non-OPEC members can assist the organization in the long run.

Chapter 4
Non-OPEC Producers, the Ever Fading "Peak Oil," and the Rise of the USA

The old order changeth, yielding place to the new.

Tennyson

Energy "Independence": Weaning Consumers Off OPEC

Since the momentous era of the 1970s when oil prices rose due to geopolitical factors such as the Arab oil embargo and more assertive national oil policies, the control of what was deemed to be limited energy resources and access to a diverse supply of oil became essential to major oil consumers like the USA, China, Europe, and Japan. While each of the above consumer countries or blocs reacted in different ways to ensure both sufficient energy flows and energy "independency," access and security of oil supply became vital elements to national security, with the USA exemplifying this policy (Cuervo 2008). However, energy "security" was also important for China, the world's second largest consumer nation, as well as for Japan and Western Europe. Gradually the issue of energy security became a global defense security concern as illustrated by tensions between countries sharing disputed mineral-rich maritime zones, like China and Japan, China and Vietnam, Russia and the Nordic countries/Canada/USA, and the UK and Argentina. More recent maritime tensions and disputes involve Israel and Lebanon/Palestine, and Turkey and Cyprus in the Mediterranean over gas field exploration rights.

It is not only access to future energy sources that is of concern to consumer nations, but that the economic growth of Europe, China, India, Japan, and the USA also depends on a *sustained* flow of crude oil at *stable* prices. Earlier in this volume, we noted the asymmetrical relations between producer and consumer nations when oil prices change, and while there is a strong desire on both sides to establish what they feel to be a "fair" oil price, more recent large fluctuations in prices have caused massive readjustments in the fiscal balances of both energy consumer and producer nations leading to government budgetary planning uncertainties.

The USA has been the most vocal in setting out an energy independence policy since *project independence* was launched in 1973 with the promise of ending energy dependency on foreign countries. This was first espoused by President *Richard Nixon* in his Address to the Nation on January 30, 1974, when he stated that "… let this be our national goal: at the end of this decade, in the year 1980, the United States will not be dependent on any other country for the energy we need" (Nixon 1974). On December 22, 1975, the USA enacted the *Energy Policy and Conservation Act* which created a 1-billion barrel strategic petroleum reserve, and President Nixon's commitment was reiterated by his successors such as *Gerald Ford, Jimmy Carter,* and *George Bush Jr.* who famously declared in 2006 that he would wean the USA away from being "addicted to oil" (Cuervo 2008, p. 5). The use of such negative and emotive words was not well taken by oil producers, especially Saudi Arabia, who see themselves as suppliers of a commodity that helps in nation's economic prosperity and should not be correlated with harmful substance addiction (Al Kadiri 2010). Based on data provided by the National Petroleum Council, the total productive capacity of the USA fell from 12.3 million bpd in 1968 to an effective zero in 1972, with industry analysts usually citing 1971 as the approximate date when full capacity utilization was achieved in the USA. According to Vietor, the 1973 oil price rises were due to several coalescing factors, with OPEC merely the catalyst, as depletion of easily accessible oil and gas deposits in the USA, masked by the drawdown of reserve inventories, was the real cause (Vietor 1984).

According to the US EIA, the SPR held 691 million barrels of both "sweet" and "sour" crude (in ratio of 40/60 %) in 2015, representing 137 days of import protection, well above the statutory 90 days mandated on the EIA by Congress. The maximum SPR reserves held were 727 million barrels on December 27, 2009; the end of year stocks held by the US SPR is illustrated in Fig. 4.1.

Fig. 4.1 End-of-year stocks in SPR

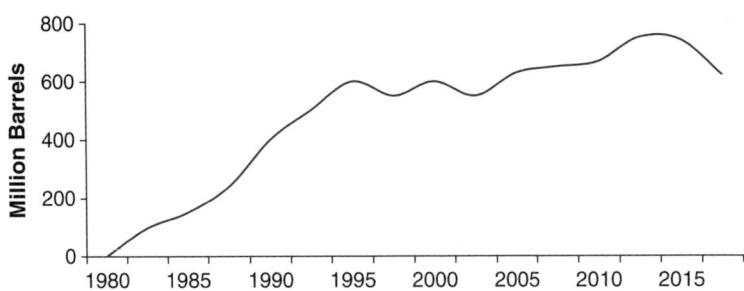

Source: US Energy Information Administration, Annual Energy Review (2011, p. 158), EIA (2015)

Similarly, President *Barack Obama*, both during his first presidential campaign and as President, has repeatedly emphasized the importance of significantly reducing US oil imports. In his speech accepting the Democratic Presidential nomination in August 2008, he promised to end US imports from the Middle East within 10 years (Al Kadiri 2010, p. 93). However, as some pointed out, energy independence is misconstrued and should not be taken as literally import-free but rather "not vulnerable" (Yergin 2011).

Given stated political sentiments, it is not surprising that the petroleum industry in the twenty-first century started to focus on production of oil and gas from unconventional sources such as heavy oils, tar sands, oil shale, and other renewables in order to ensure some viable and possible long-term energy independency. The viability of pursuing such strategies especially from the so-called shale oil "revolution" will be addressed later.

Can the world ever be "weaned" away from fossil fuel? The answer to this is that it is only possible if society at large opts to change its current way of life. This might seem impossible to accomplish very quickly, but there are indications that the global debate on the effects of climate change could bring about such changes, with compulsory home and workplace energy efficiency being the first step. The topic will be addressed in a later chapter, but whatever the world finally decides on the merits of using fossil fuel, it will involve a concerted effort by all parties—consumers, producers, and manufacturers.

"Peak Oil": The End of the Oil Era?

That the world must run out of oil someday seems obvious as there is a finite amount of oil on Earth. Assuming current oil consumption patterns and the fact that every developing nation relies on oil as a major input for its economic growth, especially the emerging giants like China and India, it stands to reason that oil demand will outstrip supply and the world reaches a "peak" in its production, with supply rapidly diminishing thereafter. The big debate concerning "peak oil" theorists is *when* such a peak will occur. Some view this as being unduly too pessimistic and argue that the world will not run out of oil at any time. This was put succinctly by the former Saudi Oil Minister Zaki Yamani when he said in 2000 that "thirty years from now there will be a huge amount of oil and no buyers. Oil will be left in the ground. The Stone Age came to an end, not because we had a lack of stones, and the oil age will come to an end not because we have a lack of oil" (Fagan 2000).

"Peak Oil" on the Back Burner?

"Peak oil," or *Hubbert's Peak,* is inextricably linked due to the publication of *Marion King Hubbert* in his seminal work that argued that world oil output was currently at or near the highest level it will ever reach and that the point of decline was nearing (Hubbert 1971). Hubbert used a statistical approach to project the kind of oil decline curve that oil fields and, by extension, countries will encounter and his research mostly concentrated on the USA. According to some who supported Hubbert's theory, the main appeal of his approach is that it represents a type of mass balance, and that given a fixed volume of oil, how long it will last will depend on its rate of extraction and consumption, with the latter being the main driving force. As long as global oil is plentiful, the effects of exponentially increasing production are not detrimental because there is ample supply to meet demand. However, as peak oil is approached, demand will overtake the ability to extract oil. As such, intuitively and mathematically, the future decline shown by Hubbert's curve, as illustrated in Fig. 4.2, appears to be the natural and inevitable result of the conflict between demand and a fixed, finite endowment (Gorelick 2010, p. 10).

Fig. 4.2 Hubbert's US 1956 prediction under different endowment scenarios

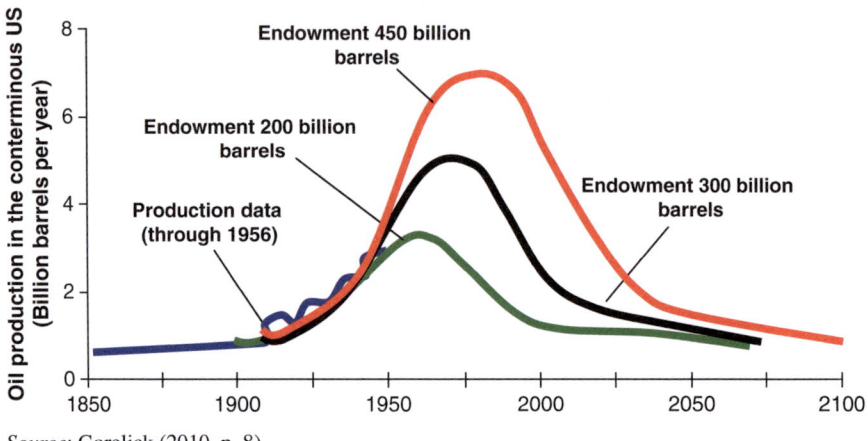

Source: Gorelick (2010, p. 8)

In the above figure, the curves are based on Hubbert's approach that matched US production data through 1956 but assuming oil endowments of 200, 300, and 450 billion barrels. The estimated US peak oil production occurred in 1971, 1981, and 1990, respectively. Between 1972 and 1976, Hubbert went on to extend his analysis to global oil depletion, with peak oil occurring in 1995, 1996, and 2000, by applying global oil endowments ranging from 1.35 to 2.1 trillion barrels. According to some analysts, Hubbert greatly underestimated the amount of oil that would be found in the USA and they highlighted that by 2010, US production was four times higher than what Hubbert had estimated at 5.9 million barrels per day vs. Hubbert's 1971 estimate of around 1.5 million barrels per day (Yergin 2011). By November 2014, US oil production stood at 9.02 million barrels per day, approaching the peak of 9.13 million barrels per day reached in September 1971 (Energy Information Administration, USA 2014[a]).

Was Hubbert Wrong?

Those analyzing Hubbert's "peak theory" forecasts point out to some key elements that were missed out by Hubbert, namely, the pace of technological progress and oil prices, and that his analysis was very much a static view of the world and the pace of change. Others also pointed out that Hubbert assumed there was a fairly accurate estimate of ultimately recoverable resources when in fact this was a constantly moving target, but the harshest criticism was laid concerning oil price factors. As some pointed out, energy activity goes up when prices go up and activity goes down when prices go down and that higher prices stimulate innovation and increase supply (Yergin 2011).

However, there are others who point out that global oil depletion has taken place despite high oil prices and technological advances and support this argument with what happened to oil-producing countries whose production "peaked" and are now in decline (Deffeyes 2001). This is illustrated in Fig. 4.3 for various oil producers, including Indonesia which was a member of OPEC, but left the organization in 2008 after a sharp reduction in its output. By 2010 Indonesia was importing oil to meet domestic consumption.

Fig. 4.3 Oil production declines in some selected producer countries

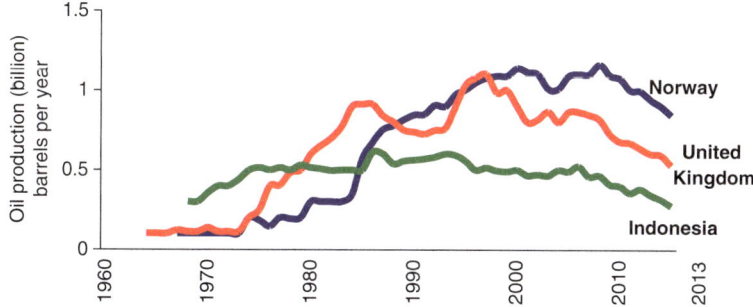

Source: BP Statistical Review (2014, 2015), Gorelick (2010)

While it was countries like the UK and Norway that contributed to global energy marginal supplies from the late 1970s when oil prices rose as explored in earlier chapters, it is the decline in the number of so-called "giant" oil fields and the oil volume over each decade of discovery from these "giants" that is pointed out to vindicate Hubbert's theories. This is illustrated in the figure that follows.

Fig. 4.4 Number of "giant" oil fields containing over 0.5 billion barrels (1860s–2000s)

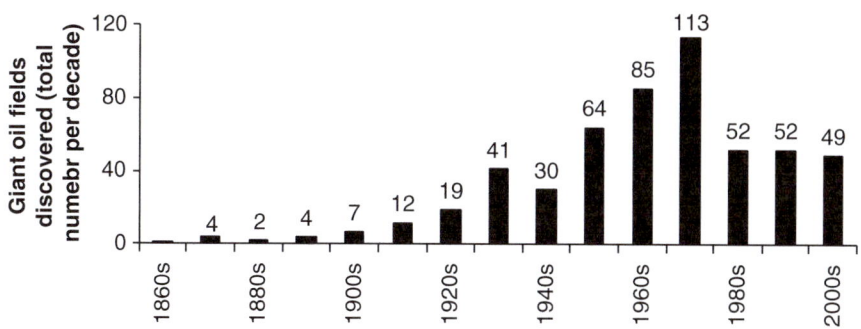

Source: Horn (2006, 2009)

From Fig. 4.4, it would be seen that the discovery of giant oil fields peaked in the 1970s and has continued to decline since then, and as Fig. 4.5 illustrates, the peak in volume of discovered oil in these giant fields seemed to have occurred before the peak in the number of new fields, reflecting the fact that the "giants" were discovered first (Horn 2009).

Fig. 4.5 Oil volume production in "giant" fields (1980s–2000s)

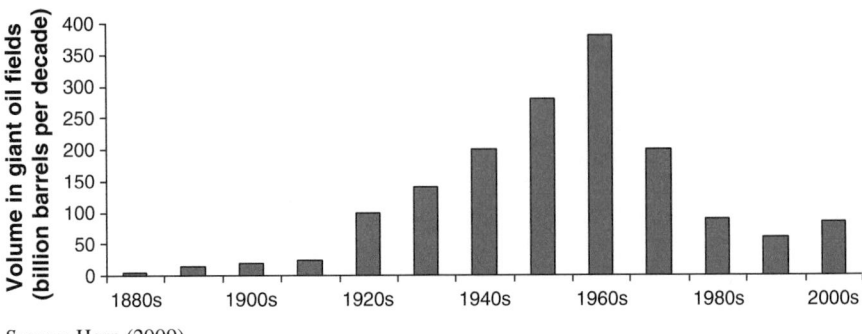

Source: Horn (2009)

Some researchers have also highlighted that production from a number of the "giants" has also declined and that compared with three decades ago when there were 15 oil fields producing more than 1 million barrels per day, by the late 2000s there were only four such fields—*Ghawar* (Saudi Arabia), the world's largest oil field with 5 million barrels per day capacity; *Kirkuk* (Iraq), discovered in 1927 still produces up to 1 million barrels per day almost half of all Iraqi oil exports; *Greater Burgan* (Kuwait), the world's second largest oil field with production capacity of 1.7 million barrels per day; and *Cantarell* (Mexico), the largest offshore development project in the world, with a total installed cost of more than $5bn and capacity of 1 million barrels a day. Of particular interest has been the Saudi giant *Ghawar*, the largest oil field in the world, accounting for over half of Saudi production and raising questions about its sustainability (Simmons 2006). The concerns about *Ghawar* are that to maintain production of around 5 million barrels per day, water is injected around the perimeter of the field to sweep the oil toward the production wells and that there is the possibility of some of this water making its way into the production wells, contaminating the oil produced. This, as some have pointed out, is not a problem as long as the so-called *water cut* or the traction of water to total liquids produced is controlled, but also point out that the "water cut" rate at *Ghawar* has been rising steadily from about 25 % to over 36 % in a short period of time from 1993 to 1999 (Nasser and Sabri 2004). Other analysts discount that *Ghawar* is in a steep decline and point to continued investment in this "super-giant" field, which according to former Saudi Aramco CEO Khalid Al Falih still remains "robust in middle age" due to the application of new technologies that continue to unlock Saudi resources (Yergin 2011, p. 238).

Are International Reserves Overinflated?

A key problem with analyzing international reserve statistics is that they are self-reported, especially from countries that have total control over national energy resources. According to some, estimates of national oil reserves "can and have been manipulated" and may overstate the amount of ready recoverable oil with three main reasons being cited. First, that geologists and geophysicists have historically been unable to provide tight estimates of reserves for many oil fields. Second, reserves estimated by international oil companies could enhance a company's competitive advantage or boost its stock price. Sometimes overexaggerating reserves come with a heavy price as illustrated when Royal Dutch Shell agreed to pay the US Securities and Exchange Commission a fine of $120 million over Shell's exaggerated reserves, which the company had to reduce by 4.5 billion barrels or 23 % of its booked reserves in 2002 (Gorelick 2010). Third, oil-producing countries could have their own motivation for claiming exaggerated reserves, either in gaining political support and protection from superpowers or, in the case of OPEC countries, to increase their export quotas based on both production and reserve values (Gorelick 2010, p. 67). The figure below illustrates the sharp upward revisions of oil reserves of some leading OPEC members.

Fig. 4.6 Revisions in oil reserves of key OPEC members

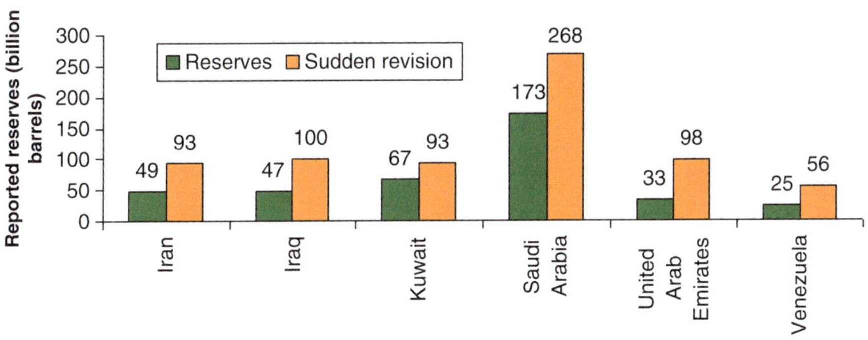

Country	Reserves	Sudden revision
Iran	49	93
Iraq	47	100
Kuwait	67	93
Saudi Arabia	173	268
United Arab Emirates	33	98
Venezuela	25	56

Source: Gorelick (2010, p. 68)

During the period 1988–1990, the six OPEC members as highlighted in Fig. 4.6 (Iran, Iraq, Kuwait, Saudi Arabia, UAE, and Venezuela) increased their collective reserves by 77 % or by 304 billion barrels, but that surprisingly, with the exception of Iran (which increased its reserves by 40 % in 2004), none of these counties has materially revised its reserve estimates since 1990 (Gorelick 2010).

Technology to the Rescue

For those in the energy sector, technology is key to the continuing success of enhanced oil exploration and production. The further advancements in technology are essential to meet future energy needs at competitive prices. While traditionally more and more oil has been found by drilling deeper, it has been the use of new technology in more challenging geological environments that has turned dormant energy resources into *viable* reserves. The exploration and production sector relies on a variety of technologies to meet its goals and the most widely used today are seismic interpretation, drilling (vertical and horizontal), logging, and stimulation, testing, and modeling. Many of these techniques, especially seismic (from 2D to 4D), were not available only several decades ago, but technological advances have made them accessible to many organizations today at competitive prices. While new technology might be very expensive, costs generally decline over time and the "*old*" new technology is eventually replaced by more advanced "*new*" new technology such as the 4D seismic which replaced 3D seismic interpretation of geological reserves without actually having to drill for oil. This expanded the industry's knowledge on reservoir management, by increasing proven reserves, upping recovery, and optimizing well locations. Such technological advances make it possible to identify and exploit additional production from the further expansion of existing proven reserves, *enhanced oil recovery* (*EOR*), unconventional sources of oil, and new discoveries both on land and offshore. This is where the confluence of technology and price meets to justify such expenditure to move lower on the pyramid resource base as illustrated in the figure below.

Fig. 4.7 Energy production: pyramid resource structure

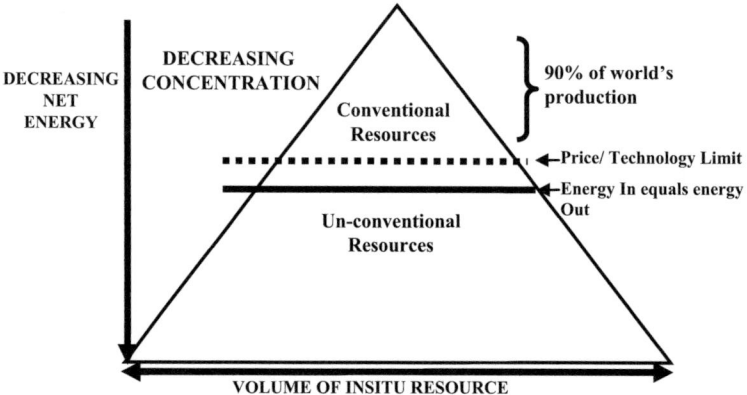

Source: Hughes (2013)

The above figure illustrates the relationship of "in situ" resource volumes to the distribution of conventional and unconventional accumulations and the generally declining net energy as well as the increasing difficulty of extraction as volumes increase lower in the pyramid. In essence, as one moves *lower* in the pyramid resource, volumes increase but quality decreases as hydrocarbons become more dispersed and more difficult to extract, and hence the energy required extracting them also increases. The dashed lines in Fig. 4.7 represent the transition from high-quality/low-cost conventional resources and lower-quality/higher-cost unconventional resources. The price/technology line reflects the fact that as prices rise, higher-cost but lower-quality resources become available. Technology once again becomes a key driver as explained earlier, and multistage hydraulic fracturing has made previously inaccessible resources such as shale oil now accessible (International Energy Administration 2014).

Barriers to Renewable Energy Technology Deployment

As noted above, the costs of many of the new renewable energy technologies have often been a major barrier to more widespread market deployment, because they have not been economically competitive with "traditional" fossil fuel energy sources

(Gould 2006, International Energy Administration 2013ª). There are a host of barriers for nonconventional energy technology to overcome before the effectiveness of their deployment can be realized. Figure 4.8 illustrates a few of these major barriers, some of which exist in some countries but not in others depending on the country's regulatory and political structure.

Fig. 4.8 Barriers to nonconventional energy technology deployment

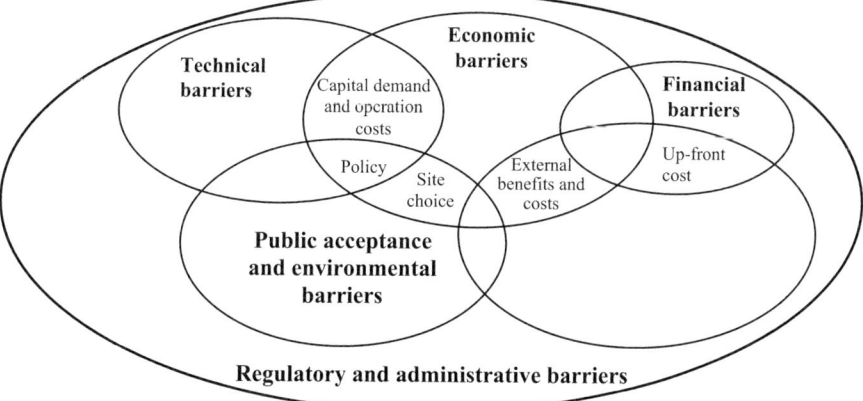

The above figure highlights both *economic* barriers and *noneconomic barriers* to energy technology deployment. In the former, if the cost of a given technology is above the cost of competing alternatives, then an economic barrier is judged to be present. Under "noneconomic" barriers are a host of often interlinked factors, some of which are addressed below. These include:

- *Regulatory and policy uncertainty barriers*: These relate to policy discontinuity and insufficient transparency of policies and legislation. As discussed later in this chapter, the boom in US shale oil and gas production has taken place in that country largely because of legal and regulatory certainties which are missing in other countries.
- *Public acceptance and environmental barriers*: These can become serious barriers, and new technology companies and those exploiting nonconventional energy have to master the art of communicating both the benefits and any social cost spillover to the general public to ensure acceptance.
- *Financial barriers*: These are associated with an absence of adequate funding opportunities for new technology breakthrough. The nonconventional energy players in the USA have been able to access substantial venture capital funds as well as raising debt through equity and debt instruments including higher risk "MEZZANINE" financing, through the US financial markets which is sometimes not available in other countries.

Besides the above, there are other barriers including the lack of sufficient numbers of skilled workers to use these new technologies, again something the USA has an advantage over other countries, as well as institutional and administrative barriers such as complicated, slow and nontransparent permitting procedures to allow the deployment of new nonconventional technology.

What Is Unconventional Oil and Gas?

There are many types of unconventional energy sources in the world. These can be broadly classified into two major groups: primary energy production and those that are converted using one fuel to produce another as illustrated below:

- Oil sands and tar sands oil
- Heavy oil
- Enhanced oil recovery (EOR)
- Oil shale
- Shale oil
- Shale gas
- Coal bed methane

Primary Energy Production

- Methane hydrates

- Coal to liquids
- Gas to liquids
- Bio fuels

Fuel conversion: one fuel into another with high energy cost

In the fuel conversion stage, there is perceived to be an upgrading of fuel quality. In geological terms, unconventional resources are defined as both conventional oil and gas trapped in shale, sandstones, siltstones, and carbonates where the permeability of the host rocks is below 0.1 *milliDarcy* (*mD*). Unconventional resources consist of shale gas, shale oil, tight gas, coal bed methane, and gas hydrates. The figure below diagrammatically sets out the technological complexity level necessary to develop unconventional resources (Fig. 4.9).

Fig. 4.9 Gas resource triangles

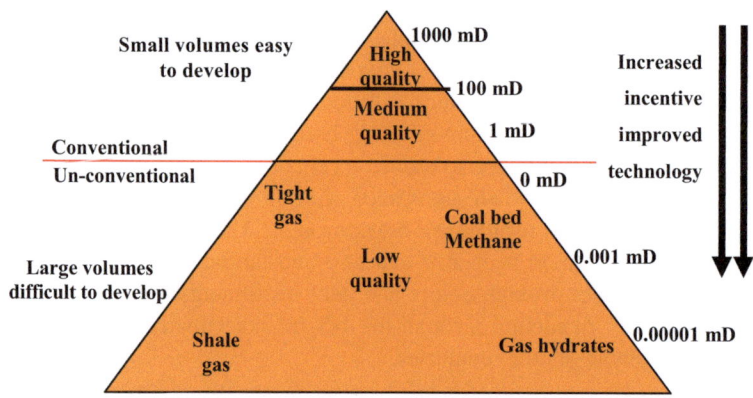

Source: Majid (2014, p. 3)

There are three stages of oil recovery, which in essence try to "squeeze" more oil from the ground given the different geological features noted above. The three stages are as follows:

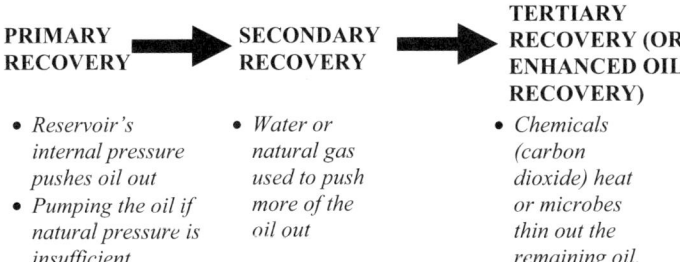

What are the characteristics of the primary energy production resources listed earlier and how significant are these estimated resources compared to conventional oil resources? The following is a brief analysis of the major resources.

- *Oil sands and tar sands*: Oil sands consist of immobile, semisolid heavy bitumen (a tar-like substance) in combination with water, sand, and clay, with bitumen comprising 1–20 % of oil sands. When near the surface, oil sands are recovered by mining, but when they are too deep to mine, the bitumen is mobilized by introducing steam or solvents and the bitumen is pumped from the ground. Canada has the largest global oil sands reserves, estimated at around 174 billion barrels of oil, but it is generally acknowledged that mining of these oil sands is expensive (between $60 and $90 per barrel) and has significant environmental impacts (CO_2 and water contamination).
- *Heavy oil*: Oil from these deposits is too viscous to extract by simple conventional technology, but progress has been made using steam injection. In its World Energy Outlook 2012 report, the IEA estimates that about 1700 billion barrels of extra heavy oil can be technically recoverable worldwide, with South America, principally Venezuela, a member of OPEC, holding major reserves of around 1.2–1.36 trillion barrels of reserves (EIA 2013[b]).
- *Shale oil or "tight" oil production*: The presence of hydrocarbons in deep impermeable shale formations has been known for a long time, but it is only in the last 15–20 years that first for gas and then for oil that US entrepreneurs combined hydraulic "fracturing" and horizontal drilling to extract hydrocarbons. Tight oil production has grown impressively and made up about 20 % of US oil production or 2.3 million bpd in 2013.
- *Shale gas*: As early as the late 1990s, the hydrocarbon industry recognized that the USA possessed many formations from which gas could be extracted (like tight oil) through "fracking" and US production of gas from shale formations started earlier and grew much faster than oil production from tight oil formations. According to the Energy Information Agency, the total remaining reserves and underdeveloped shale gas resources in 2013 amounted to 1161 trillion cubic feet in the USA (EIA 2013[a]).

With such a variety of nonconventional resources available, should OPEC be concerned at this looming competition?

How Much Global Energy Resources Are Available?

To answer the above question, one must assess global oil resource for both conventional and nonconventional resources irrespective of the source. Table 4.1 sets out estimated global oil reserves for both conventional and nonconventional resources, while Fig. 4.10 estimates the relative price ranges necessary for the profitable extraction of these various forms of energy sources.

Table 4.1 Estimated global initial oil-in-place resources (trillion barrels)

	Middle East OPEC	Other OPEC	USA	Other non-OPEC	Global total beyond 2030
Conventional oil and condensate	2.6	2.6	0.9	2.9	9.0
Natural gas plant liquids	0.3	0.3	0.2	0.4	1.2
Heavy oil	0.0	2.3	0.0	0.0	2.3
Oil sands	0.0	0.0	0.0	2.4	2.4
Shale oil	0.0	0.0	2.1	0.7	2.8
Source rock	0.9	0.9	0.3	1.0	3.1
Total (corrected for rounding)	*3.8*	*6.0*	*3.4*	*7.4*	*20.6*

Source: EIA (2013a), Sweetnam (2008)

From the above table, it is apparent that the combined Middle East and non-Middle East OPEC countries are still a force to be reckoned with in the future, as they hold substantial conventional and heavy oil, the latter primarily located in Venezuela. The non-OPEC producers dominate in nonconventional resources such

Fig. 4.10 Relative price ranges (in 2007 US$) for profitable extraction of energy resources

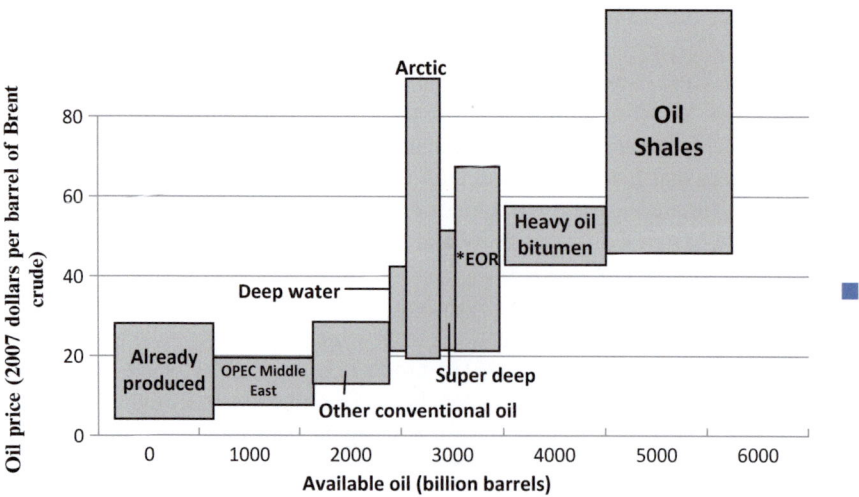

Source: Gorelick (2010, p. 164). *EOR* enhanced oil recovery

as oil sands and shale oil. However, as discussed earlier, the confluence of both technology and price is important to ensure that these forecasted global reserves come onto the market, and Fig. 4.10 examines the availability of different types of resource reserves at different price ranges.

As the above figure illustrates, greater quantities of both conventional and nonconventional oils can only be profitably recovered as prices increase in real term taking into account cost inflation. Assuming a constant price range at or above $80 per barrel in 2007 $ terms, then it is predicted that over 5.5 trillion barrels of oil can be profitably recovered. What Fig. 4.10 also highlights is the relative cost production advantage that Middle East OPEC members have over other high-cost nonconventional resource producers, but that EOR has an important role to play in extracting resources at lower oil prices due to technology advances in this field, compared to the more expensive deepwater, arctic, and oil shales.

With oil prices losing over half of their value during the latter part of 2014 to around $50 dollars a barrel and remaining weak during 2015, such price developments, if sustained, will put high-cost reserve extraction under pressure, as well as on the more financially stressed OPEC members. Despite uncertain oil prices from 2014, the amount of proved conventional oil reserves continued to rise, with the Middle East region accounting for the lion's share, as illustrated in Fig. 4.11.

Fig. 4.11 Distribution of proved reserves in 1993 and 2013 (%)

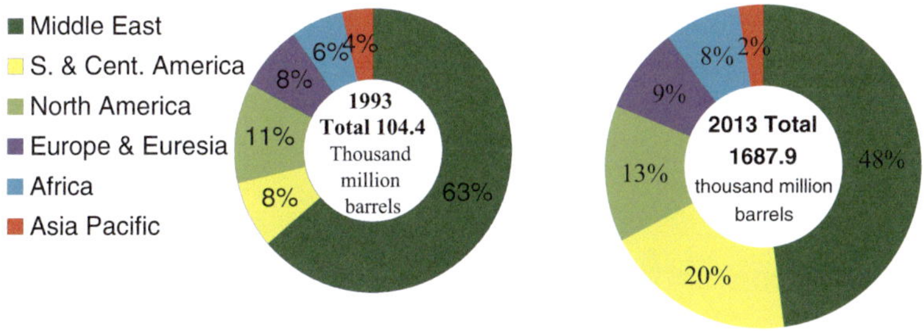

Source: BP Statistical Review of World Energy (2014, p. 7)

As the figure highlights, over a 20-year period from 1993 to 2013, total proved oil reserves rose from 1041 billion barrels to 1687 billion barrels, with the Middle East's share at just under 50 % in 2013 compared with around 64 % two decades earlier. What has taken place is that EOR and development of existing reserves, aided by advances in new technology, have added to reserves from North America and South and Central America. Taking a long-term view of world oil production, this pattern of exploiting existing capacity will continue as illustrated in the next diagram.

Figure 4.12 predicts that existing oil capacity will experience a decline over the next 15 years, but that new discovery, especially in offshore fields and unconventional oil, will provide new resources. Key growth will be technology-driven development of existing reserves and EOR, as these two will provide the majority of production. To the extent that technology can reduce costs, its efficient application can extend the life of an existing field and, for financially stressed OPEC producers,

Fig. 4.12 Prediction of world oil production to 2030

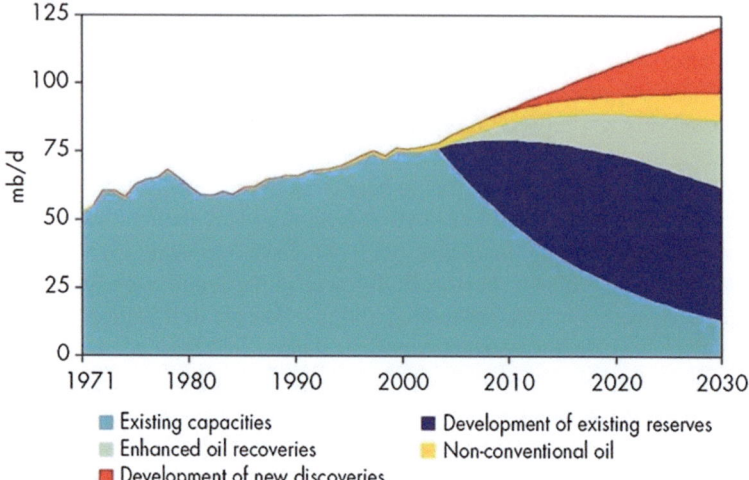

Source: Gould (2006, p. 179)

make smaller fields economical and can even enable the redevelopment of fields that were abandoned. As noted earlier, the number of new "giant" field finds has diminished and is mostly concentrated with a few OPEC oil producers. The smaller OPEC producers will stand to benefit more from EOR application. However, as noted from Fig. 4.11, unconventional oil will have a role to play in future global energy production, and the question is then: who are the major unconventional oil and gas world players?

The New Unconventional Energy "Stars"

The tables below list the top ten countries with technically recoverable shale and gas resources according to the latest US Energy Information Agency estimates (EIA 2013[a]).

The EIA in its 2013 report also highlights the rapid upward revision in technically recoverable shale oil and gas from 2011 data, indicating world total of shale gas at 6622 trillion cubic feet and shale oil at 32 billion barrels. However, when considering the market implications of abundant shale resources and which countries have these resources, it is important to distinguish between a *technically* recoverable resource and an *economically* recoverable resource. Technically recoverable resources represent volumes of gas and oil that can be produced with current technology, regardless of oil and natural gas prices and production costs. Economically recoverable resources are those that can be profitably produced under current market conditions with three key factors being cost of drilling and completion of wells, the amount produced from an average well over its lifetime, and prices received. These distinguishing features

Table 4.2 Top ten countries with technically recoverable shale oil and gas resources

Rank	Country	Shale oil (Billion barrels)
Shale oil resources		
1	Russia	75
2	USA	58
3	China	32
4	Argentina	27
5	[a]Libya	26
6	[a]Venezuela	13
7	Mexico	13
8	Pakistan	9
9	Canada	9
10	Indonesia	8
	World total	*345*

Rank	Country	Shale gas (trillion cubic feet)
Shale gas resources		
1	China	1115
2	Argentina	802
3	[a]Algeria	707
4	USA	665
5	Canada	573
6	Mexico	545
7	Australia	437
8	South Africa	390
9	Russia	285
10	Brazil	245
	World total	*7299*

Source: U.S. Energy Information Administration (2013[a], p. 10)
[a]OPEC members

are important as will be discussed later. Following from the depressed market oil prices in 2014, the issue of economically recoverable resources has taken center stage. At the same time, technically recoverable resources assume that all players have the same access to technology and few of the constraints noted earlier in Fig. 4.7 concerning barriers to nonconventional energy technology deployment.

Table 4.2 highlights few OPEC members have substantial shale oil and gas resources. The most notable exception is Algeria, ranked third in terms of technically recoverable shale gas with around 707 trillion cubic feet and ahead of the USA's 685 trillion cubic feet although another estimate by the *Advanced Resources International (ARI)* puts US shale gas technically available resources at 1163 trillion cubic feet (EIA 2013[a]). China tops the ten country list with 1115 trillion cubic feet. However, while Algeria might rank third ahead of the USA, extraction of shale gas requires injection of water for hydraulic fracturing to move out the resource.

Like almost all countries of the Middle East and North Africa (MENA) region, Algeria faces a water shortage and water security is a major challenge especially for the Gulf Cooperation Council (GCC) states where there is increased reliance on energy-intensive desalinated water (Saif et al. 2014). The majority of the Algerian shale gas fields are in the drier, southern part of the country, making an economic extraction costlier compared to other water-abundant shale gas countries, with a "composite success rate" recovery factor set at 20 % for most of the Algerian shale gas basins by *ARI* (EIA 2013[a], p. 42). The Chinese technically recoverable shale gas resources at 1115 trillion cubic feet places that country at the top of the world league. The question is whether China can harness the current technological advances made by the USA in shale gas fracking to enable the Chinese to reduce their energy dependency quickly is not easy to predict. The long-term implication to major OPEC producers exporting oil and gas to China is that this could be something of a concern should the Chinese become less energy dependent on key OPEC countries in the future.

Analysis of the top ten countries with technically recoverable shale oil resources places two OPEC members—Libya and Venezuela—in this league but with Russia dominating at 75 billion barrels, almost one-third of reported Saudi conventional oil resources. The implication is clear: Russia will continue to play an important role in both conventional and nonconventional oil supplies in the foreseeable future, and how OPEC manages this relationship with a key non-OPEC oil producer will be a litmus test of future oil producer cooperation.

A key uncertainty regarding non-OPEC supply is the potential diffusion and mass use of shale technology outside the USA, in countries like Russia. Leaving the issue of the embargo of high-technology end use by the USA on Russian oil companies following the Ukraine-Crimea events of 2014, there is some doubt on whether the US tight oil revolution could be replicated in other countries like Russia. The shale oil development in the USA has come at a great cost by drilling many expensive experimental wells and the accumulation of large amounts of debt, which in periods of low oil prices could and has put such shale operators under strain as will be discussed later. The US "expensive" model might not be easily replicated in other countries, especially those facing fiscal pressure like Russia following on from the fall in oil prices and Western government-led sanctions. Ignoring external-driven political barriers, the Russian energy sector is different from the USA, whereby in Russia the government owns underground reserves, the corporate landscape is dominated by a few large vertically integrated companies, the service sector is "weak," there is low availability of rigs, the tax system is in need of reform, and the capital markets are thin (Fattouh 2014[a]).

The US Shale Revolution: *Drill, Baby, Drill…*

It has been called the "energy revolution" with shale gas being dubbed a "game changer" in leading this new energy revolution with implications for both OPEC and non-OPEC conventional energy producers (Dagher 2014; Fattouh 2014[a, b], Westphal et al. 2014).

It is now assumed that recent advances in this fossil fuel production will herald the long-awaited age of energy abundance and "energy independence" for the USA, with many arguing that the period of US energy decline is now over (Donilon 2013; Dunn and McClelland 2013; Rosenberg 2014). However, others caution against such a long-term scenario of US "energy independence" and continue to point out that the USA's future energy demand growth will still be reliant on foreign imports (Emerson and Winner 2014, Hughes 2013). Some analysts, while conceding that indeed an energy or more precisely a shale technology revolution has taken place, predict that the long-term impact on such major OPEC producers like Saudi Arabia will not be as serious to their energy market share as believed (Jadwa Investment 2013). This is based on the premise that despite the rhetoric, the USA which has been leading this shale revolution is highly unlikely to become energy independent unless the rate of its energy consumption is radically reduced (Hughes 2013).

The rhetoric was very much in evidence since the famous slogan *Drill, Baby, Drill* in 2099 was first used at the US Republican Party's National Convention meeting by Michael Steele, who was later elected Chairman of the Republican National Committee (Hughes 2008). The phrase proved to be popular and gained further prominence after it was used by Republican Vice Presidential nominee *Sarah Palin* in her debate with *Joe Biden*. Sometimes slogans can become an embarrassment as *Drill, Baby, Drill* turned to *Spill, Baby, Spill* following the 2010 *Deepwater Horizon* oil spill at a British Petroleum offshore drilling rig in the Gulf of Mexico (Corrigan 2010).

There is no disputing that the size of the US oil and gas supply shock has been phenomenal, with tight oil production rising from around 1 million bpd in 2010 to more than 3.5 million bpd in the second half of 2014 and shale gas rising from negligible amounts in 2006 to 11,400 billion cubic feet by 2013. The growth in both US oil and US shale gas production is illustrated in the figures below.

Fig. 4.13 US crude oil and liquid fuel production (2007–2014)

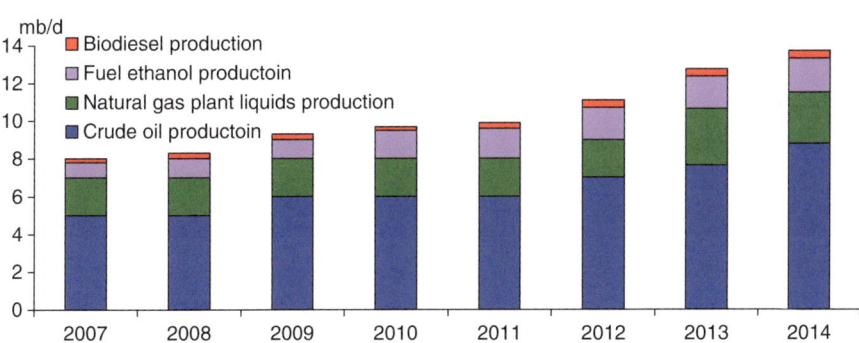

Source: Fattouh (2014[a])

From Fig. 4.13, US crude and liquid fuel production increased from around 7.3 million bpd in 2007 to above 11 million bpd in 2013 and 12.6 million bpd in 2014, with the USA adding more than 1 million bpd of liquid production in 2013, with only Saudi Arabia having a bigger increase than the USA in the same year, but that was because of the Kingdom's ability to tap into spare capacity and unlike the

USA's "organic growth" (Fattouh 2014[a], p. 15). While US oil production has been impressive, the limelight has been stolen by the growth in US shale gas, as illustrated below.

Fig. 4.14 (**a**) US gas production: conventional and shale (1970–2013) (billion cubic feet/day). (**b**) US shale production 2006–2013

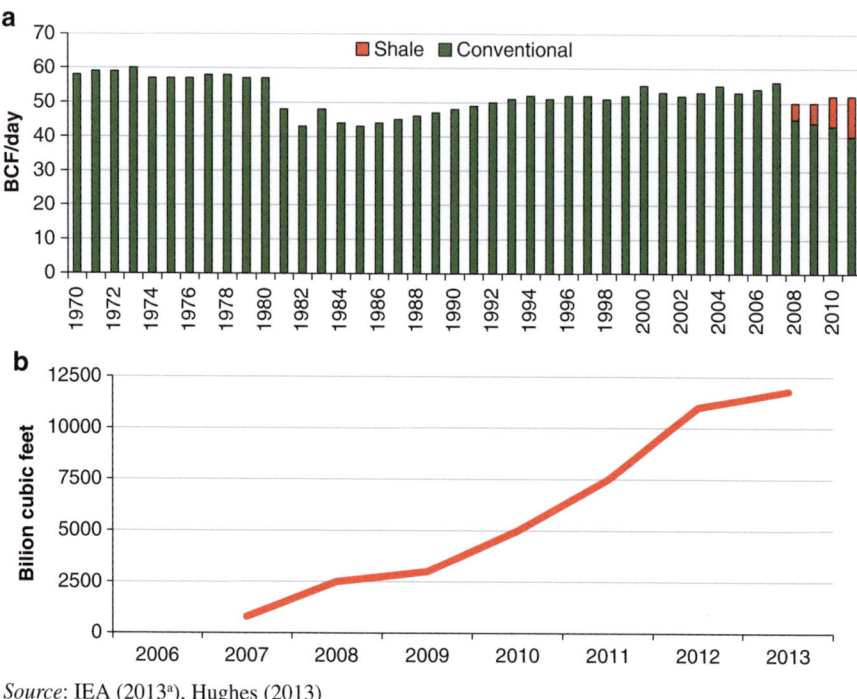

Source: IEA (2013[a]), Hughes (2013)

Figure 4.14 highlights the sharp rise in US shale production from 2006 to 2007 to around 11.5 billion cubic feet per day by 2013, with shale gas increasing its overall share in US gas production. Both conventional and shale US gas productions have now surpassed the peaks in US conventional gas production of the early 1970s.

The surge in US oil and gas production has implications for both the volume and origins of North American imports with drops forecasted by the IEA by 2018, as illustrated in Fig. 4.15.

From the above figures, the share of the Middle East to North American imports is estimated to decline to 1.7 million bpd by 2018 compared with 2.7 million bpd in 2012, with the sharpest decline in imports coming from Africa and a marginal decline from Latin America. According to Saudi Arabia, the Kingdom exported around 521 million barrels of crude oil to North America in 2012, up from 442 million barrels in 2010 (SAMA 2013, p. 150). The Saudi share of the North American imports for 2012 represented around 53 % of all imports. The main issue facing Saudi Arabia is whether it will see the Kingdom's share of exports erode further as forecasted by 2018, or will the USA cut back from other Middle East suppliers, like Libya or Iraq due to continued political and military conflicts and uncertainties in

Fig. 4.15 Origin of North American crude imports in 2012 and forecast for 2018 (million barrels per day)

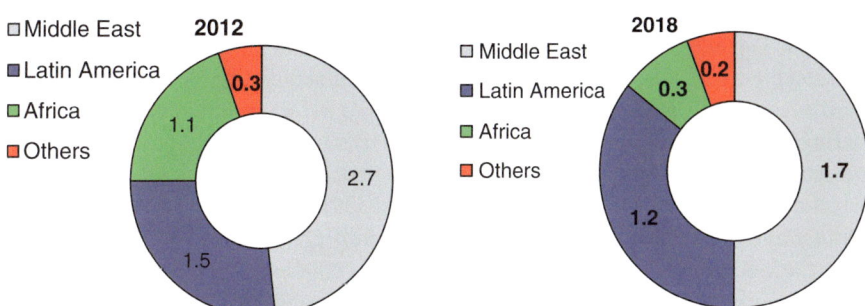

Source: International Energy Agency (IEA). "Medium term market outlook report. Market trends and projections to 2018." Platts 6th Annual Crude Oil Summit. London. May 14, 2013

these OPEC producers. By 2014, the Saudi crude exports to North American had dropped to 457 million barrels, a decline of around 13 % over the 2012 exports (SAMA 2015, p. 143). In February 2015, President Barack Obama reiterated once again the importance of energy security to the USA's national security strategy, stating that the USA "must promote diversification of energy fuels, sources and routes as well as encourage indigenous sources of energy supply. Greater energy security and independence within the Americas is central to these efforts" (White House 2015[b], p. 16). Is this achievable?

The USA in the Driver's Seat: A New "North American OPEC"?

US politicians spent much of the past decade discussing the break-up of OPEC and the US House of Representatives even passed *Act HR 6074* entitled *No Oil Producing and Exporting Cartels Act of 2008*, or *NOPEC*, allowing the Department of Justice to pursue cases against OPEC member states on antitrust grounds via amendments to the *US Sherman Act*. President Bush did not approve it. The original act was reintroduced in the House of Representatives as *HR 1674* in June 29, 2012, but again the US Administration did not approve it (Learsy 2012). The *NOPEC* Acts illustrate the continuing belief that OPEC behaves like a true cartel in setting prices and controlling production allocation among its members, when in earlier chapters it was highlighted how OPEC in reality did not act as a cartel, but pursued individual member's interests when it suited them.

Despite this reality, the OPEC "cartel" myth persists, but it becomes ironic to note that the USA might now be tempted to form its own, albeit, unofficial North American OPEC club when the energy ministers of the USA, Canada, and Mexico gathered for their first high-profile talks in Washington in December 2014, with the aim of "defining areas for energy cooperation moving forward and address a strategic vision for the North American energy community" (Critchlow 2014).

Combined, these three North American nations have a production capacity of around 20 million bpd and could, theoretically if they act in concert, pose a credible threat to OPEC's dominance of the market with production of around 30 million bpd. The gap between the two groups is narrowing as the US and Canadian nonconventional energy petro-nexus rises sharply as discussed earlier. While the tripartite North American discussions could be a reaction to the sharp fall in world oil prices during the latter part of 2014, putting some US and other shale producers under pressure, it could certainly smack of political hypocrisy if the USA pursues such an unofficial North American cartel, after decrying OPEC's power over the years. The momentum for further North American countries' cooperation is gaining ground with the US House of Representatives passing the controversial Canadian *Keystone* pipeline bill in January 2015, which the US senate approved in February but with President Barack Obama threatening to veto it (BBC 2015ª). In the end, the US senate's 62–37 vote in March 2015 fell short of the 67 votes needed to push past President Obama's veto (Litvan 2015). The *Keystone* project is a 1179 miles extension of an existing pipeline that would bring additional oil from the tar sands of Alberta, Canada, to refining facilities near the Gulf of Mexico and is a $5.4 billion project that was first initiated in 2008.

The presidential veto is based on the controversy that the tar sand oil that the pipelines transport would be more polluting than other types of oil, as well as to ongoing litigation in the State of Nebraska. However, in January 2015 the Supreme Court of Nebraska dismissed the lawsuit on which President Obama's veto threat was partly based, putting pressure on the President to accept the bill (BBC 2015ª). While transporting Canadian tar sands might or might not introduce environmental concerns, the successful implementation of the project will put further pressure on Saudi crude oil exports to the USA, as Canadian tar sands oil is similar in density (heavy crude) as Saudi oil exports to the USA, intensifying the Kingdom's search for alternative Asian export markets.

As part of this new-found North American energy tripartite cooperation, the USA is also actively considering light crude oil exports to Mexico, representing a potential dramatic shift in the oil trade between the two countries. According to reports, the Mexican national oil company *PEMEX* said it wanted to import as much as 100,000 bpd of light crude from the USA, following the meeting between President Obama and Mexican President *Enrique Peña Nieto* (Quinn 2015). Since 1970, US laws have prohibited exporting crude, though the American government can authorize exceptions as it did with Canada (Quinn 2015). What Mexico preferred was to "swap" its heavy crude with US lighter crude oil to increase Mexico's gasoline production and improve refining and see it as benefiting both countries (Navarro and Murtaugh 2015).

Most commentators seem to agree that lifting the US crude oil export ban would be beneficial to the American economy with *IHS*, an international energy analyst group, estimating that such a move would create an additional 964,000 US jobs by 2018 and increase US crude production to 11.2 million bpd, as well as adding investments of nearly $750 billion (IHS 2015). Such prospects have led the US

Administration to put aside some of its reservation that global warming could be exacerbated by leaks of methane, a potent greenhouse gas, from shale sites if further production is encouraged by approving many pending requests to export processed ultralight oil (Reuters 2014). Apparently some 20 pending requests were approved by the US Department of Commerce which also issued the first formal guidelines and definitions for what constitutes exportable crude oil, including clarifying the degree of distillation required (Reuters 2014).

Others see such a reversal of policy in terms of bigger geopolitical strategies, both domestically and internationally. Domestically, ending the export ban would encourage more US shale investment or at least offset the negative effects of low oil prices and encourage US refiners to process more American crude. US exports would help American producer's profits by closing a gap between the *Brent* crude price and the US crude price *West Texas Intermediate*. Internationally, such a move could dampen the effect of lower oil prices due to OPEC's stance on not cutting production and forcing international oil prices to fall sharply during 2014. Additional US exports would increase prices in certain categories of crude, like light oil (Financial Times 2015).

US Shale Revolution: Key Success Drivers

The US shale revolution did not take place in isolation but a host of accommodating enablers helped to propel this energy sector forward, stamping it as a "unique American model" which might be difficult to replicate in other shale-rich countries. The legal and regulatory framework for the development of unconventional resources in the USA is a mixture of laws, statutes, and regulations at the federal, state, regional, and local levels. Most of these rules apply to all oil and gas generally and were in place before unconventional resource development took off (IEA, Special Report 2012[a], p. 104). The regulations cover virtually all phases of an unconventional resource development, from exploration through to site restoration, and include provisions for environmental protection and management of air, land, waste, and water. Within this diverse structure, a major challenge for the USA is to maintain consistency of regulation among different states, closing regulatory gaps where necessary, especially in environmental regulations, but doing this in a way that encourages best practice and responds to changes in production technology (IEA 2012[b]).

As some have rightly pointed out, there would *not* have been a US shale revolution without high oil prices and innovation in hydraulic fracturing and horizontal drilling which were key enabling technologies. For example, *eight horizontal wells drilled from only one pad can access the same reservoir volume as 16 vertical wells*. By using multi-well pads, it can significantly reduce the overall number of well pads, access roads, pipeline routes, and production facilities required, thus minimizing the impact to the public and reduce growing public opposition on environment degradation (Majid 2014).

Besides technology advances, other US-specific enabling factors include:

- Favorable mineral rights with land ownership that allows landowners to have rights not only on the surface of their property but to everything below the surface
- Dynamic exploration and production industry with many independent operators
- Tax and royalty incentives from federal and state governments to the operators
- Strong logistics and service provider industry
- Huge number of rigs availability
- Deep financial markets and relatively cheap credit
- A liquid futures market allowing producers to hedge production forward and reduce risk
- High associated condensate production within the shale, paired with the robust condensate pricing that can make shale gas reservoirs economic despite lower prices relative to the late 2000s.
- A large existing amount of well control from historical production, with 4 million drilled wells in the USA compared with 1.5 million drilled wells in the rest of the world
- An extensive network of pipelines and other support infrastructure (Fattouh 2014[b], p. 51; Majid 2014, p. 9)

From the above environment, a distinct pattern has emerged in the USA in the evolution of operationalizing tight energy fields or *plays*. They seem to follow the following pattern:

(a) A "play" is identified and leasing bids ensues.
(b) A drilling boom to hold leases follows, as lease agreements typically include "held-by-production" arrangements which mandate drilling, with leases typically having terms of between 3 and 5 years.
(c) The first wave of drilling defines the so-called *sweet spots* or areas of highest productivity, as well as the extent of the play or field. Large leaseholders than cash out of their worst/least productive wells by selling to other hopeful would-be producers.
(d) The drilling boom causes production to rise rapidly with drilling focusing on the *sweet spots*.
(e) Application of better/newer technology, such as longer horizontal laterals with more hydraulic fracturing stages, serves to maintain production even as drilling moves away from sweet spots to lower-quality parts of a play.
(f) Eventually even better/new technology cannot make up for lesser quality geology, as noted earlier in the chapter and the incremental production of new wells declines. In this scenario, as the rate of incremental production decreases, more wells are required to offset the overall field declines, and without massive amounts of new drilling, the "plays" go into decline, some terminally (EIA 2013[a]).

While the above US success enabling factors have given North American players a distinct advantage over other potential shale producers in other countries, there are also problems facing the tight energy industry which should not be underestimated

Success Comes at a Price…

Shale oil activities bring out strong public reaction, especially concerning the environmental impact of shale production. As pioneers of large-scale unconventional energy development, all stakeholders—policy makers, regulators, producers, and the general public in the USA—have been the first to face the question of how to evaluate and minimize associated environmental risks.

The effect of the public debate in the USA should not be underestimated internationally, as other countries have reacted to their own domestic constituents, with France banning hydraulic fracturing entirely, and Germany has put a de facto moratorium in place. In the UK there has been vocal opposition to this new technology, but the industry has not been banned despite some public unease following reported mini earthquakes in fracking zones the latest being in Kent in May 2015. In the UK, the level of approvals is much tighter than in the USA, and in order to engage in hydraulic fracking in the UK, a company must achieve permission from several regulatory bodies according to the Department of Energy and Climate Change—DECC (Department of Energy 2013). First it will need to acquire a Petroleum Exploration and Development License from the DECC; then they need to gain planning permission from the Mineral Planning Authority and consent from whoever owns the land where drilling is to take place. After that, permits from the Environment Agencies will be required and finally the drilling operation needs to be signed off by the Health and Safety Executive. Once drilling has commenced, the Environment Agencies and the Health and Safety Executive will both oversee the process to ensure adherence to the terms of the permits and this will involve regular monitoring and inspections.

This cautious UK approach contrasts with the approach taken by the US authorities, especially in trying to understand the causes of some earthquakes generated by hydraulic fracturing. Research in the USA into the cause of these quakes has found that the majority were caused by the reinjection of waste water into disposal bores rather than the hydraulic fracturing or drilling operations. This practice is not allowed in the UK (Department of Energy 2013). Other ways in which the UK regulation regime is much stricter than in the USA include the type of fluids that can be used in the hydraulic fracturing process and the ways in which waste material can be disposed, as there is a wide variety of fracking fluids used in the USA, including highly toxic chemicals, unlike in the UK where there are also stricter controls over the disposal of waste materials. In the USA there are regulations covering the disposal into lakes and other waterways, but the disposal of waste over land is unmonitored by the authorities (Department of Energy 2013). The US authorities are awarae of these concerns and new regulations around the disposal of wastewater from fracking could as early as 2016 affect discharge costs, and oil output for companies (Harvey, 2015b). Despite the widespread use of fracking in the oil and gas industry in the USA, many municipal treatment plants are not designed to remove all water constitutes associated with shale gas/oil extraction and disposal of water is often done using deep injection wells, onsite recycling, or reuse (Majid 2014, p. 10).

North American Shale "Plays": Location, Location, Location

Despite the environmental issues listed above, shale oil and gas production has grown explosively in North America. However, the major producing areas are few in number with around 80 % of the shale gas production coming from five *plays*, several of which are in decline as will be analyzed later, while in tight oil production, more than 80 % of shale oil produced comes from two unique "plays": the *Bakken* in North Dakota and Montana and the *Eagle Ford* in Southern Texas. Figure 4.16 illustrates the major North American shale resource locations.

Fig. 4.16 Location of major unconventional gas resources in North America

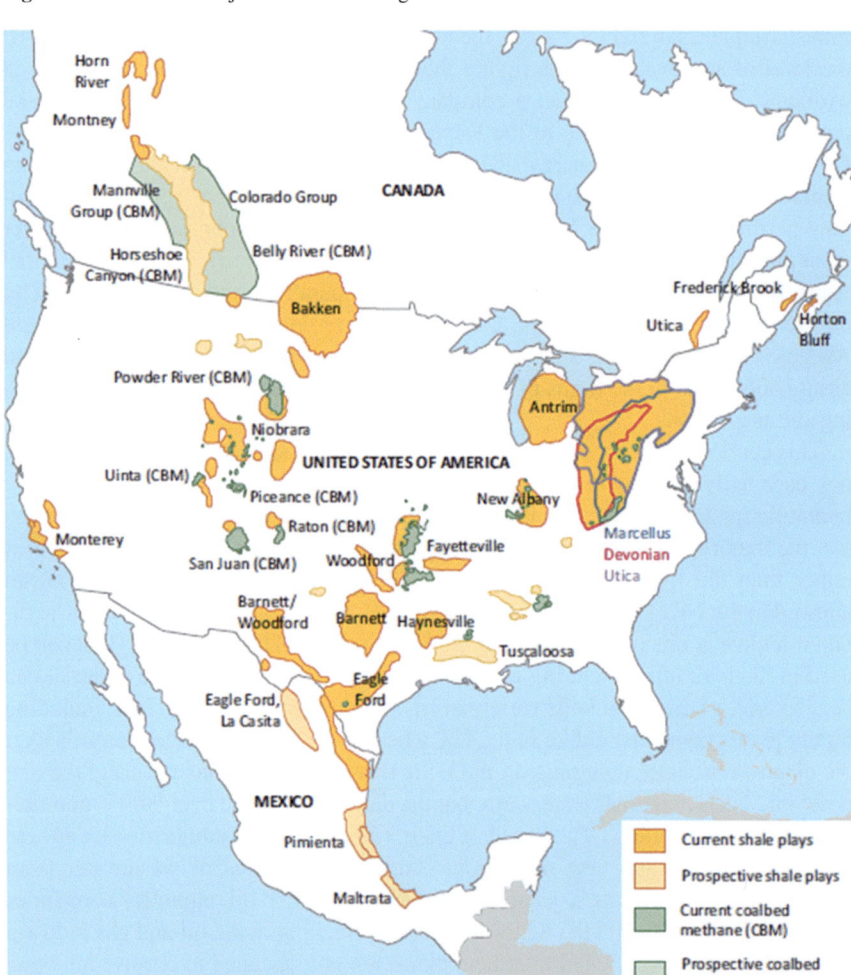

Source: International Energy Agency (2012) "Golden rules for a golden age of gas: World Energy Outlook. Special Report on Unconventional Gas," IEA, p. 103

Some of these unconventional resource "plays" have become famous, and along with *Bakken* and *Eagle Ford*, there are *Antrim, Barnett, Marcellus,* and *Haynesville*. What the above map reveals is that unconventional resources in North America are widespread in the continent and that most states have benefited from the tight resources boom. However, high productivity shale plays are not widespread and relatively small "sweet spots" within plays often offer the most potential, but as explained further below, both oil and gas shale productions are characterized by high decline rates and consequently require continuous inputs of capital to maintain the same level of productivity. It has been estimated that it required around $42 billion per year to drill more than 7000 wells in the shale gas plays in order to maintain production in 2012, while in comparison the value of shale gas produced in the same year was $32.5 billion. For tight oil, it is also estimated that more than 6000 wells at a cost of $35 billion annually are required to maintain production, of which 1542 wells annually at a cost of $14 billion are needed in the *Eagle Ford* and *Bakken* plays alone to offset declines (Hughes 2013). These are heady numbers by all account and underpin once again two important factors: that such a "new" energy revolution cannot be easily replicated and, critically, the importance of a *sustained* level of high oil prices being maintained to ensure the profitability of tight energy production.

A Heady Cocktail of Productivity, New Finds, and Price

In Chap. 5, we assess OPEC's fiscal breakeven pricing as being driven by a confluence of *T, E, and P* whereby Technology, Economics, and Politics/Policy are the key drivers affecting individual OPEC countries' breakeven pricing levels. Concerning tight shale oil and gas production, the confluence is *PPP* or *productivity, prices,* and *plays* as illustrated in the figure below (Fig. 4.17).

Fig. 4.17 Shale breakeven: "PPP" confluence of productivity, prices, and plays

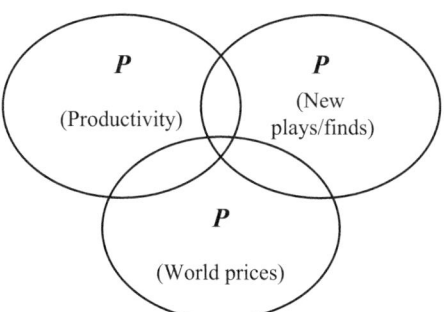

Despite advancement in "new" fracking technology, it has become apparent that the rate of depletion and hence of productivity of many of the US shale plays is steep. This in turn calls for new plays to maintain production at current levels. This can only be achieved if world prices make such investments worthwhile. The next set of figures and tables illustrates the interlinked factors of productivity and new plays and dramatically highlights the options faced by tight energy producers.

As noted, not all shale plays are ubiquitous with the same high productivity characteristics. Figure 4.18 sets out the production of the top 30 shale plays as of 2012, indicating that the majority of production is concentrated in a few plays.

Fig. 4.18 US shale gas production by play (May 2012)

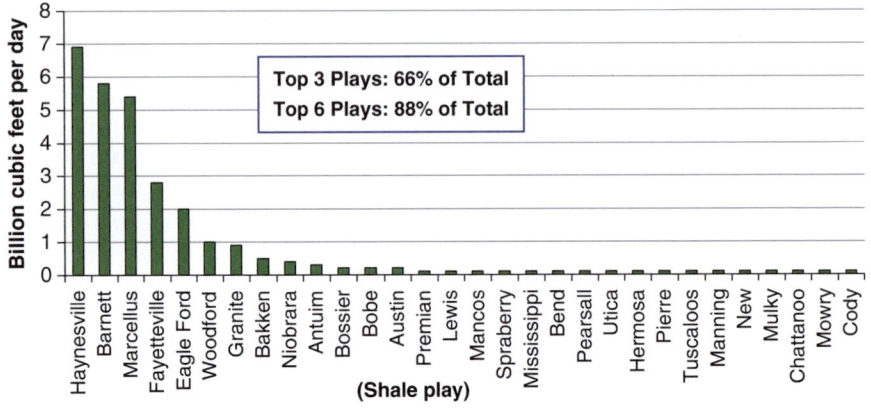

Source: Hughes. Post Carbon Institute, Feb. 2013

From Fig. 4.18, it can be seen that the top three plays—*Haynesville, Barnett, and Marcellus*—account for 66 % of total production and the top six plays that include *Fayetteville and Eagle Ford* account for nearly 90 % of production. The bottom 17 plays collectively contribute just over 1 % of production. However, it is more illustrative to put the growth of energy production in a longer historical perspective to assess average productivity trends and the incremental new capacity required to maintain and increase production. These are illustrated in Figs. 4.19 and 4.20, which take into account both conventional and nonconventional energy productions.

Fig. 4.19 US natural gas productions vs. annual drilling rates of successful gas wells (1990–2012)

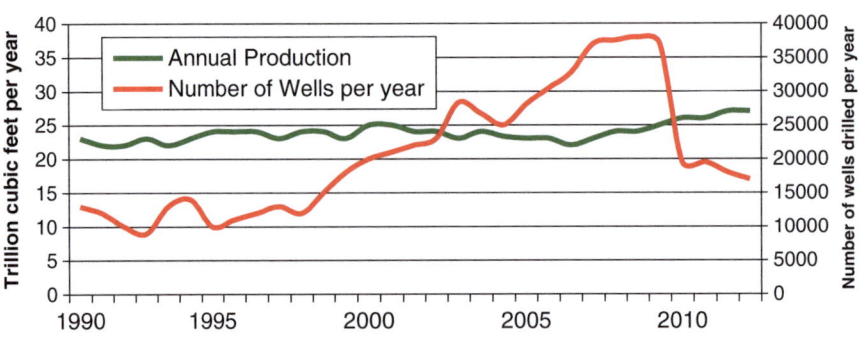

Source: Hughes. Post Carbon Institute, Feb. 2013

Fig. 4.20 US operating wells vs. average well production (1970–2010)

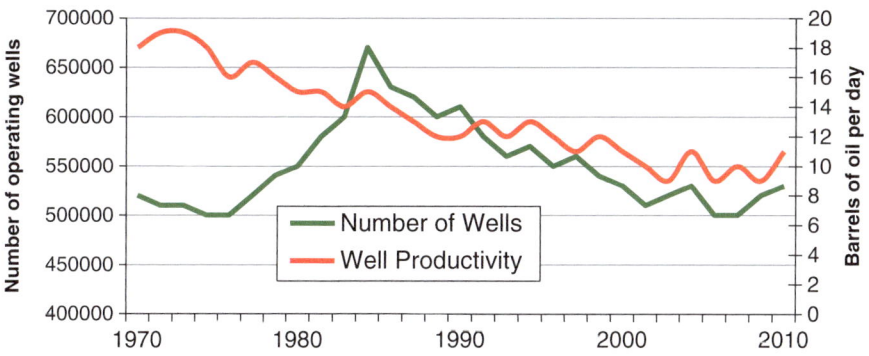

Source: Hughes. Post Carbon Institute, Feb. 2013

The magnitude of this private sector driven US energy growth is sometimes difficult to grasp in terms of its size. To put it in a global context, the number of operating wells in the USA was close to 500,000 in 2008 compared with about 900,000 in the entire world, with all the OPEC nations combined having about 38,000 wells, and Saudi Arabia the largest OPEC producer and holding the largest conventional oil reserves in the world having only 1560 producing wells (Gorelick 2010). By 2010, the situation had not changed but with the number of operating US wells dropping from a peak of around 650,000 in 1985 as per Fig. 4.19. However, in terms of productivity per well, there has been an overall decline from the 1972 peak of an average 18 barrels per day to around 10/11 barrels per day by 2010, with average well productivity falling by around 44 % over the past four decades. The number of new wells drilled per year is also on a staggering scale, reaching almost 40,000 by 2008 before falling to around 15,000 a year in 2012, but with annual production remaining relatively stable. The US energy sector seems to almost possess the characteristics of a traditional "cottage industry."

Some petroleum engineering research has also highlighted that recent drilling operations have tended to be more productive, although evidencing the same sharp rate of declines later on. The key is that the somewhat "hit-and-miss" type of well drilling and exploration is gradually being replaced by more sophisticated modern seismic technology and geology evaluation which pinpoints with more accuracy where the most productive "sweet spots" are available to enable drilling with lower overhead costs. This is illustrated in Fig. 4.21 for monthly average well production for the *Barnett* shale plays for three different drilling years (2004–2006).

Fig. 4.21 Average monthly well production—Barnett shale (2004–2006)

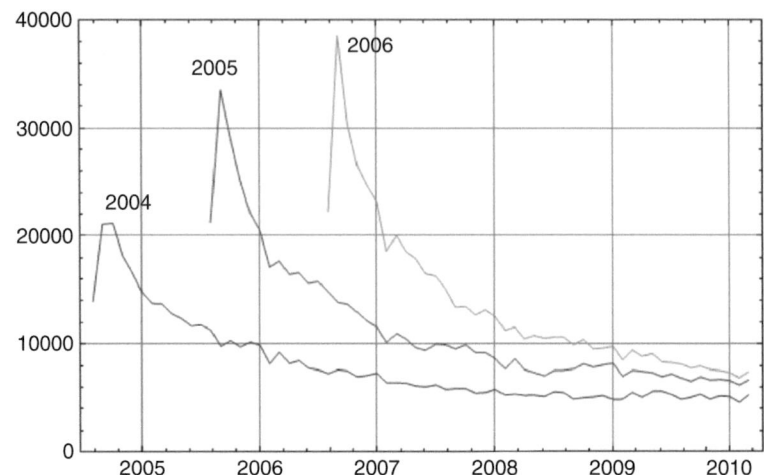

Source: Valko and Lee (2010, p. 4)

A better understanding of reservoir geology has led to "pinpoint" drilling being made which, as the above diagram illustrates, has led to higher initial production from the same "play" than previous drilling and production rates. Once again, the use of technology has helped to ensure that production diminishing rates are extended.

Forecasts by the US Energy Information Administration and the *ARI* estimate significant remaining shale resources from the key US plays as set out in Table 4.3, with remaining shale gas reserves and undeveloped resources put at 1161 trillion cubic feet and shale oil at 47.7 billion barrels.

Table 4.3 US remaining shale gas reserves and undeveloped resources

	Shale gas resources		Shale oil resources	
	Distinct plays (#)	Remaining reserves and undeveloped resources (*Tcf*)	Distinct play (#)	Remaining reserves and undeveloped resources (billion barrels)
1. *Northeast*				
Marcellus	8	369	2	0.8
Utica	3	111	2	2.5
Others	3	29	–	–
2. *Southeast*				
Haynesville	4	161	–	–
Bossier	2	57	–	–
Fayetteville	4	48	–	–

Table 4.3 (continued)

	Shale gas resources		Shale oil resources	
	Distinct plays (#)	Remaining reserves and undeveloped resources (*Tcf*)	Distinct play (#)	Remaining reserves and undeveloped resources (billion barrels)
3. *Midcontinent*				
Woodford	9	77	5	1.9
Antrim	1	5	–	–
New Albany	1	2	–	–
4. *Texas*				
Eagle Ford	6	119	4	13.6
Barnett	5	72	2	0.4
Permian	9	34	9	9.7
5. *Rockies/Great Plains*				
Niobrara	8	57	6	4.1
Lewis	1	1	–	–
Bakken/Three Forks	6	19	5	14.7
Total	*70*	*1161*	*35*	*47.7*

These estimates are based for the top US shale players, but other analysts stress the uneven future progress rates for these key fields, with varied estimates for the number of new wells needed to be drilled in order to maintain current productivity and offset declines. Table 4.4 summarizes some of these productivity growth concerns.

Table 4.4 Prognosis for future production in the top nine shale gas plays in the USA (May 2012)

Field	Rank	Number of wells needed annually to offset decline	Wells added for most recent year	October 2012 rig count	Prognosis
Haynesville	1	774	810	80	*Decline*
Barnett	2	1507	1112	42	*Decline*
Marcellus	3	561	1244	110	**Growth**
Fayetteville	4	707	679	15	*Decline*
Eagle Ford	5	945	1983	274	**Growth**
Woodford	6	222	170	61	*Decline*
Granite Wash	7	239	205	n/a	*Decline*
Bakken	8	699	1500	186	**Growth**
Niobrara	9	1111	1178	−60	*Flat*

Source: Hughes, Post Carbon Institute, Feb. 2013

From the above table we note the uneven prospects for growth in the top nine US plays. Only three plays—*Marcellus*, *Eagle Ford*, and *Bakken*—show attractive economics since these also produce *Natural Gas Liquids* (*NGL*) and other condensates and with *Marcellus* and *Eagle Ford* also connected to pipeline infrastructure which was earlier pointed out as one of the key success enabling factors for tight resource exploitation. These top nine US plays accounted for 95 % of US shale gas production as of mid 2012, but Table 4.4 also shows that the number of rigs and wells recently drilled are below what is needed to keep to 2011 production levels, with the exception of the more attractive *Marcellus, Bakken,* and *Eagle Ford* plays which were still able to attract investments necessary for production growth.

Even those most optimistic about the growth potential of the most productive US plays forecasts that in the long run and unless significant new technological advances are made, there will be an eventual decline in the production of tight resources in the USA. This is illustrated in Fig. 4.22 for *Eagle Ford and Bakken* tight oil production projections to 2025.

Fig. 4.22 Projection of tight oil production by plays in the USA to 2025

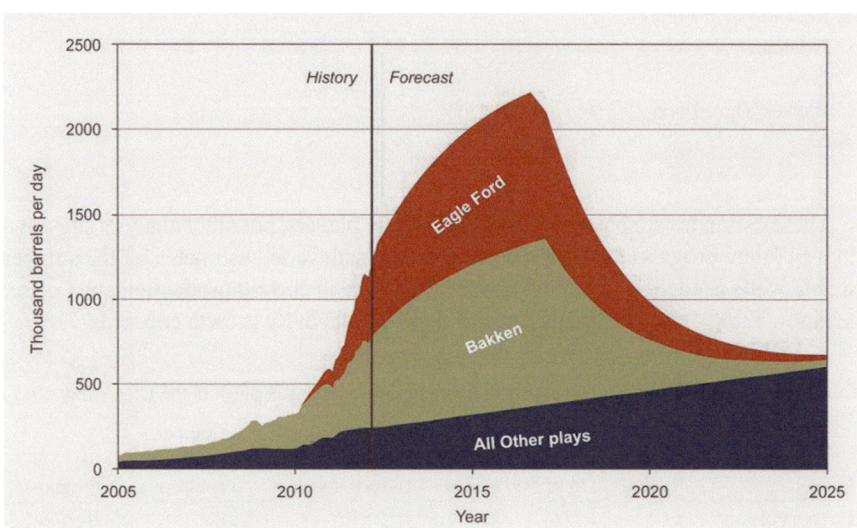

Source: Hughes, Post Carbon Institute, Feb. 2013

This long-term projection may be unduly pessimistic, as it is based on *current understanding of the shale formations and fracking technology* as the above figure shows a "peaking" in tight oil production from these major plays by around 2017/2018, while the EIA in its "2013 International Energy Outlook" predicts a US tight oil production of about 2.5 million bpd in 2025—or three times more than the projection above—and declining slowly to 2 million bpd in 2040 (EIA 2013[b], p. 42). While there are differences of opinion on long-term production and growth estimates, it is worthwhile examining in more detail the actual decline rates and productivity levels of some of the key US plays to get a better understanding of how the post-2014 low oil price levels might have a bearing on these trends. The next set of figures illustrates these for *Bakken* for tight oil and *Haynesville* for shale gas plays.

Fig. 4.23 (**a**) Typical decline curve for *Bakken* tight oil wells (months on production). (**b**) *Bakken* tight oil production and number of producing wells (2000 to mid-2012)

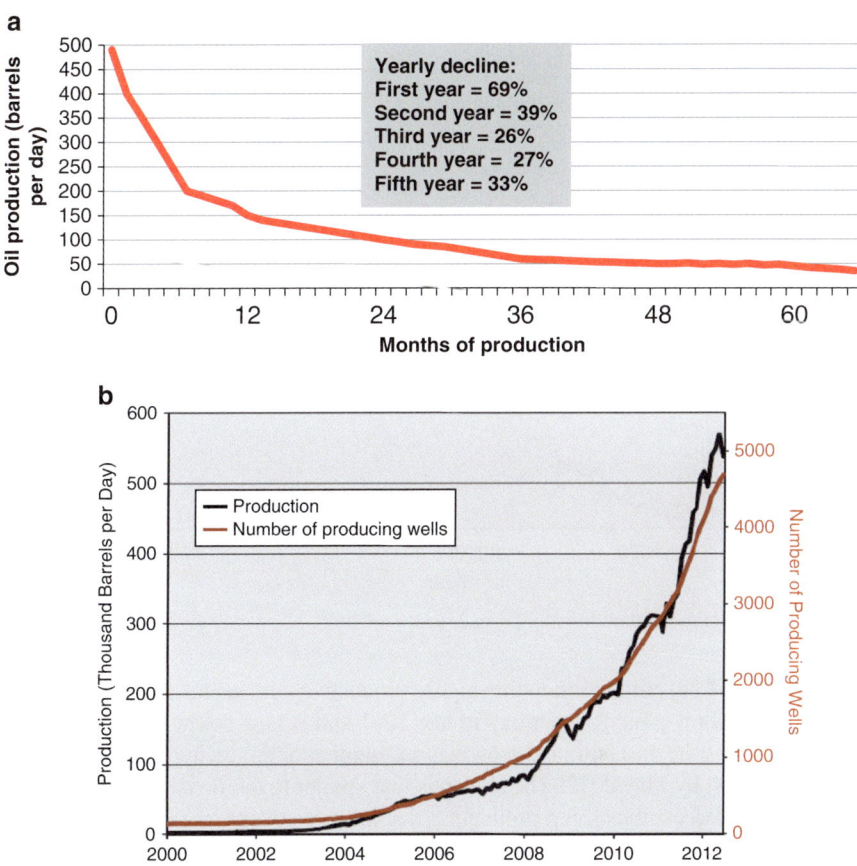

Source: Hughes, Post Carbon Institute, Feb. 2013

From Fig. 4.23a, the pattern of sharp declines from almost the first year of production at a rate of 60 % is immediate. To keep average productivity and production levels even, one needs to dig more and more wells, let alone increase production, thus putting into perspective the slogan *Drill, Baby, Drill*. Figure 4.23b highlights the increasing numbers of producing wells for the *Bakken* play.

The situation for tight shale gas plays is no different as illustrated in Fig. 4.24 for *Haynesville*, one of the top ranked shale plays and the most productive shale gas field in the US as of May 2012, accounting for almost 26 % of the US shale gas production (Jadwa Investment 2013, p. 38).

Fig. 4.24 *Haynesville*—shale gas production and number of producing wells

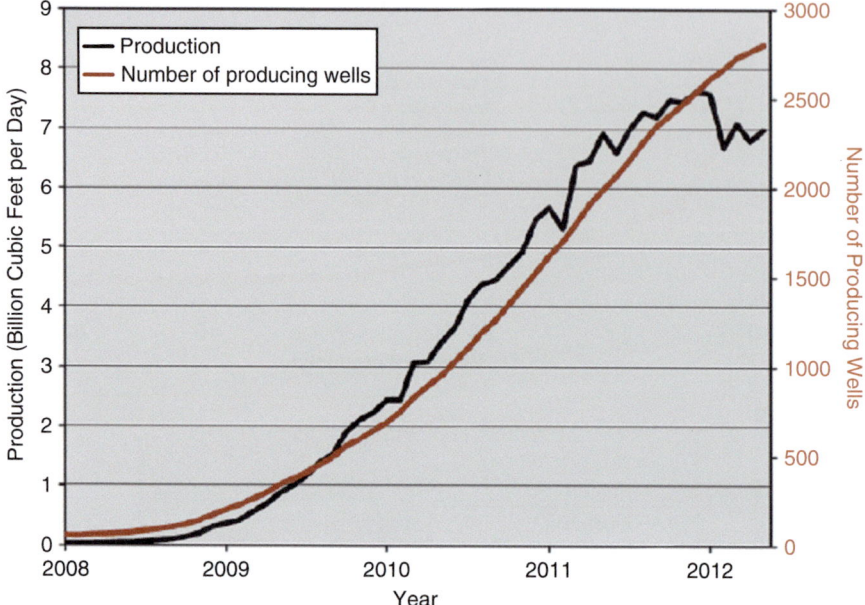

Source: Hughes, Post Carbon Institute, Feb. 2013

From Fig. 4.24, production in this highly productive field seems to have peaked at around 7.2 billion cubic feet per day in late 2011 and is now below 7.0 billion cubic feet per day despite the continued growth in the number of producing fields which rose to around 2800 by May 2012. The reason is that similar to the *Bakken* tight oil fields, *Haynesville* also exhibits steep production decline rates over time, averaging around 68 % in the first year, 50 % in the second, third, and fourth years. Given such productivity trends, necessitating a continuous addition of new wells to maintain production, what is the effect on US shale producers from the recent sharp fall in oil prices?

The November 2014 Oil Price Shock: What Future for the Shale Oil Boom?

The sudden and persistent fall in oil prices toward the latter half of 2014 from $110 a barrel ranges earlier in the year to under $50 a barrel was a shock to many in both OPEC and non-OPEC producers, and the effect of these price drops on OPEC members' breakeven fiscal sustainability will be examined later. Concerning US producers, the fall in prices has come at a price, threatening to derail the "golden boom" shale revolution and turn it to "gloom." This touches upon two of the *3 P's*—productivity and new plays as being important elements for shale breakeven. The third *P*—price—has now entered the equation as being a very critical element on whether the shale boom can continue in its *present format*.

Fig. 4.25 (**a**) US land rig count (2013–2015). (**b**) USA: rig count vs. total crude production (DOE, Baker Hughes)

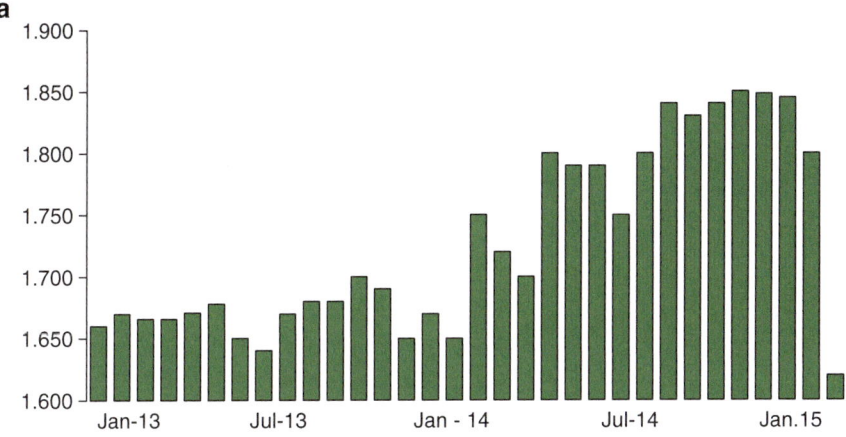

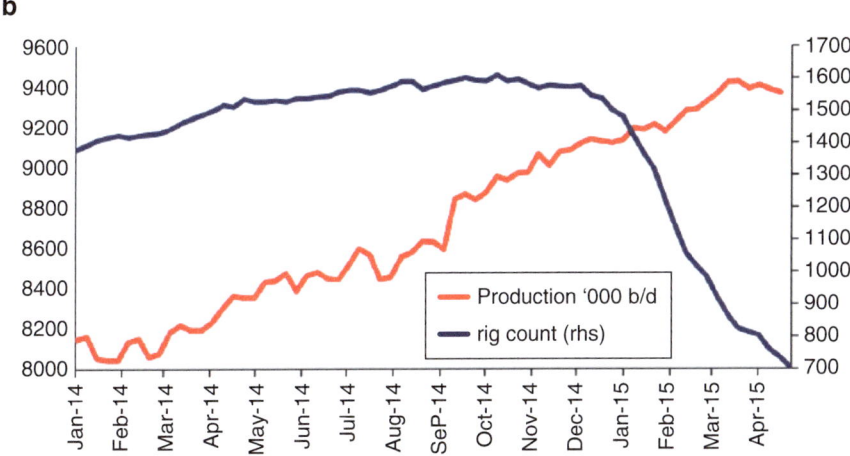

Source: Jadwa (2015a, b, p. 8), SAMBA (2015, p. 3)

By all accounts, some US energy producers are retrenching, with the latest rig counts showing a sharp decline compared with mid-2014 levels as illustrated in Fig. 4.25.

From a peak of around 1850 land rigs, the US rig count has sharply fallen to just over 1600 rigs in January 2015 and according to latest estimates stood at 1397 land rigs as of mid-February 2015 (Baker Hughes 2015). According to Baker Hughes, the *combined land, inland water, and offshore* US rig count has fallen to 1456 rigs in February 2015 compared with 1771 a year earlier (Baker Hughes 2015) with the decline continuing sharply to reach just over 700 rigs in April 2015 as illustrated in Fig. 4.25b. In analyzing the major US plays by rig count, it was noticeable that, with the exception of three plays (*Fayetteville, Haynesville, and Utica*), all other plays registered rig count declines as illustrated in Table 4.5.

Table 4.5 US rig count—major plays February 2014 vs. February 2015

Play	Feb. 2014	Feb. 2015	Change
Woodford	56	50	(6)
Barnett	31	19	(12)
Niobrara	55	48	(7)
Eagle Ford	216	168	(48)
Fayetteville	9	9	0
Granite Wash	54	39	(15)
Haynesville	43	43	0
Marcellus	81	71	(10)
Mississippian	78	53	(25)
Permian	483	417	(66)
Utica	41	41	0
Williston	177	137	(40)

Source: Baker Hughes (2015, online)

The fall in rig numbers was the first sign that a serious rethink was happening in the shale sector and some oil producers started to bail out of long-term contracts for drilling rigs when crude prices fell below $50 a barrel. According to reports, companies such as *Helmerich & Payne, Inc.*, the biggest rig operators in the USA, said it had received early termination notices for rigs and similar reports came from other US contract drillers like *Pioneer Services Corp.* with Bloomberg analysts predicting another 50–60 agreements may be cut short by producers, with companies paying to cancel rigs rather than keep drilling in the face of a 55 % fall in oil prices (Doan 2015). According to analysts, it sometimes made economic sense to terminate rig contracts as oftentimes a rig termination clause is equal to the cost of one well and is cheaper, compared to drilling $10 million wells multiple times (Doan 2015). However, one should be cautious in predicting that a fall in rig count translates immediately into a drop in oil output, as both the time lag and the extent to which a declining rig figure translates into lower production are tricky to project. It is also important to assess the number of well completions per month. Delayed completions could result in both a bigger and quicker decline in output compared to what would be seen by simply looking at the rig count.

The result of the sharp fall in oil prices is that dangerous and geological difficult oil fields that looked very promising at $100–$110 a barrel prices had now turned unprofitable, not only for producers, but put at risk dividend payouts for investors who had backed this sector as explained earlier, under the "held-by-production" lease agreements marketed to investors. However, it was not only the small players that were being squeezed by the fall in oil prices but also the biggest explorers and IOC's, with analysts estimating that for every $10 in price drop erases some $2.8 billion in annual cash flow from companies such as Exxon Mobil and for those IOC's which are more crude-dependent like Chevron, this translated to a $3.85 billion cash flow reduction (Carroll 2015). The drop in oil prices also saw cancellation of some mega energy projects worldwide with France's *Total SA* putting its C$11 billion *Joslyn* joint venture project with *Suncor Energy Inc.* on hold. This had followed a 2013 cancellation of their C$12 billion *Voyageur* oil sands upgrade. In September 2014, Norway's *Statoil ASA* delayed work on the 40,000 barrel a day *Corner* oil sands development in Canada, while in Brazil *Petrobras* said that it needs a minimum price of about $45 a barrel to justify tapping into the mega offshore discoveries. According to some estimates, the cutback on capital expenditure of around 1–2 % globally could take 1 million bpd of oil off the market in 1 year and help to balance the market (Carpenter 2015).

A key factor causing global uncertainty for producers is to what price level oil prices could yet fall. Some believe that crude could trade in the $40 a barrel range for 2015 and even 2016 which might be close to an absolute price "floor level" but feel that the plunge in oil prices are not sustainable in the long term but that some US shale plays could still be attractive even at $15 a barrel at the wellhead according to the North Dakota Department of Mineral Resources (Bloomberg 2015[k]). Others estimate breakeven costs for plays such as *Bakken* to be in the range of $29–$41 a barrel but that costs of operating the fields may be reduced as engineering and procurement management seek additional efficiency gains and cost concessions from suppliers which might sustain output levels for longer at lower oil prices and render breakeven thresholds even more difficult to estimate (Bloomberg 2015[n]).

Some of the better known personalities in the shale business in the USA such as *Continental Resources Inc.* CEO Harold Hamm, the largest leaseholder and producers in the *Bakken* shale play, was quoted as saying that continental can whether low crude prices "forever" as it idles wells, cuts cots, awaits a rebound or goes into a cash conservation mode (Weber 2015). According to Hamm, whose net worth was estimated at around $11.3 billion, well completion costs should come down in the future as producers save more, and if more producers take this approach, then the biggest drop in oil prices in 5 years will "quickly be a memory" (Weber 2015). But not all shale producers are like *Continental Resources*, with its relatively low debt-to-equity ratio and a healthy credit line to allow it to ride out the sharp drop in oil prices.

According to analysts, the 2014/2015 oil price plunge, the biggest since 2008, is prompting bond traders to treat some $27 billion of investment grade energy debt as junk, amid concern that those companies will have to cut spending to conserve cash and investors are now demanding more yield premiums to own the debt of high-trade companies such as *Transocean Ltd.*, *Noble Corp.*, *and Continental Resources Inc.* and then the average for bonds with the highest junk rating (Natarajan 2015). Figure 4.26 illustrates the fall in US energy corporate junk bonds in comparison with the S&P 500 Index.

Fig. 4.26 US energy corporate junk bonds vs. S&P 500

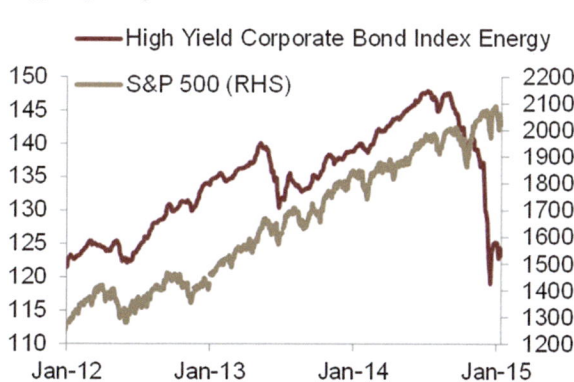

Source: Jadwa (2013, p. 8)

According to analysts, the US shale oil industry took advantage of the low interest environment and higher risk appetite in recent years in the USA to obtain financing via the high-yield corporate bond market, but that as prices dropped, the fear of default has risen resulting in a sharp sell-off and drop in value as illustrated in the above figure (Jadwa Investment 2015[a, b], p. 9). The consequences of this decline is that for the smaller shale producers new financing will become limited and expensive which will increase breakeven costs for shale production further and impact new drilling which was noted earlier by the reduction in rig numbers. According to *Bloomberg*, the average spread on bonds with a Bloomberg composite rating of BB+ is 361 basis points over government bonds with yield premiums for the larger producers such as *Transocean, Noble, Weatherford International, and Continental Resources* exceeding that level as some of these companies carry significant amounts of debt. For example, *Transocean* has $9.1 billion of obligations, while *Continental Resources* has $5.8 billion of borrowings (Natarajan 2015). However, not all is pessimistic for those investing in oil bonds as in April 2015 it was reported that the holders of three of the biggest US exchange traded funds (ETFs) that follow oil had withdrawn almost $300 million after oil prices rebounded to $55 pb from a low of $43.5 in March, but some of the ETF outflows could also be due to traders cutting some of their losses (Zhou 2015[b]). Some analysts such as New York Hedge Fund manager *David Einhorn* (who shot to fame after exposing illiquid real estate investments at Lehman Brothers months before its 2008 collapse) predict that the debt-led shale oil boom, "basically a spending binge fuelled on cheap credit," could become the next big financial catastrophe waiting to happen. He furthermore added that if oil prices remain in the ranges of under $60 a barrel, that the cost of acquiring, fracking, and developing a barrel of oil has now outstripped internal cash flows made from selling the actual oil and that large US oil frackers have spent $80 bn more than they have received from selling oil since 2006 and as a result share prices of the fastest growing US energy companies such as *Pioneer, Concho Resources, and EOG Resources* have grown completely out of range with their actual return on capital (Williams 2015).

OPEC and the Shale Producers: Who Blinks First?

The shale energy revolution is a reality and here to stay. Conventional producers and specifically the major OPEC producers have to learn to adapt to this new reality. The options seem to center on a new strategy that Saudi Arabia imposed on other fellow OPEC members following the November 2014 OPEC meeting: hold firm with no production cuts and let market demand and supply determine oil prices, even if this means financial pain in the interim until oil prices recover once higher-priced marginal shale producers and other non-OPEC members cut production. If OPEC had hoped that official US government policy might intervene to try and promote some cooperation on production and prices between US suppliers and OPEC, they soon realized that the US government would not intervene and that the official US policy was to let "the market" decide what happens according to *Amos Hochstein*, US Special Envoy and Coordinator for International Affairs at the State Department's

Bureau of Energy Resources (Dipaola 2015[a]). With this impasse, and given that the US shale oil producers are a multitude of private firms with little or no US government influence on their production decisions, unlike OPEC's sovereign producers, the question then arises on who blinks first and cuts back on production?

To answer this question, one must assess the position of both protagonists for market share from short and long run perspectives. In the short run, there are indications that there is some cutback in actual and planned production and new well drilling, but the results are uneven across all plays, with some of the major shale plays confident that they can weather even lower oil prices from around $45 a barrel. At the official level, OPEC cut its forecasts for global oil supply growth for 2015 and lowered its estimates for non-OPEC supply by about 400,000 barrels a day, led by a reduction of around 130,000 barrels a day in the USA (Smith 2015[a]). The organization's own supply estimates indicated an increased amount of crude of around 29.2 million barrels a day for 2015, still 1 million barrels a day below the 2014 output (Smith 2015[b]).

The basic fact is that there are few who are able to predict with precision on where market prices might fall to or how quickly and to what levels they will rebound, and the situation is not helped by contradictory policies by some leading OPEC members. The firm "no production cut" policy has been endorsed by OPEC's largest producers—Saudi Arabia, Kuwait, and the UAE, but it also comes as a surprise to note UAE's announcement that, despite low prices in 2015, the country will stick to its plan to *boost* crude production capacity to 3.5 million barrels a day in 2017 even in the face of global oversupply, compared with 2.7 million barrels in January 2015 with a capacity of 3 million barrels a day (Dipaola 2015[b]).

This UAE's planned increase in production seems to fly in the face of the November 2014 OPEC agreement when OPEC confirmed that the organization was effectively entering a battle for market share with both non-OPEC and US shale producers and which prompted the Venezuelan Foreign Minister *Rafael Ramirez* to state bluntly that "OPEC is always fighting with the US because the US has declared it is always against OPEC …. Shale oil is a disaster as a method of production, the fracking. But it is also too expensive and there we are going to see what will happen with production" (Arab News, Nov. 29, 2014[a]). Global energy politics does not operate in a vacuum and OPEC knows that some form of cooperation or mutual understanding has to be another option to try, at least with major non-OPEC conventional producers.

In January 2015, there were reported meetings between Saudi Oil Minister Ali Al Naimi with the Ambassadors of Russia, Norway, as well as Finland, ostensibly to discuss joint investment in the petroleum sectors of these countries (Tuttle 2015) as well as separate meetings between Russia and Saudi Arabia (Razzouk 2015). It was not only the UAE that decided that the time was opportune to announce an increase in its future production capacity. In February 2015, Kuwait announced that it would increase its capacity to 3.15 million barrels a day in 2016, up from around 3 million barrels a day in 2015 and increase its drilling rigs to 120 from 80 based on higher oil prices of $60 a barrel in mid-February, which some members viewed as a vindication of their policy (Mahdi and MacDonald 2015), while Saudi Arabian oil production registered a new peak of 10.3 million bpd in March 2015 with the Kingdom vowing to maintain production at the 10 million bpd levels from now on, according to its oil minister (Reuters 2015[b]).

Having a better understanding of Saudi Arabia's intentions, both in the short and long term, is a key issue in understanding how OPEC's strategy continues to unfold. In the short term, OPEC's largest producer Saudi Arabia despite holding substantial financial reserves was not immune from trimming on its planned capital investments in the face of the sharp fall in oil prices. Saudi Aramco announced in early 2015 that its capital investment will be lower than the company's own target for 2015, but still "higher than 2013" (Carey and Dipaola 2015), with Saudi Aramco also stating that it saw other forms of renewable energy such as solar power being attractive. While the Saudi company opted for rationalization of some of its planned capital expenditure, it took the market by surprise by announcing that Saudi Arabia would be investing substantial amounts to develop its own shale projects with an additional $7 billion marked for developing unconventional gas, according to former Saudi Aramco CEO Khalid Al Falih (Arab News, Jan. 28, 2015). This planned expenditure is over and above what Aramco had already invested in unconventional gas to the tune of $3 billion according to the company CEO, who also stated that the US shale innovation had led the way for Saudi Arabia to pursue similar techniques and that the Kingdom "will be the next frontier after the US where shale and unconventional will make a contribution to our energy mix, especially gas" (Arab News, ibid). In line with what the Saudi Oil Minister had been stating, the Aramco CEO also stressed that the Kingdom will not single-handedly balance the global oil market and that Saudi Arabia "got spoiled with $100 oil and ...were focused on building capacity and we lost focus on fiscal disciple" (Arab News, ibid). An implication of this comment is that should oil prices remain depressed over a long period of time and fall to $20 levels, then Saudi Arabia might reconsider its current policy of maintaining an idle and expensive spare capacity of 2.5–3.0 million barrels a day.

The Kingdom's sudden and new-found enthusiasm to become the next shale frontier after the USA is also somewhat of a surprise given previous official pronouncements that Saudi Arabia was "unconcerned" by US shale output. It also leaves unanswered the question of the additional desalinated water supplies and energy input required to extract shale gas. Prince Abdulaziz bin Salman, then Assistant Deputy Minister of Petroleum and Minerals Resources, said in November 2013 that Saudi Arabia remained unconcerned by surging US shale output and that "we need to make sure that the world economy comes out decisively on a growth pattern, and if that can be established ... the world economic growth will be able to handle growth from all sorts, shale oil, shale gas, tight oil and including renewable" and, furthermore, that the Kingdom "welcomes new resources of energy supplies as they are needed" (Arab News, Nov. 21, 2013). What a difference 1 year can later make given the depressed global oil prices in November 2014. The prince was promoted to Deputy Minister in the February 2015 government reshuffle by the new *Saudi King Salman bin Abdulaziz*.

Should Saudi Arabia have reasons to be concerned over US shale and potential loss of Saudi and OPEC market share? According to BP's 2015 statistical review of World Energy, the USA passed both Saudi Arabia and Russia to become the world's largest oil producer for the first time since 1975 (BP 2015). Apparently US oil production increased by 1.6 million bpd in 2014, the first time the country had increased production by more than 1 million bpd for 3 consecutive years (ibid). To answer the question, one has to be either a conventional resource optimist or a pessimist. While undoubtedly there has been a tight oil production revolution brought about by a fortuitous confluence of technology breakthrough and unique enabling factors in the USA, it is doubtful that tight oil production will represent more than 3–5 % of total liquid supply in the long run and that the growth in oil production from this source in the USA or elsewhere will materially affect Saudi Arabia's long-term oil position.

A key factor in this calculation is the future price of oil. As relatively cheaper conventional sources of oil are depleted, the production of oil from relatively more expensive unconventional sources will increase, but this also *assumes that no new conventional energy sources are exploited*. As noted in this volume, deep water and heavy oil resources whether offshore in Brazil, Venezuela, or Canada can be developed as technology and oil prices justify.

Is the US shale sector ready to assume the role of "swing producer," given that in the face of the decline in oil prices from $115 to $60 pb nearly 3000 wells went off the market and producers cut billions in spending? If the US shale is to act as a "swing producer," recoveries will become more volatile as it will take a large rise in oil prices to bring back idle shale production, as operators will have to wait for hydraulic fracturing fluids and completion rigs and laid off workers to come back. These cannot be as easily mobilized by shale operators as OPEC countries like Saudi Arabia with its large spare capacity and its demonstrated ability to add significant new output as it did during March 2015. While US shale producers' responses might be less consistent than those of Saudi Arabia, they are solely driven by market conditions and US shale producers may, in fact, inadvertently be now serving as a *check* on the rest of the world when oil prices rise, but it will take time.

In the short term, the production of tight US oil could narrow the price differential between heavy and light crude which could bring about a restructuring of the downstream refining sector. While Saudi and other OPEC producers of conventional oil might face a less uncertain future, the situation is not the same for *OPEC's gas producers*. The large production of shale gas in the USA and the potential for this to be replicated in other major world economies such as China could bring about a large production of cheap NGL by-products that could also affect not only OPEC NGL producers like Qatar, but also OPEC members' petrochemical industries.

The US shale gas revolution and supply reduced NGL prices and have forced Qatar to shelve its plans to export NGL to the USA and seek other Asian markets. At the same time, cheaper US feedstock is reducing the comparative profitability of Saudi Arabia's large petrochemical sector and inducing the sector to consider overseas expansion, particularly in the USA, to profit from a new source of cheap feedstock. Saudi Arabia's giant petrochemical basic industries—*SABIC*—has already moved in this direction with joint ventures in the USA, as illustrated in Table 4.6, that set out proposed US ethylene capacity additions to 2020.

Table 4.6 Proposed US ethylene capacity additions—2013–2020 (Million tons per year)

Company	Location	Proposed capacity
Chevron Phillips	Baytown, TX	1.5
Exxon Mobil	Baytown, TX	1.5
Sasol	Lake Charles, LA	1.4
Dow	Freeport, TX	1.4
Shell	Beaver Co., PA	1.3
Formosa	Point Comfort, TX	0.8
Occidental/Mexichem	Ingleside, TX	0.5
Dow	St. Charles, LA	0.4
LyondellBasell	Laporte, TX	0.4
Aither Chemicals	Kanawha, WV	0.3
Williams/SABIC	Geismar, LA	0.2
Ineos	Alvin, TX	0.2
Westlake	Lake Charles, LA	0.2
Williams/SABIC	Geismar, LA	0.1
Total		10.1

Source: Jadwa (2013, p. 52), EIA Annual Energy Outlook (2013[b])

These new capacity plans would add about 10.1 million tons per year which would increase US ethylene capacity by 40 % and the world's capacity by around 7 %. Saudi Arabia has already started on its own mega petrochemical projects—the $10 billion *SATORP* (Saudi Aramco Total Refinery) and the $20 billion *SADARA* (Saudi Aramco Dow Refinery) which are expected to add 1 million metric tons p.a. (and 400,000 barrels of refined fuel per year) and 3 million metric tons per year, respectively, underscoring Saudi Arabia's ambitious plans for exploiting the Kingdom's own shale gas as highlighted earlier. According to analysts, the "indirect" impact on the Arab Gulf petrochemical plans could be more severe than the direct impact from oil production by shale producers. The large increase in US NGL production has helped the country to cut its LPG (Liquefied Petrol Gas) imports by more than half, while its own exports of NGL have increased sharply, and as a result, LPG prices have declined and countries like Saudi Arabia have lost market share to US producers in Central and Latin America (Al Hajji 2014[b]). The key to the competitiveness of the petrochemical industry is the low cost of feedstock and energy and the US shale revolution has made the US petrochemical industry very competitive by providing cheap ethane and fuel. With ethane prices declining by about 70 % over the period 2012–2014, resulting it being cheaper than natural gas, the lower price of ethane increased the profitability of the US petroleum sector. Given the tight natural gas market in Saudi Arabia and the abundance of tight natural gas in the USA, the cost of US ethylene will be the lowest in the world, hence underlining SABIC's joint venture projects in the USA. To sum up, according to analysts, the US shale revolution will "make global markets for petroleum products, NGL's and petrochemical more competitive and more efficient, by reducing the role of government-owned companies and increasing the role of the private sector.... This increase in competitiveness is the most important impact of the shale revolution" (Al Hajji 2014[a]).

Shale Oil and OPEC: A Threat or a Stabilizer?

The Bahraini Energy Minister *Abdulhussein Mirza* stated at a conference of oil industry professionals in Manama on March 8, 2015, that the shale oil revolution "had the most profound impact on the market since many years." The pressing questions now are how does OPEC view the shale oil revolution? Is it a threat or a source of stability to the market and OPEC's position? Do OPEC officials have deep understanding of the shale oil revolution? For OPEC, shale oil is a stabilizer for the market in the long run as it can help meet an expected increase in world demand (El-Badri March 2015), but in the short term, it is viewed as a source of threat and stability for oil prices. Supplies of US shale oil contributed to market stability in 2013 when the market experienced disruptions from some OPEC members, but once supply from OPEC countries started to recover, shale oil started to be viewed as a short-term competitor for OPEC's light crude and a threat for its global market share. Shale oil, in the view of OPEC, was also identified as a high-cost producer that contributed to the excess supply in the market that exerted pressure on prices starting from the mid of 2014.

This was a radical change in the way Saudi policymakers viewed the impact of the technological advances of the "shale oil revolution" on the market and on prices. At first Saudi policymakers argued that the development cost of unconventional oil, a group of oil resources that requires special and costly methods of recovery, will set a floor for oil prices as they are considered to be high-cost developments (Al Naimi June 2014; Abdulaziz bin Salman Sept. 2014). The Saudi Oil Minister Ali Al Naimi considered the high-cost shale oil developments to have many benefits for the market as they will set a high floor for prices that will enhance long-term investments in the industry, provide a sense of security to customers, and add depth to markets and help in stabilizing them (Al Naimi 2014). Saudi Deputy Oil Minister Prince Abdulaziz bin Salman argued that unconventional resources in general and not only shale oil will have an impact on the price and market balance relation in the long run. For the petroleum market to be in balance, oil prices have to stay high as all the new developments are costly, and oil prices in the long term will remain high supported by the high cost of the marginal barrel of unconventional resources (Abdulaziz bin Salman Sept. 2014).

It was clear that OPEC was depending on shale oil to be a marginal producer that can set a high floor for oil prices as the group lost that pricing power. The main reason why it was welcomed at the beginning is because it was, first, a substitute for OPEC's shut-in capacity and, second, a high-cost producer. As these two conditions no longer are the case, there was a *change of heart* at OPEC toward shale oil. The shale oil revolution developed rapidly to the extent that it left OPEC as a whole and its individual member countries unable to fully breakdown and grasp its different elements. The different views that member countries held on shale oil is a testimony for that inability to catch up with the rapid technological and economic developments in the production of shale oil. For some who had warned OPEC in 2012 that shale oil "is here to stay" but with the advice falling on deaf ears, it must have come as bittersweet for Conoco Phillips CEO *Ryan Lance* when he repeated the message during OPEC's June 2015 meeting. His message again was that shale oil had not only transformed the global energy industry, but had proved far more resilient to lower prices than most had expected and the increase in shale oil production of 3.3 million bpd—more than the output of the UAE—over the 3-year period from when he had made his first warning, put this growth in context (Blas 2015ᵇ).

Have such warnings been taken on board and what have been the public pronouncements of key OPEC officials concerning their position on the emerging challenges from US shale oil? Nowhere is that found than during a gathering in Abu Dhabi in December 2014, less than a month after the historic November OPEC 166th meeting, where oil officials convened for the tenth Arab Energy Club. In one panel session, both the heads of OPEC and IEA along with Saudi oil minister were unable to come up with a single view on shale oil breakeven cost. The quotations below indicate that there is still a wide discrepancy in viewpoints with often contradictory statements and differences on the critical issue of shale oil breakeven pricing:

- Saudi Oil Minister:

  "Shale oil rocks are not homogenous. Some areas have higher porosity than others. Sweet spots areas that have good porosity can produce at $20–$30/B. But some places need $80–$90/B to make it."

- OPEC Secretary-General:

  "$70–$80 is average cost of tight oil. No single study can give decisive answers on cost of shale oil."

- IEA Executive Director:

  "Cost of shale is difficult to identify as it varies from field to field. Costs of production differ from well to well and according to efficiency and technology."

A few months later, in March 2015, the OPEC Secretary-General appeared in Manama, Bahrain, stating that production of shale oil needs oil prices to remain above $100 to be sustainable over the long run as $70, $80, or $90 cannot make shale oil production feasible (El-Badri March 2015). He also confirmed that OPEC still welcomes shale oil, but he argued that "OPEC cannot subsidize another source of energy—if we reduce (production) in November we will reduce in January. We will reduce in December. We will reduce maybe for another four to five years" (Ibid). It is very clear to notice from the above statements that when OPEC officials entered the closed meeting on November 27, 2014, they had not a unified single view on shale oil and its impact. They apparently did not have a clear understanding at what price shale oil production will start to slow down. This was evident in the words of Iran's Oil Minister *Bijan Zanganeh* who told Bloomberg in an interview in Tehran 2 days after the November 2014 meeting: "there is no fact or figure to say that shale production will definitely decrease. With these prices that we see, it's not reason enough to say that definitely in the next 4–5 months shale production will decrease by 1 or 2 million barrels" (Motevalli Nov. 2014).

For most participants at future OPEC meetings, the question is now no longer at what prices shale producers would go bankrupt, but at what prices *they start drilling again*, boosting output further and putting into question OPEC's production target, which in effect is a free-for-all policy.

Given this scenario, should OPEC and, principally its largest member, Saudi Arabia feel that they can face the future with more confidence against the US shale with more confidence against the US shale producers and what can they do to explore new opportunities? For OPEC to feel more secure, it will be important that it closely monitors what happens in two key countries: China as a major consumer of OPEC energy and potential nonconventional shale gas producer and Russia as a major conventional and potential nonconventional oil producer.

Enter the Dragon But Beware of the Year of the Goat

The rapid economic growth of China has generated an equal mix of anxiety and opportunity both for China and the rest of the world. In the political sphere, there is anxiety by some, especially the USA and South East Asian countries like Japan, South Korea, and Vietnam, that Chinese economic growth could translate into potential disputes over natural resources. For the Chinese, assurance of sufficient energy supplies is a national imperative and crucial to energy security to ensure that shortages of energy do not constrain the country's economic growth necessary to reduce poverty and avoid domestic political discontent (Yergin 2011, p. 193). For oil producers, especially those in OPEC with spare capacity, China represents an increasingly important element in their energy supply equation, in a market that has become oversupplied by marginal non-OPEC producers. One now looks to China, along with other emerging Asian economies like India, as a market of decisive importance with significant impact on global supply and demand and hence on world prices. It was a combination of increased non-OPEC marginal oil supplies and a weaker Asian, especially Chinese demand during the latter part of 2014, which combined to reduce world oil prices. By 1993, the Chinese were no longer self-sufficient in domestic energy and imports began to assume a greater share as illustrated in Fig. 4.27.

Fig. 4.27 China's overall oil production and surplus (1980–2011)

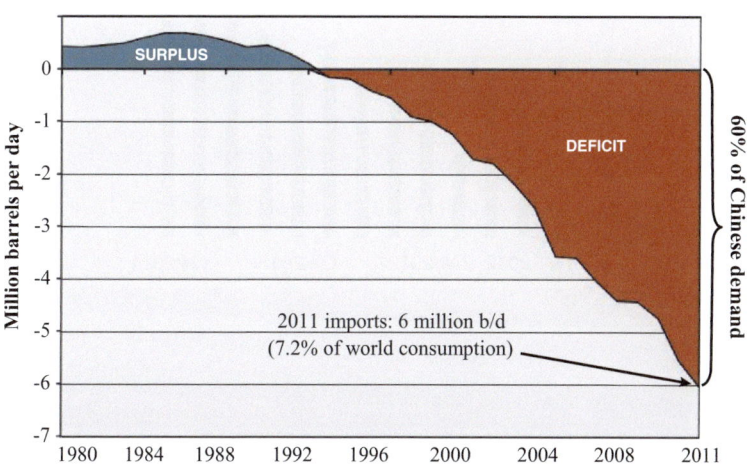

Source: IEA (2013)

By 2011, total Chinese oil imports had reached 6 million bpd representing around 7.2 % of world consumption. In 2013, according to BP, China held 18.1 billion barrels of oil reserves or less than 1 % of world reserves while domestic production totaled 4.180 million bpd and consumption at 10.756 million bpd, or 5 % and 12.1 % of world total, respectively (BP 2014, pp. 6–8). Given its spectacular economic growth, China may seek to secure more energy deals from the Middle East and elsewhere, putting it on a possible collision course with other countries' energy interests (Salameh 2012, p. 107).

The Chinese embrace a policy of mutual respect in the affairs of other countries and place great store on the Chinese Zodiac calendar, or *Shengxiao,* which repeats itself every 12 years. The year 2015 is the Chinese *Year of the Goat*, which lasts until February 2016 and is supposed to bring peace and prosperity, but four previous visits of this auspicious animal have brought chaos and instability to the Middle East region with some of the effects continuing (Lee 2015). The four previous visits were: *1967* (the Six-Day War between the Arabs and Israel and the continuing conflict); *1979* (the Islamic revolution in Iran, with Iranian oil output falling from 6 million bpd to around 3 million since then); *1991* (Operation Desert Storm to expel Iraqi forces from Kuwait and the destruction of Kuwaiti oil wells—and dumping of up to 6 million barrels of oil into the Arabian Gulf—with sanctions on Iraqi oil exports until 1996 before easing); and *2003* (US-led invasion of Iraq and continuing instability in the country, with Iraqi oil production falling to 300,000 barrels a day from 7.5 million barrels a day, and only recently rising). The 2014/2015 fall in oil prices is not an auspicious start to the new Chinese Year of the Goat for oil producers of the Middle East, if past history is anything to go by.

Fig. 4.28 Chinese oil imports

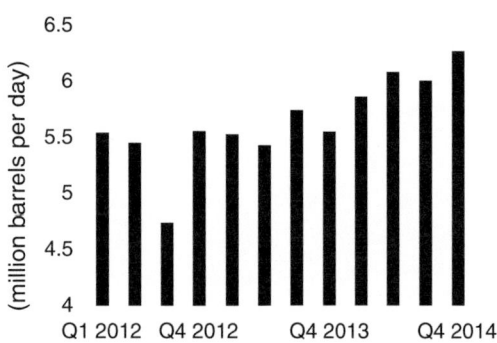

Source: Jadwa (2015[a], p. 3)

The pace of economic growth in China is of crucial importance to many oil-exporting countries, both OPEC and non-OPEC, and the level of Chinese oil imports has fluctuated from 4.6 million barrels a day to 6.3 million barrels a day as Fig. 4.28 illustrates.

The fluctuations in Chinese oil imports, especially in periods of low oil prices, are a cause of some concern for producers aiming to capture some of this Asian market, while at the same time trying to assess China's domestic energy supplies from biofuels and coal to liquids. China has invested significant amounts in the development and initial commercialization of coal-to-liquids technology in order to produce liquids which can be used for transport and petrochemicals (Andrews 2010, p. 15). According to analysts, the Chinese talk about development of this sector more as an "insurance policy" against imported energy disruptions rather than large-scale substitution (Yergin 2011, p. 214). Fueled by trade, China's growth has strongly supported diesel demand, and this has been boosted by the Chinese government's mandate requiring all trucks to be fuelled by diesel since 2010. Gasoline demand has been supported primarily by consumers, with the transportation (non-truck) sector accounting for nearly 50 % of total gasoline consumption, given the astonishing rise in private car ownership with total passenger vehicles growing from just under 13 million in 1997 to over 100 million vehicles in 2012 (Fattouh and Sen 2014).

China's Nonconventional Gas Reserves: Large But Difficult to Exploit

As noted earlier, China is believed to posses the largest shale gas resources in the world, surpassing even those of the USA with an estimated 1115 trillion cubic feet according to the EIA (2013[a]). These technically recoverable basins are located in seven areas—*Sichuan, Tarim, Junggar, Yangtze, Song Liao, Jiangnan, and Subei* as illustrated in Fig. 4.29a that follows, while Fig. 4.29b illustrates the projected production of Chinese gas by the year 2035.

Fig. 4.29 (**a**) Major unconventional natural gas resources in China. (**b**) Natural gas balance in China (*The sum of production and net imports represents total demand*)

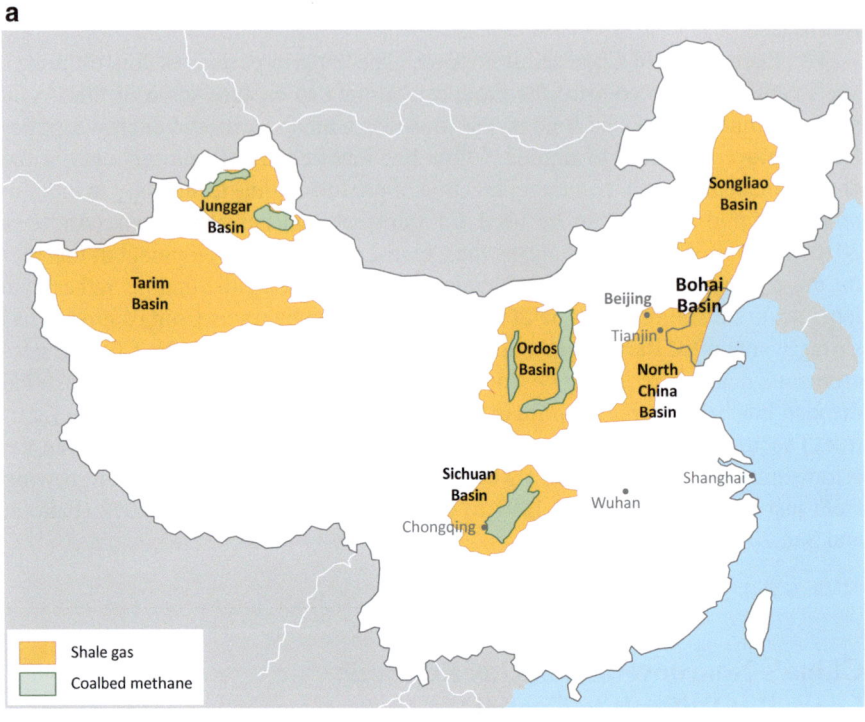

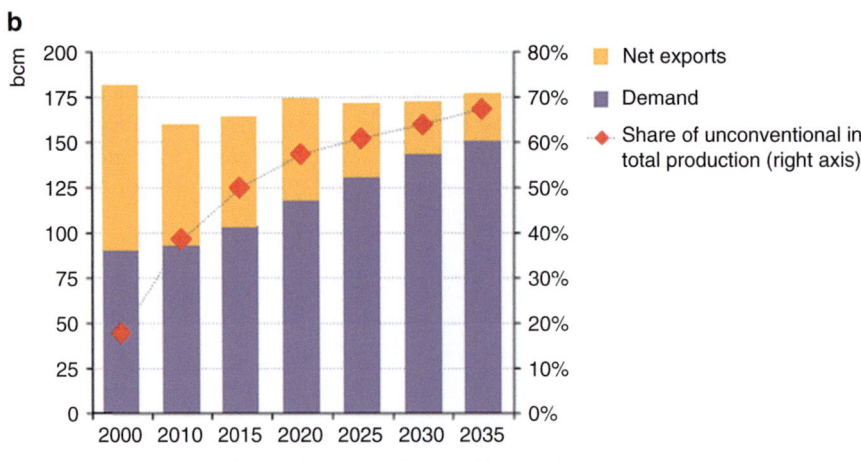

Source: OECD/IEA (2012[b], p. 116), OECD/IEA (2012[b], p. 115)

According to geologists, the largest and most promising basins are in *Sichuan*, which is a highly populated area with about 56 % of the total estimated shale gas reserve, and the other in a more remote and water stressed region of *Tarim*. The exploitation of China's huge shale reserves has been somewhat difficult at best, and despite government subsidies, it has not progressed to meet the 2015 target of 6.5 billion cubic meters and 60–100 billion cubic meters target by 2020, as initial wells confirmed that some formations have a high clay content making them more pliable and less apt to fracture and are deeper compared to the US shale gas wells, causing Shell and its local partner *China National Petroleum Corporation* (*CNPC*) to report significant aboveground challenges (Jadwa 2013, p. 46). These included inadequate power and road infrastructure, forcing the number of wells that could be drilled from a single site to shrink, as well as issues concerning water pollution (Jadwa, ibid). As such, it would seem that large-scale exploitation of China's shale gas reserves, and reducing the country's external energy dependency like in the USA, is still at an early stage.

Maintaining Producer's Market Share: Discount and More Discounts

The relations between OPEC and non-OPEC producers and the Chinese government have been critical to the development of energy relations, especially in the case of countries with large oil resources such as Saudi Arabia, Russia, Iran, Iraq, and Kuwait, with cordial relations often being cemented at the highest political levels of both oil-exporting countries and Chinese government officials (Andrews 2010). Such cordial relations have not stopped leading oil producers from engaging in some sharp price discounting in order to protect their diminished market share in both the Asian and other major export markets as illustrated in Figs. 4.29 and 4.30, illustrating Saudi Arabia's Official Selling Price, or *OSP*, discounts and the competition for Chinese oil exports.

Fig. 4.30 Saudi OSP's discount

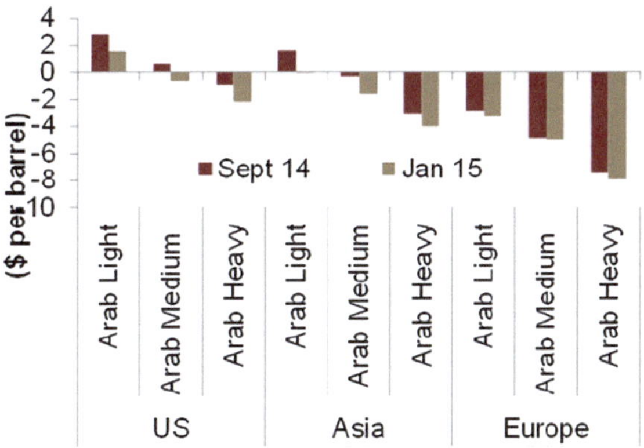

Source: Jadwa (2015[a, b], p. 6)

In Fig. 4.29, Saudi Arabia's response to the sharp fall in oil prices from the second half of 2014 has been to decrease its *OSP* for virtually all regions of the world, whether Europe, America, or Asia, with the heaviest *OSP* discounts in Europe, by as much as $8 a barrel in January 2015 for Arab Heavy crude. In the USA, Saudi

Fig. 4.31 Chinese oil imports. *Asterisk*: Year to November

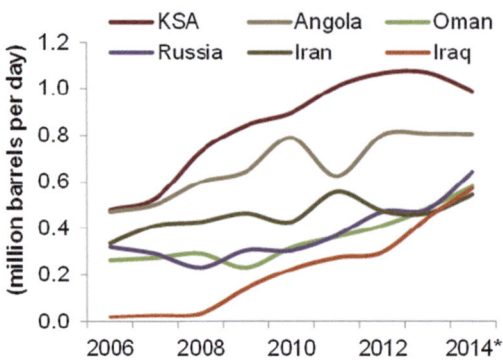

Source: Jadwa (2015[a, b], p. 6)

Arabia's export of its heavy crude came under pressure from Canadian heavy crude imports to the USA, underscoring the potential deterioration in Saudi Arabia's future heavy crude exports to the USA should the Canadian *Keystone* pipeline project be finally approved as discussed earlier.

Figure 4.31 illustrates the varied sources of Chinese oil imports from both OPEC and non-OPEC members like Russia and Oman and indicates that Saudi crude sales to China had lost out to Iraq, Iran, and Russia, forcing Saudi Arabia to offer even bigger discounts for future oil cargoes to Asia (Cho and Zhu 2015). Despite the sharp fall in prices, Saudi market share fell by 5.7 % in 2014, led by a decline in China, its biggest customer, which cut Saudi crude imports by 7.9 % in 2014 while it increased its imports from Iran, Iraq, Kuwait, Angola, and the UAE (Mahdi 2015[a]). The decision to cut *OSPs* by Saudi Arabia and its insistence that OPEC does not cut back on production based on the November 2014 OPEC decision indicates that, in a competitive global oil market, prices were not a priority in the short term but the maintenance of market share being the primary objective. The problem though for the major oil exporters is that with relatively abundant non-OPEC supplies still on the market, the level of competitive *OSP* discounting by different countries could very well depress global prices further, fueling a viscous cycle of downward oil prices.

The Russians Are Coming: The Bear Awakens

The Russians are coming back in the global energy sector and in 2013 Russian output stood at 10.7 million barrels per day or nearly 13 % of the total world output (BP 2014). While Russian proven oil reserves of 93 billion barrels are not as large as other major OPEC members like Saudi Arabia, Iraq, or Iran, yet current Russian oil production exceeded those of Saudi Arabia's average 9.2 million barrels a day for 2013 (BP 2014). The figure below illustrates the remarkable resurgence in Russian oil production over the period 1994–2013 (Fig. 4.32).

Fig. 4.32 Resurgent Russian oil production (1994–2013)

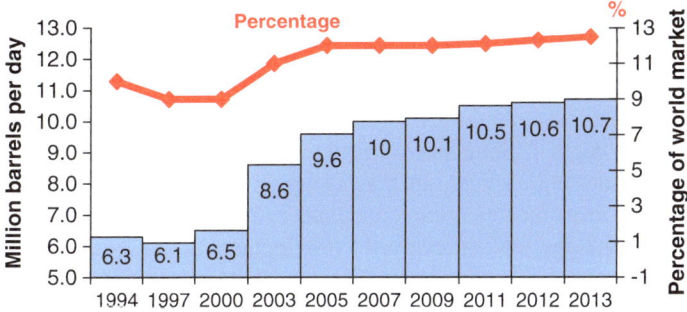

Source: BP (2014)

Three-quarters of Russian oil has been produced from the Western Siberian Basin, a region that contains some of the richest petroleum accumulation on Earth. Spanning 2.2 million square kilometers, it is the largest petroleum basin in the world (Gorelick 2010, p. 143). As noted earlier, Russia is reputedly the top country with 75 billion barrels of technically recoverable shale oil resources, but there are other estimates of the country's oil reserves. These indicate that so far, 144 billion barrels of oil have been discovered and that the estimate of *undiscovered* oil is another 55 billion barrels, which, if taken together with the unconventional shale oil reserves, would place Russia in the top league of oil reserve nations, enough for Russia to support around 20 years of current global-energy equivalent oil production (Umishek 2003). With such potential energy reserves, Russia is a critical player in the world's energy market, and how OPEC and Russia interact on production and prices becomes an important matter for both. It is certainly of some concern to OPEC that Russian oil imports have been rising despite the 2014 fall in prices, as illustrated in Fig. 4.33.

Fig. 4.33 Russian oil exports

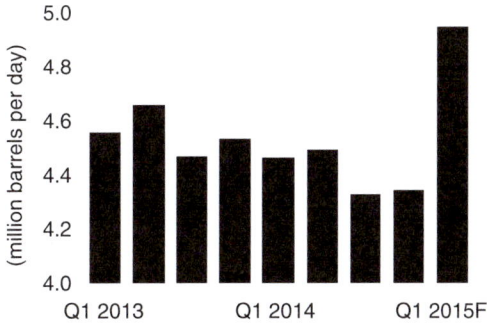

Source: Jadwa (2015[b], p. 5)

According to analysts, Q4 2014 exports from Russia totaled 4.3 million bpd, down 4 % in 2014, but that this drop was mainly due to exporters delaying crude shipments to take advantage of the new tax regime introduced by the Russian government in early 2015, which resulted in almost a 40 % drop in duties compared to 2014. Analysts are forecasting a gradual rise in Russian oil exports to around 5 million bpd in 2015, with these levels sustained (Jadwa Investment 2015[a, b], p. 6). The centrality of oil and gas to the Russian economy has highlighted an important question faced by increased economic and trade sanctions by Western governments against Russia due to the Crimea/Ukraine events: would Russia be able to maintain its level of oil output or will it witness a decline? Some have argued that Russia would not be able to sustain production without big changes, such as a step-up in new investment, a tax regime that encouraged investment (similar to the January 2015 duties reduction), augmentation of new technology, and, crucially, the development of the so-called next generation of oil and gas fields, particularly offshore in the Arctic region. All these will be challenging and costly (Yergin 2011, p. 41). To do this requires both Russian and foreign capital and Western expertise.

While Russia has tremendous long-term prospects, the imposition of Western sanctions in 2014 has added complications on both its drilling technologies and external debt financing in foreign currency. In and of itself, the sharp oil price decline is not expected to directly hit Russian oil production due to the Russian *ruble* depreciation as illustrated below (Fig. 4.34).

Fig. 4.34 The Russian ruble's decline (2014) (rubles per US dollar)

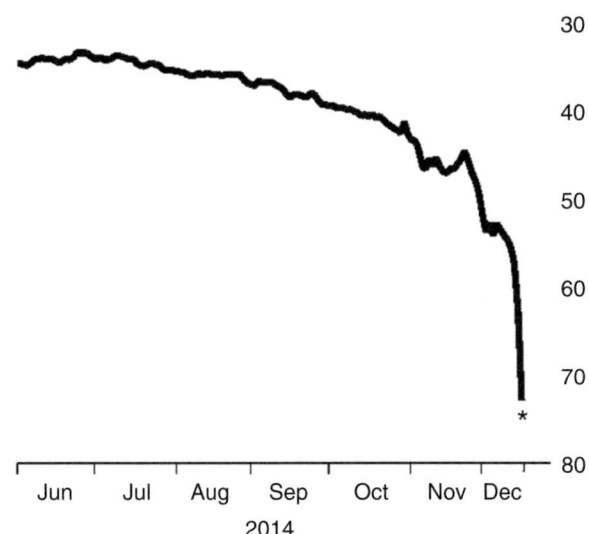

Source: Thompson Reuters, Bloomberg

The reason is that the *ruble*-based costs of a large number of Russian oil firms acts as a counterweight compared with external debt financing which is more of a serious issue for Russian firms such as *Rosneft* which has large short-term US dollar debt commitments, and ruble-based debt issuance has precipitated further ruble depreciation in anticipation that the cash raised will be used to buy US dollars (Citi Research 2015, p. 8).

While John D. Rockefeller was attributed to have used the phrase "a good sweating will be healthy for them" to describe his company Standard Oil, it is still an open issue on whether the current Russian financial predicament can be phrased in the same terms as Rockefeller's. However, past history has indicated that Russia's energy-related capital expenditure has indeed suffered due to a shortage in funding as occurred during the 2008/2009 global financial crisis and credit rationing. This is illustrated in Fig. 4.35.

Fig. 4.35 Russia: significant drops in capex in the past can occur again

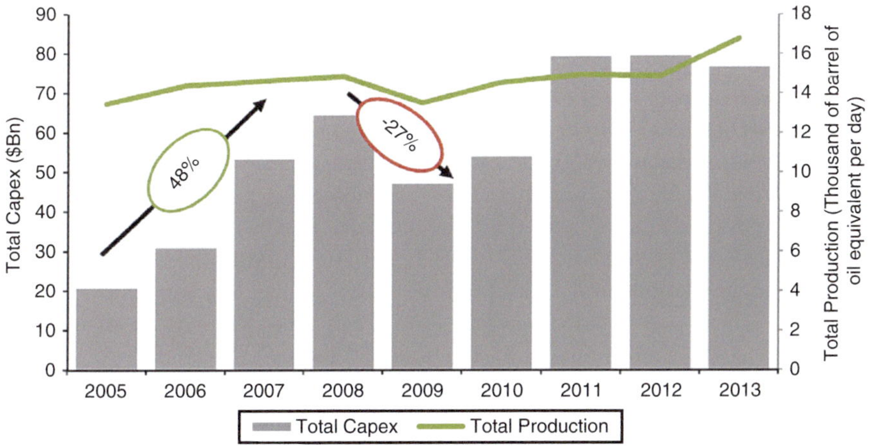

Source: Bernstein Research (2014, p. 14)

During the 2008/2009 financial crisis, there was almost a 30 % drop in overall Russian capital expenditure (*capex*, leading to a reduction in production, but this was quickly reversed when the financial markets stabilized and Russian energy risk was perceived to be acceptable. This however was in a period of "normal" geopolitical relations compared to the post 2014 sanctions era. It is estimated that Russian *capex* declines during 2015 will be severe, with a reduction of 20 % at average prices of $65 a barrel for 2015 (Bernstein Research 2014, p. 1). The fact that both Moody's and Standard & Poor's downgraded Russia's debt to junk bond status in 2015 will not help Russian companies access international capital markets at favorable rates and can only hasten the fall in forecasted capital expenditure (Guardian 2015).

Russia and OPEC

The initial OPEC approach to Russia to discuss possible mutual production cuts prior to the OPEC November 2014 meeting was rebuffed by the Russians with no deal made (Bloomberg 2014^e). This however did not stop OPEC's largest member

Saudi Arabia from holding some cordial discussions with Russia on possible cooperation and market stability (Tuttle 2015), which was later followed by a high-profile meeting between Saudi Oil Minister Ali Al Naimi with *Gazprom's* Chairman and Presidential envoy *Viktor Zubkov* in February 2015 (Razzouk 2015). During the annual Davos World Economic Forum in January 2015, the OPEC Secretary-General Abdullah El-Badri went to great pain to deny that OPEC was targeting other countries by stating that

> the organization "… is not targeting. We are not targeting the United States shale oil. And we are not targeting Russia. We are not targeting anybody. It is a pure economic decision (not to cut production) for the interest of all member countries" (Bloomberg 2015[1]).

While there might be no targeting of other oil producers, one of the problems between OPEC and non-OPEC producers is that there is not much familiarity about their respective energy sectors and future plans, as illustrated when *MEES Magazine* interviewed Saudi Oil Minister Al Naimi and asked him on whether he saw Russian oil supply going down by 500,000 barrels per day as a result of lower oil prices and sanctions and the Minister admitting that "I cannot say…. that is unknown because I do not know all Russia's fields (and) the only area I know is West Siberian fields, but I do not know their other fields" (MEES 2014). For those who have researched OPEC-Russia relations, this lack of mutual understanding has not come as a surprise, and they have doubted whether Russia can join OPEC as a full member (Simoniya 2010). They point out to structural differences between the two, arguing that while OPEC moved toward a "cartelization" process and quota mechanism, Russia proceeded in the opposite direction by the disbanding of the old USSR era of a strict centralized energy sector, resulting in the privatization of the state oil assets, with only *Rosneft* remaining in State hands. Production sharing arrangements with foreign companies such as Exxon, Shell, and Total was also another difference with OPEC (Simoniya 2010, p. 176).

Under such divergent ownership structures, even if any Russian government decides to formally join OPEC in the future, the heterogeneous nature of the Russian oil industry will conspire, due to vested private sector ownership and lobbying, to try and maintain their interests, and which might be in conflict with the State's wishes based on tactical or geopolitical reasons. OPEC, or those members of the organization with large financial resources, can cooperate with the Russian energy sector in joint investment and development of the oil and gas provinces either in the oil-producing countries themselves (like the Russian company *Lukoil's* activities in exploration in Saudi Arabia and Iraq) or in East Siberia, the Russian Far East, and the Arctic, with the new Russian focus on simultaneous construction of oil processing, petrochemicals, gas chemical, and gas liquefaction plants, all of which are long- term projects and do not constitute short-term competitive issues for OPEC. Following a high profile visit to Russia in June 2015 by the Saudi Second Deputy Crown Prince Mohammed bin Salman to discuss cooperation in various sectors, Russia's Novak agreed to sign a cooperation pact and Lukoil held talks with Saudi Aramco to increase the company's stake in the Russian - Saudi Luksar exploration company joint venture to 50 % (Bierman, 2015).

Chapter 5
Facing Realities: OPEC Fiscal Stress and Break-Even Pricing

In life, as in chess, forethought wins.

Charles Buxton

Introduction

As explored in earlier chapters, OPEC was unable to coordinate production quotas and increase prices throughout the 1990s, and the situation from 2014 seemed to echo this collective impasse. Earlier periods of collusion were short-lived, based on some production cuts that were undertaken, notably by Saudi Arabia. As noted, OPEC overwhelmingly controls the proven world's reserves of oil, with nearly 80 % of reserves concentrated in Saudi Arabia, Iraq, Iran, and Venezuela. Over the years, countries like Saudi Arabia, Venezuela, Kuwait, and the UAE have amassed significant financial reserves, while some have seen their fortunes, notably Iraq and Iran, affected by either war or embargoes. Toward the end of 2014, all OPEC countries, including those who had amassed large financial surpluses in the past, faced the same questions: for how long could they cope with weak oil prices and what was their "break-even" oil price level to avoid deepening fiscal stress?

OPEC: A Phantom Menace?

From the above, some further questions arise: what possible action can OPEC collectively take and, if not, what options do individual OPEC countries have to postpone the impact of worsening fiscal stress on their national economies and social fabric? The earlier analysis of the workings of a cartel indicated that OPEC did not strictly conform to such a designation as the organization rarely if ever influenced its individual members' oil production rate, but some politicians still believe that it was a cartel, creating the illusion that OPEC was a "phantom menace" (Colgan 2014). This however does not preclude a market perception that while *OPEC as a whole* seems powerless to influence current weak oil prices, some *individual* members, notably Saudi Arabia, do have market influence, which explains why market participants pay very close attention to Saudi Arabia for signals about present and future pricing behavior. However, the sharp fall in prices toward the end of 2014 and continuing weak prices during 2015, has had a far greater impact than previous falls like in 2009 which was attributed to the fallout from the global financial crisis of 2008. This time around there is a higher degree of uncertainty and a change in market psychology, with the oil markets also unsettled by geopolitical events in the Ukraine, the advances of the Islamic State in the Middle East, and, above all, by the unprecedented North American shale oil-led production boom brought about by hydraulic fracturing. According to some analysts, a $40 per barrel price of oil now seems possible (Bloomberg 2014^e) and in fact oil prices fell below $40 a barrel in August 2015 in face of weakening Chinese economic prospects and a fall in global stock markets.

Fiscal Challenges: Addressing Subsidies First

Assessing fiscal challenges and break-even price bands for OPEC countries would not be complete without understanding the current state of energy subsidies in these countries and their cost to the national economies. According to a joint report conducted by the IEA, OPEC, OECD, and the World Bank in 2011, curbing growth in energy demand via subsidy reform would have several important energy security implications for both energy-importing and energy security implications (IEA 2011^b). For net importing countries, lower energy demand would reduce import dependence and consequently spending on imports. For net exporting countries, removing subsidies would boost export capacity and earnings for energy-related products but could potentially have a negative impact on nonenergy sectors, mainly a significant increase in domestic inflation and negative effects on the competitiveness of the manufacturing sector, resulting in a loss of jobs and a rise in unemployment (IEA 2011^b, p. 8).

The above effects illustrate some of the sensitivities and challenges that governments in OPEC face with regard to phasing out energy subsidies, especially in the aftermath of the so-called Arab Spring and an increase in government social welfare and subsidy programs in the major GCC economies (Ramady 2013). However, some oil producers have attempted to phase out subsidies and direct these to needy sections of society rather than a blanket subsidy for all segments of the population. This was the approach adopted by Iran under President Mahmoud Ahmadinejad who introduced an energy subsidy reform program in 2010 with the aim of saving an initial $20 billion per annum, which would have meant increasing prices for subsidized goods on average between 2.5 and 4.0 times their current levels (Harris 2010). These measures were a bold move by a president who was elected on a populist platform. A key element of the reform was that 50 % of the revenue gained would be given back to "needy" Iranians in direct cash transfers or indirect welfare benefits and, which removes the top 30 % of earners from the subsidy payment system. To effectively implement this reform program, Iran needed to have an advanced tax system to clearly identify top earners which caused some initial confusion in the implementations of the program. However, Iran came up with a novel method by identifying "top earners" as those who owned luxury cars and made frequent foreign trips (Etebari 2013). Figure 5.1 illustrates the effect of the Iranian subsidy reform program on domestic energy prices, pre and post the reforms.

Fig. 5.1 Selected pre-reform and post-reform energy prices in Iran (in 2012 US$)

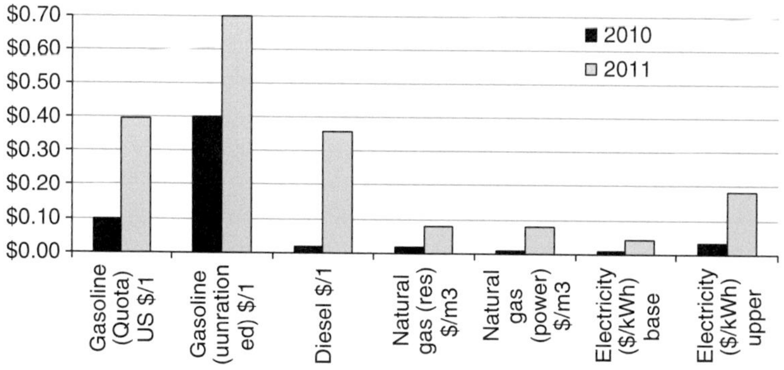

Source: Baker Institute (2014, p. 5)

Fossil fuel subsidies have allowed many energy-exporting countries to distribute revenue often based on a so-called rentier social contract between rulers and those governed (Luciani 1987), with some arguing that such a political system can hinder democracy (Rose 2011; Herb 2005; Haber and Menaldo 2011). While reforms of state benefits can be politically dangerous, the Iran experience illustrates that subsidies can be rolled back, albeit in a complicated process, without undermining government legitimacy. Despite "ruling bargains," underpriced energy has encouraged energy demand to the point that some major OPEC countries like Saudi Arabia face future oil revenue declines if current domestic energy demand continues to rise (Gately and Al Yousef 2012). Table 5.1 sets out electricity and gasoline prices for selected OPEC members, compared with non-OPEC energy exporters Russia and Norway and unsubsidized but relatively low-tax USA.

Table 5.1 Electricity and gasoline prices in selected oil-exporting countries, in comparison with the USA (2012)

Nation	Avg. residential electricity price (US cents per kWh)	Avg. gasoline price (US$ per liter)
(A) *OPEC members*		
Kuwait	0.7	0.23
Saudi Arabia	1.3	0.16
Iran	2.7	0.33
Venezuela	3.1	0.023
Angola	4.2	0.63
Algeria	3.9	0.29
Nigeria	7	0.62
Ecuador	9.6	0.58
(B) *Non-OPEC members*		
Russia	11	0.99
Norway	14.9	2.53
(C) *USA*		
USA	11.8	0.97

Source: Baker Institute (2014, p. 1)

As noted from Table 5.1, OPEC members rank among the lowest in the world in charging residential electricity prices, with Kuwaiti residents able to purchase electricity at seven tenths of a US cent per kilowatt hour since 1966 compared with Norway's 14.9 and 11.8 cents for US consumers. It is as if some states are encouraging their citizens to waste national funds and resources.

Economic Cost of Subsidies

The economic cost of subsidies should not be underestimated. Studies have identified some key elements (Baker Institute 2014; Kaplow and Kretzmann 2010). These include:

- Emissions of carbon linked to climate change are exacerbated.
- Energy exporters' bottom line is affected.
- Reduced national revenues detract from reinvestment in infrastructure and production which is required to maintain exports.
- Retail subsidies sometimes encourage smuggling to higher cost countries, with Iran, Venezuela, Algeria, and Saudi Arabia losing large amounts of subsidized fuel to neighboring countries.
- Subsidies undermine national industry competitiveness.
- Subsidies can be a significant share of gross domestic product (GDP).

Table 5.2 sets out energy subsidies in major OPEC and non-OPEC countries in relation to their GDP in 2011 as well as energy subsidy per capita.

Table 5.2 Energy subsidies in major oil exporters in billions of US$ and percent of 2011 GDP

Nation	Oil subsidy (US$bn)	Gas	Electricity	Total subsidy ($bn) 2011	Total subsidy as share of GDP (%)	Popul. (millions)	Subsidy per capita
(A) *OPEC members*							
Iran	41.4	23.4	17.4	82.2	16	77.9	**$1055.0**
Saudi Arabia	46.1	0	14.8	60.9	9	28.0	**$2175.0**
Venezuela	22	0.9	3.2	27.1	9	29.3	**$924.0**
Iraq	20.4	0.3	1.6	22.2	12	32.3	**$687.0**
UAE	3.9	11.5	6.4	21.8	6	8.0	**$2725.0**
Ecuador	5.4	0	0.1	5.6	7	15.0	**$373.0**
Nigeria	3.6	0	0.7	4.3	2	162.0	**$27.0**
Libya	2.3	0.2	0.7	3.1	5	6.2	**$500.0**
Algeria	11.3	0	2.1	13.4	7	36.0	**$372.0**
Kuwait	4.3	2.1	4.7	11.1	7	3.0	**$3700.0**
Qatar	2	1.9	2.1	6	3	1.8	**$3333.0**
Angola	1.1	0	0.3	1.3	1	2.0	**$650.0**
(B) *Non-OPEC members*							
Russia	0	21.9	18.3	40.2	2	143.0	**$281.0**
Indonesia	15.7	0	5.6	21.3	3	238.0	**$89.0**
Mexico	15.9	0	0	15.9	1	115.0	**$138.0**
Azerbaijan	0.6	0.8	0.5	1.9	3	9.2	**$206.0**
Malaysia	5.4	0.9	0.9	7.2	2	29.0	**$248.0**
Kazakhstan	3.2	0.3	1.7	5.8	3	16.6	**$349.0**
Turkmenistan	0.8	4.4	0.7	5.8	20	5.1	**$1137.0**

Source: Baker Institute (2014, p. 2)
GDP figures (in current US$) from World Bank, World Development Indicators 2014, Author's estimates for per capita subsidies

From the below table, OPEC members spend a significant amount of subsidies per capita, with Kuwait and Qatar spending the most, followed by the UAE and Saudi Arabia. By comparison, non-OPEC oil exporters, including ex-OPEC member Indonesia, spend far less on subsidies, with the exception of Turkmenistan which has the characteristics of a "rentier state" political governance system. Some OPEC members are cognizant of high subsidy expenditures and, in lieu of tackling energy subsidies head on, try to tackle other subsidies, especially those provided to expatriates as the per capita data in Table 5.2 includes both nationals and expatriates. In 2015, Saudi Arabia announced that all foreign workers and their dependents must have a mandatory health insurance cover to ease increasing health public sector expenditure (Saudi Gazette 2015[a]).

Despite such measures, the widespread use of energy subsidies continues to be widely defended on the basis of social safety and ensuring energy access. Some have argued that energy subsidies are largely inequitable as they naturally accrue most to the largest users—energy-intensive industries and medium- to high-income households—and that energy subsidies could otherwise be invested into other socially productive channels like free public health and education, infrastructure improvements, and alternative tax reductions to small- and medium-sized enterprises (El Katiri and Fattouh 2015). The Iranian example is an important one, as it has demonstrated that a reform of domestic energy pricing can be economically and politically feasible for large oil and gas producers. The recent fall in the oil price and its possible persistence for a few more years increase the urgency to adjust oil producers' spending patterns and reform energy subsidy which necessitates a well-planned and executed public relations campaign to accompany the reforms. In July 2015, the UAE, OPEC's third largest producer, announced the removal of domestic fuel subsidies and would now link gasoline and diesel prices to global oil markets helping to ease on the UAE's first fiscal budget defiit since 2009 (Capenter and Khan, 2015). This will put pressure on other GCC Arab oil producers to consider similar measures.

Break-Even Prices: Different Pain Thresholds

There is agreement among those following the fortunes of OPEC that break-even pricing thresholds are important factors that could shed some light on how OPEC collectively and individually might react to changes in oil prices (Davis et al. 2003; Aissaoui 2010; Deutsche Bank 2014[b]; Sfakianakis 2014; Jadwa 2014[a]). It is not often that OPEC members would call for a price restraint in periods of sharply rising oil prices, and yet this is precisely what Saudi Arabia did when oil prices reached a record $139 per barrel in June 2008. In that same year, King Abdullah of Saudi Arabia said that in the Kingdom's view, "$75 per barrel would be a fair price (and that) our budgets are not based on earlier high price but on a lower one" (Aissaoui 2010, p. 110). By making such a statement, Saudi Arabia's oil price preference was established as a target price band around $75 a barrel. The 2008 stated price band preference is in stark contrast to the sharp oil price fall during the latter part of 2014 when the Kingdom did not state what its preferred "fair" price should be, but rather allowed "market forces" to determine prices as discussed in Chap. 1.

Can a "fair" price ever be determined, one that meets the demand and supply security needs of both oil-consuming and oil-producing countries? What are the major confluences of factors that could lead to such price cohesion and acceptance? According to some analysts, these confluences are based on technology, policy, and economics as illustrated in the figure below.

Fig. 5.2 The confluence of key factors as determinants of oil prices

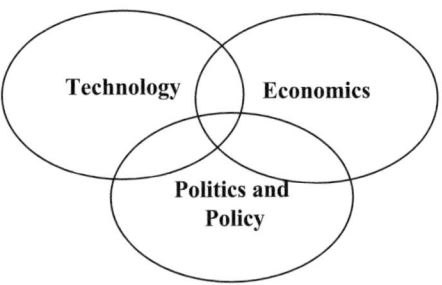

Source: Aissaoui (2014[a])

From Fig. 5.2, the *technology* element describes prospective developments in the fields of geology, engineering, and processes that benefit exploration and development. The *economics* element concerns the viability of different upstream projects under prevailing and forecasted geological, technological, and environmental and credit market risk. The *politics and policy* element refers to sovereign political decision making with regard to legislation, regulation, and fiscal/tax regimes. The sharp fall in oil prices during the latter half of 2014 highlighted the importance of these confluences but in different magnitudes, with reduced energy demand and the impact of shale and other unconventional oil sources due to technology breakthrough adding to overall uncertainty for price stability and an agreeable "fair" price band.

Uneven Pain and Gain

While falling oil prices are the cause of pain for oil producers, they are, by contrast, a source of gain for oil importers, and the hope of many OPEC members is that reduced energy import bills would trigger an economic takeoff in some sluggish economies and help stabilize falling prices and eventually raise them. Table 5.3 sets out the asymmetrical macroeconomic relationship between oil-consuming and oil-producing countries.

Table 5.3 Macroeconomic asymmetry between oil-consuming countries and oil-producing countries (2008 estimates)

Main indicators	Unit	IEA countries	OPEC countries
Average oil price	US$/bbl	97.19 (*CIF*)	94.45 (*FOB*)
Share of energy imports in total imports	%	20.5	1.4
Share of energy exports in total exports	%	10.1	84.5
Share of energy trade in GDP	%	6.8	43.7
Share of petroleum taxes in budget revenues	%	7.1	72.3

Source: Aissaoui (2010, p. 113)
Notes: *CIF* cost, insurance, freight; *FOB* free-on-board

The above table illustrates this asymmetry based on average 2008 oil prices of just under $100 a barrel. The asymmetry becomes even sharper if the comparison between oil-consuming and oil-producing countries uses $50 per barrel as happened when prices collapsed in 2014. The most significant impact will be on OPEC countries' share of petroleum taxes in budget revenues which accounted for around 73 % in 2008 and could conceivably account for lesser than 40 % in 2015 at depressed oil prices of around $50–$60 per barrel. The USA's EIA (Energy Information Administration) has estimated that the 12 OPEC members will lose around $257 billion in lost revenues in 2015. By comparison, the share of petroleum taxes in the budget revenues of IEA oil-consuming countries will only fall slightly due to the relative inelasticity of demand for energy, even if import prices fall, but with significant gains made in the share of energy imports in total imports for IEA members.

Given the asymmetrical pains and gains from fluctuating oil prices, which OPEC members stand to face high pain, and, above all, what are the estimated threshold levels for "break-even" pain? This can provide an indication on how OPEC as a whole and individual OPEC members can withstand current depressed oil prices without putting the organization under acrimonious stress and possible fracture.

There are many assessments on OPEC "break-even" prices, mostly from investment banks like Deutsche Bank, Goldman Sachs, and Barclays to name but a few. In the Middle East and North Africa (MENA) region, local banks and investment houses like SAMBA, National Bank of Qatar, NCB, National Bank of Kuwait, and Jadwa Investment have been active in this respect, especially concerning GCC OPEC members. The results obtained sometimes offer a diverse range of views on what constitutes either OPEC or individual members' "break-even" pricing levels. *This mostly depends on what fiscal indicators have been included or excluded, like the analysis on the level of OPEC members' subsidy programs, the level of exploration technology complexity, and the level of national debt to maintain current and capital expenditures.* A major issue in trying to estimate the level of break-even pricing is time inconsistency factors, as this type of analysis depends on when the analysis was made and oil price levels at the time. Despite the above, most analysts agree on one major aspect: some OPEC members, especially Venezuela, Iran, and Nigeria, have higher break-even prices than others like Saudi Arabia, Kuwait, the UAE, and Qatar (Deutsche Bank 2014[b], 2015). Figure 5.3 sets out some key OPEC members' break-even prices and their relative output of total OPEC production.

Fig. 5.3 OPEC break-even prices and output (2014)

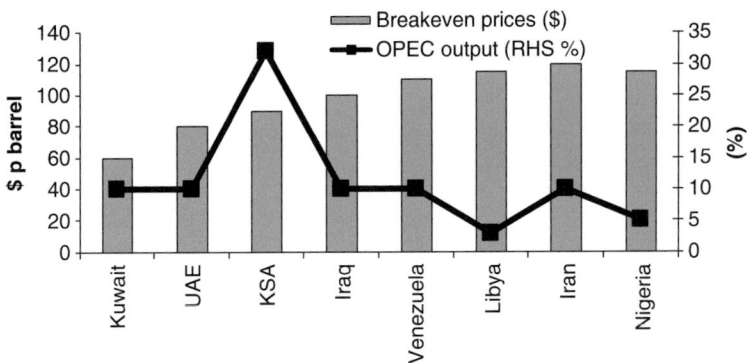

Source: Jadwa (2014[b], p. 6), Deutsche Bank (2014[b], p. 1)

What is immediately noticeable from the above figure is the output dominance of the GCC's OPEC members like Kuwait, the UAE, and Saudi Arabia followed by Iraq, Venezuela, and Iran. However, when it comes to break-even pricing, it is the GCC's OPEC Arab bloc that has the lowest break-even pricing levels, making it no coincidence that it was these three countries that stood firm in the face for calls to cut back on production during the November 2014 OPEC conference meeting.

From the above figure, it was apparent that at oil price hovering at the $50–$55 per barrel range, all OPEC countries faced higher break-even price thresholds as was beginning to happen in August 2015, when oil prices tumbled to below $ 40 a barrel to reach lowest levels for the first time in six years. As mentioned earlier, the effect of a fall in energy prices has an asymmetrical impact on energy consumers and producers, and this is illustrated in Fig. 5.4.

Fig. 5.4 A $20 per barrel decline on GDP: effect on global energy consumers and producers

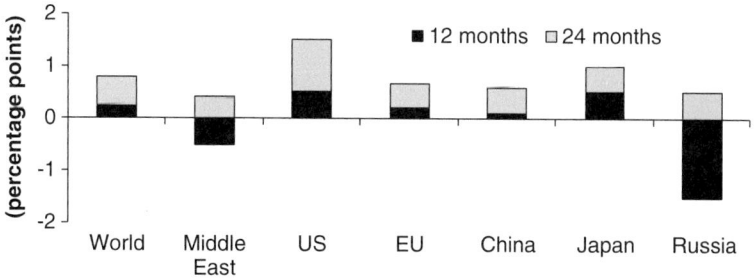

Source: Jadwa (2014[b], p. 8)

The forecasted drop of $20 per barrel over the next 24 months (2014–2016) indicates a significant decline in Russia's GDP and a more modest decline in Middle East oil producers. However, it is not only Russia which is most at risk from falling oil prices, but also OPEC members such as Venezuela and Iran to a lesser extent, as shown in Table 5.4.

Table 5.4 Uneven fiscal stress: Venezuela under pressure (2014)

	Russia	Iran	Venezuela
GDP (nominal US$bn)	2118	366.3	374
GDP (real change, %)	1.3	(1.7)	1
Net debt (% of GDP)	12	1.8	50
Fiscal break-even price (US$ per barrel)	107	127	120
Oil revenue (of total govt. revenue, %)	47	75	50
Foreign reserves (months of imports)	16.4	14	4
Govt. expenditure (5 years avg. % of GDP)	37.4	19.7	36.2

Source: Jadwa (2014[b], p. 7)

From the above table, Venezuela has no meaningful cushion of financial reserves to absorb the shock of continued falling oil prices, and the level of its national debt is also a cause of concern. *GMA*, a data provider owned by *McGraw Hill Financial Inc.*, forecasted that Venezuela has a 93 % chance of defaulting on its debt over the next 5 years but with Venezuelan President Nicolas Maduro stating that "there is no possibility of default" and that the country has the capacity to obtain the financing it needs (Bloomberg 2015c). Compared to Venezuela's debt burden, the dominant Arab OPEC members, Saudi Arabia, Kuwait, and the UAE, have significantly low debt to GDP as illustrated in Fig. 5.5.

Fig. 5.5 Debt to GDP: GCC energy-exporting countries (2012–2014). *Note: asterisk* OPEC members

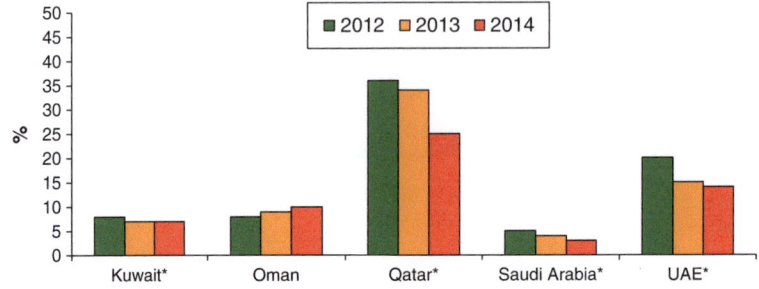

Source: Skafaniakis (2014, p. 2)

The only exception is Qatar's high debt level, but that is mostly due to the ongoing infrastructure investment for the 2022 World Cup preparation and backed by the country's large sovereign wealth fund (SWF) holdings (Ramady 2013).

While Saudi Arabia has the lowest debt–to–GDP ratio among all OPEC and GCC countries, with latest estimates putting this at around 2.0 % for 2014 (Jadwa 2014[a], p. 12), the Kingdom also has the highest break-even pricing compared to the other GCC OPEC members as illustrated earlier in Fig. 5.5. However, Saudi Arabia has amassed substantial financial reserves which has given the country the ability to enforce its policy of maintaining current oil production levels on other OPEC members.

Figure 5.6 highlights the significant foreign asset buffer that Saudi Arabia has built up since 2006.

Fig. 5.6 Saudi Arabia foreign assets (1996–2014) $ billion

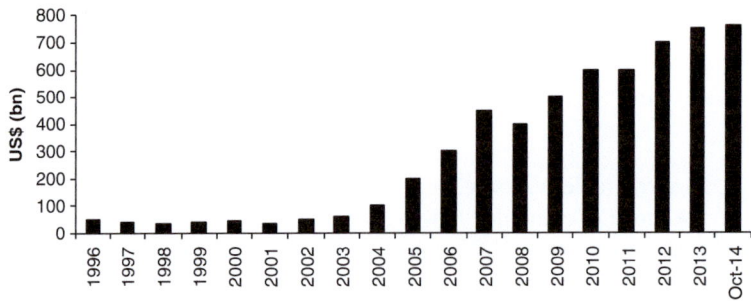

Source: SAMA, Skafaniakis (2014, p. 2)

With such reserves, Saudi Arabia can sustain high spending levels, and it would take a decade and a half to deplete its foreign assets at current expenditure levels. According to latest Saudi Arabian Monetary Agency (SAMA) figures, Saudi foreign reserves have dropped by around $50 billion at the end of April 2015 due to increased nonbudgeted government expenditures (Saudi Gazette 2015c). According to SAMA, the country's foreign assets fell further by 672 bn, the lowest since April 2013 (Reuters, 2015e), forcing the Saudi Government to start reissuing domestic bonds worth SR 15 bn (26.6 bn) by year end 2015. The main risk for Saudi Arabia is for a prolonged downsizing in oil prices to $30 or below levels *over many years*. Due to oil price uncertainties and Saudi drawdown on reserves and new sovereign borrowings, Fitch changed the Kingdom's Issuer Default Rating to 'negative', but still at AA- (Armental, 2015). This seems highly unlikely as will be analyzed later, but Fig. 5.7 assesses the impact on Saudi Arabia's fiscal balances assuming very low, low, base, and high oil prices (*$20, $79, $85, and $100 per barrel*) over the forecasted period 2015–2016.

Fig. 5.7 Saudi Arabian fiscal balances under different oil price scenarios (budget balance)

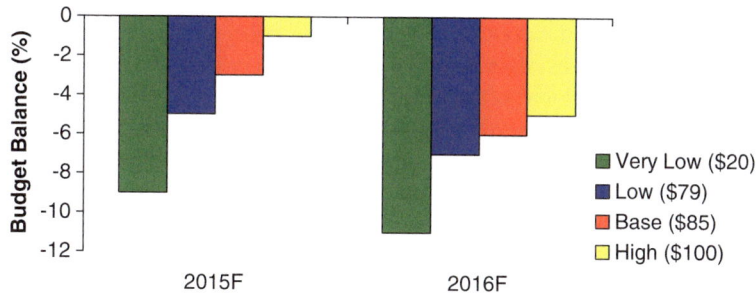

Source: Jadwa (2014b, p. 10)

From the above figure, using a baseline oil price of $85 a barrel, the projected Saudi fiscal deficits were 2.7 and 5.7 % of GDP for 2015 and 2016, but with oil prices at a "low" of $79 a barrel, the fiscal deficits would rise to 4.6 and 7 % for the 2 years. In January 2015, oil prices tumbled to $47 a barrel at one stage, indicating that the above "low price" scenario was somewhat optimistic given the bearish market mood. If such low oil price ranges of around $20–$30 remain for a sustained period, then Saudi fiscal deficits could well reach 10 and 12 % levels for 2015 and 2016. Given the size of Saudi financial reserves and on the premise that such low prices remain for the forecasted 2-year period before firming again, then Saudi Arabia can afford to see through its inflexible "no production cut" policy. For Saudi Arabia, it would seem that a "buffer" has been built and the country can withstand low prices for some years, and the financial buffer can be extended by using the country's borrowing capacity as the Kingdom has neither a dedicated SWF nor a fiscal stabilization fund. Despite Standard & Poor's cutting Saudi Arabia's credit grade outlook to negative at AA, in February 2015, the Kingdom is still an attractive emerging market investment at AA compared to Venezuela which had its credit rating reduced to CCC or eight levels below on investment grade with a negative outlook (Xie and Popina 2015). The question, though, from Venezuela's stressed fiscal position is whether other OPEC countries are willing to follow through Saudi Arabia's "wait-and-see" or more precisely "wait-and-be-firm" no production cut policy, however long it takes.

What Happens to OPEC Energy Investments?

While in the short to medium term, oil sands and other nonconventional oil surges onto the market diminish OPEC's market share, there is another important long-term question. Unlike "demand destruction" arising from vagaries in the economic fortunes of nations that depend on domestic fiscal, monetary, or geopolitical factors, such as the sluggish Eurozone economies of the post-2007 financial crisis, "supply destruction" due to a slowdown in exploration, reserve capacity building, and closure of uneconomic and costly oil fields takes a far longer period of time to readjust.

Countries might have underground proven reserves, but if they are financially stressed and begin to disinvest or postpone new energy investments because of sharply reduced oil prices, then in the long run when demand for oil rises as economies recover, higher global oil demand will be met by *reduced* global supplies, leading to higher oil prices, reversing the "gain and pain" situation of oil-consuming and oil-producing nations to the advantage of the latter. As addressed earlier, the OPEC countries account for the overwhelming share of proven energy reserves, and several OPEC members will still play a leading role in any future output increase as illustrated in Fig. 5.8.

Fig. 5.8 OPEC members' likely output, 2013–2020

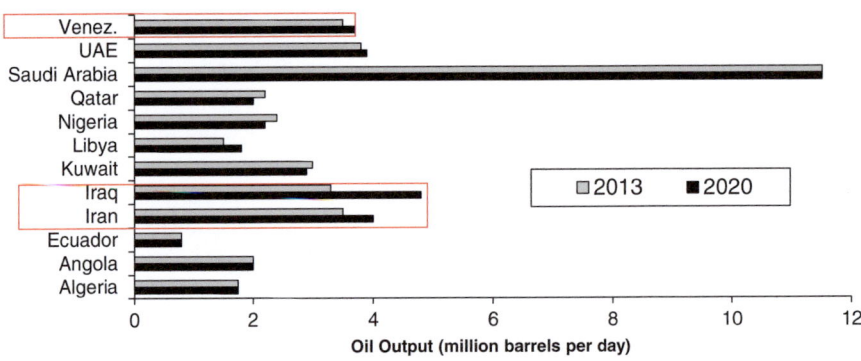

Source: Aissaoui (2014[a], p. 29)

The below figure indicates that, besides Saudi Arabia, the other major OPEC countries who have the highest potential for production capacity growth are Iran, Iraq, and Venezuela. A key factor though is whether under depressed oil prices they will be able to sustain current production *and* add to new capacity investments.

Such future uncertainties will undoubtedly hamper OPEC's decision making and long-term investment policies, leading to a seemingly perpetual OPEC dilemma as illustrated in the figure below.

Figure 5.9 encapsulates OPEC's dilemma: in periods of *high* oil prices, the more incentive there is for OPEC to invest in new production capacity, but high oil prices will induce non-OPEC marginal suppliers to come into the market, leading to a *lower* call on OPEC's supply. This was the situation during the period leading to the increase in nonconventional oil supply in 2013/2014. Conversely, in periods of *low* oil prices, there will be less desire to invest in new capacity by OPEC, an exit of marginal non-OPEC producers, and a *higher* call on OPEC supply.

Fig. 5.9 OPEC's perennial dilemma: investment, production, and market share

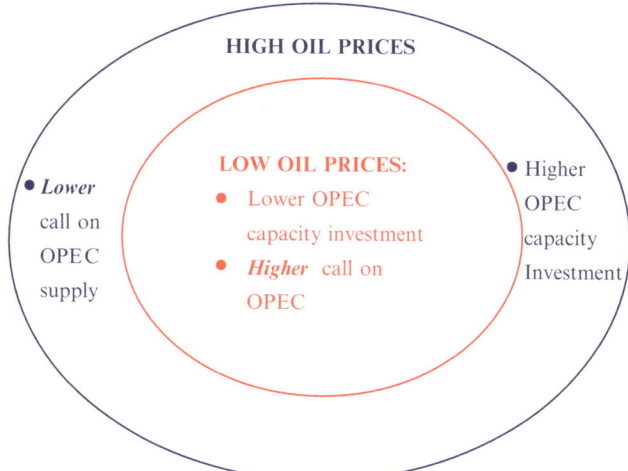

While the above figure indicates that in times of high oil prices, OPEC countries increase their capacity investments, this might not necessarily be true for *all OPEC countries*. National oil companies in some of these countries might face serious investment constraints for different reasons. Some analysts have pointed out that efficiency and competitiveness of national oil companies might not be the key drivers of national governments, but rather the maximization of oil revenue which can be used to achieve broad socioeconomic objectives. As such, the capital expenditure budget of national oil companies might be set by the finance ministry, oil ministry, or other political authority, and capital budgets in the oil sector end up being determined, not according to availability of investment opportunities in the energy sector, but subject to general government budgetary requirements (Johany 1980, p. 83; Mabro 2006, p. 106).

This is unlike privately owned IOCs (international oil companies) that can access capital markets to raise debt and equity financing based on investment opportunities and returns. It was private capital that led investment in new marginal excess capacity in the USA, Africa, and other Latin American non-OPEC countries (Gorelick 2010, p. 143). According to the IEA, based on an estimated annual growth of demand in energy, the total amount of investment required for energy supply and infrastructure was $16 trillion over the period 2001–2030, assuming cost inflation does not escalate. In the MENA region, it has been estimated that around $755 billion is required for energy capital investment over the period 2015–2019 and that internal financing could tighten if the price of oil stays below OPEC's average fiscal break-even price of around $105 per barrel, with external financing becoming more challenging in a low-oil-price environment (Aissaoui 2014^c).

Besides financial constraints, political factors also have direct implications on investment decisions, such as the turbulence following the onset of the "Arab Spring" and the security breakdown in some OPEC oil-producing countries like Libya and Iraq, which curtailed production and export capacity. These security concerns introduce additional costs and risks that reduce the attractiveness of investment opportunities to outside investors. Given the large financial commitments and long time span for upstream oil projects, political stability becomes important, and it is not only MENA OPEC countries that have been affected by political instability, but also other OPEC countries like Nigeria and Venezuela due to domestic unrest (Deutsche Bank 2015).

The above fiscal uncertainties facing long-term capacity investment raise a question for fiscal policymakers in OPEC to decide on how their expenditures can be insulated from oil revenue shocks as occurred in late 2014. This involves questions on which reserves should be saved for future generations, with *oil funds* being suggested as stabilization mechanisms, especially when there are strong political pressures to embark on populist spending programs. "Oil funds" are principally long-term funds for future generations whose primary objective is to minimize the transmission of oil price volatility fiscal policy by smoothing budgetary oil revenues. However, some argue that such oil funds often do little to improve the conduct of fiscal policy and entail the risk of fragmenting fiscal policy and asset management, creating "dual budgets, reducing transparency and accountability" (Davis et al. 2003, p. 7). Some point to the Norwegian petroleum fund as a model, as the fund is viewed as a tool to enhance transparency in the use of oil wealth and is fully integrated into the national budget and has flexible operating rules. Analysts agree however that given the high degree of consensus, transparency, and accountability of the Norwegian fund, this model might not easily be applied to many other oil producers (Davis et al. 2003).

Within OPEC, the only country to have established a long-term future generation fund is Kuwait, but even this has been dogged by disputes between parliament and the government on investment decisions which has hampered its effectiveness and independence, compounded by frequent government changes, often paralyzing executive decision making especially since the 2011 "Arab Spring" movement (Ramady 2013).

Who Blinks First?

The sharp fall in oil prices from the second half of 2014 was a wake-up call for OPEC in that the old way of doing things has fundamentally changed. The old adage that the key factor was consumer demand security was suddenly transformed to producer and, specifically, OPEC producers' supply security. Market analysts differed on whether OPEC, but principally Saudi Arabia which had the largest excess capacity, would or would not allow the free fall in oil prices to continue. Those who thought that the Kingdom would intervene and cut back on its production to stabilize prices did so on the premise that Saudi Arabia would once again play the role of a swing producer like in the 1980s and 1990s, which is explored more fully in the opening chapter. Those who disagreed pointed to the numerous statements made by Saudi oil officials at the highest level that the Kingdom had now abandoned a swing producer role and that market forces would henceforth determine oil prices even "if they fell to $20 a barrel" as stated by Saudi Oil Minister Al Naimi. As noted earlier, with its massive financial resources and production capacity, the Kingdom could very well afford to back its words. However, the main issue for countries like Saudi Arabia, Kuwait, and the UAE—the "OPEC trio" who are the proponents of this new market-driven oil pricing policy—is that they realize that the longer the price falls take, the greater the pressure on other financially stressed OPEC countries to break rank by offering ever deeper oil price discounts.

Some now believe that Saudi Arabia wants to see a quick end to the oil price slide by tacitly allowing even further sharp price falls to take place to drive the marginal high-cost non-OPEC producers, especially US shale oil, out of the market. This "shock-and-awe" oil price strategy seems to have succeeded in some measure, as in mid-2015 cracks began to appear in the US shale oil producer market. This was not the first time that US drillers have been caught up in a battle with OPEC's largest producer, as in 1986 the Kingdom increased production which sparked a 4-month and almost 67 % plunge in prices that took oil to just over $10 a barrel. The result was that the US oil sector nearly collapsed, triggering almost a quarter century of US oil production declines until the most recent shale oil technology-driven boom of the 2000 period and the more recent fall in US rig counts discussed earlier.

Whether all this signals the gradual turnaround in world oil prices to $80 plus price levels very soon is still too early to tell, but what the "OPEC trio" have done is to add another element of *uncertainty* to an already volatile market where balancing long-term energy demand and supply still seems illusive. However, whatever the final outcome, the actions of Saudi Arabia and the change in its oil policy have indicated that the Kingdom has come of age in a complex multipolar world order.

Chapter 6
Environmental Obligations and Climate Change Politics

> *When a diplomat says yes, he means perhaps: when he says perhaps, he means no; when he says no, he is no diplomat.*
>
> Anonymous

Introduction: Looming Challenges

The issue of climate change and environmental obligations of energy consumer and producer nations is an emotive one with significant long-term consequences for both in terms of the quality of life and impact on economic growth. It was only a few decades ago that discussion on the effect of climate change was firmly the domain of so-called "fringe" environmentalist pressure groups like *Greenpeace*. By the beginning of the twenty-first century, the debate had entered mainstream economics and politics, and governments throughout the world, to varying degree of enthusiasm, have subscribed to the notion that not doing anything on global emission was not an option any more. The key is how to obtain not only consensus on emission targets but on actual national and voluntary programs to translate intentions into reality. This could prove to be hard to accomplish, but sometimes even the impossible at the time turns out to be a reality in the future. This is best illustrated by the late US *President Kennedy's* bold vision of winning the space race against Russia by placing a man on the moon within a decade when he said, "its hazards are hostile to us all. Its conquest deserves the best of all mankind, and its opportunity for peaceful cooperation may never come again …. We choose to go to the moon in this decade and do the other things, not because they are easy, but because they are hard" (Kennedy 1962).

Like President Kennedy's bold vision and challenge to achieve what seems the impossible, obtaining a global consensus might not be as impossible if the major industrial consumer countries and those producing fossil fuels see that it is to their mutual benefit and achieve compromise. Some nations seem to have converted from being an inflexible opponent to climate change control to supporters, even if not totally enthusiastic about the time frame for implementation, and the USA is now in this League of Nations. From strong objections under *President Bush Jr.*, the USA under President *Obama* has placed confronting climate change at the center of its National Security Strategy, on par with energy independence and cybersecurity as outlined in its 2015 National Security Strategy document which stated that, "building on the progress made in Copenhagen .... We are working towards an ambitious new global climate change agreement to shape standards for prevention, preparedness and response over the next decade" (White House 2015[b], p. 12). The 2015 US National Security strategy goes on to state that the world's two largest emitters, the USA and China, had reached landmark agreements to take significant action to reduce carbon pollution and that substantial contributions have been pledged to the *Green Climate Fund* (GCF) to help the most vulnerable developing nations deal with climate change, reduce their carbon pollution, and invest in clean energy (White House, ibid. p. 12). With such seeming goodwill all around, what are the obstacles for more vigorous implementation of climate change and emission reduction programs? The answer is in the perception that these programs provide to those countries that feel most threatened.

What Future for Fossils? Oil Producers' Fears

Oil producers see themselves as part of the international community and responsible stakeholders who are providing energy and well-being to the world to increase living standards. It is no coincidence that OPEC's largest producer, Saudi Arabia, has stressed its mission as providing energy to all and states in its objectives that it "… plays a key role in the global economy by maintaining substantial spare crude oil production capacity to contribute stability to worldwide oil prices. Whenever the Kingdom or global markets have called, Saudi Aramco has delivered to communities and consumers around the world" (Saudi Aramco 2014). Maintaining such an expensive spare capacity is something that Saudi Arabia and other OPEC members like Kuwait, Abu Dhabi, and others, to a lesser extent, are bearing. However, in periods of low oil prices as from late 2014 and in the face of increased pressure to reduce carbon emission from fossil fuel, these countries may begin to question the wisdom of carrying expensive spare capacity. OPEC members are willing to cooperate and participate in efforts to reduce risks associated with climate change phenomenon but only in accordance with a *fair share* of responsibilities, as economic development of most of these single-resource economies is their first and foremost objective (Mabro 2006, p. 14).

While there are some differences of opinion on the scientific evidence of global warming discussed later in the chapter, there seems to be a global consensus to agree on measures to reduce emissions. With this "fact," oil producers, and specifically OPEC, have two choices: either to ignore scientific evidence and global consensus or to try and seek fair terms for themselves given their limited economic base, bearing in mind that the objective of these oil producers is to establish the basis of a *sustainable* economic growth model in the long run. Some, like Saudi Arabia, have positioned themselves as providers of mutually beneficial energy suppliers to both consumers and producers and have taken a broader picture of their own interest and of others by highlighting oil market stability and "fair" prices although what now constitutes market-led "fair prices" is more difficult to predict, following the sharp oil price drops of 2014. At the same time, irrespective of volatility in oil prices and export capacity, many of the oil producers understand that their level of revenue growth might not be continually sufficient to cover the needs of their growing population and social expectations. This is especially true for Arab oil producer countries that have shielded themselves to a certain extent against the fallout from the so-called Arab Spring of 2011, by using previously amassed financial reserves to shore up domestic support (Ramady 2013; Abdullah 2012). Some have questioned whether this type of social contract in so-called rentier economies can be sustained (Beblawi and Luciani 1987) and argue that diversification of one-resource economies is the only viable long-term option. It is rare to hear voices of criticism from officials in the Arabian Gulf, and comments on the lack of meaningful diversification of the GCC economies by Qatar Central Bank's director of research and monetary policy, *Khalid Al Khater*, were indeed a rarity when he stated that "the GCC countries have failed in diversifying their economies away from oil revenues so far" and some may eventually face political challenges as a result (Arab News 2015). As noted in the volume, OPEC oil producers are not a homogeneous group, with some having small but rapidly growing populations without significant oil reserves, while others have large populations, but relatively large oil resources and little economic diversification and public sector employment predominating. Some argue that the "oil curse" theory, whereby oil reduces the incentive to introduce necessary and painful reforms, is not only a feature of oil resource-based economies but also in other non-oil-developing countries (Mabro 2006).

The Scientific Debate

The debate concerning global warming and its effects has ranged from cautious scientific analysis to extreme doomsday scenarios of very rapid rising global temperature, melting of glaciers, and threatened coastal cities. Despite such emotive divisions, there seems to be an evolving consensus that greenhouse gases or *GHGs* and CO_2 levels are gradually concentrating in the atmosphere, and a major cause seems to be *anthropogenic* or man-made emissions, mainly the use of fossil fuels (Muller 2006). The 2007 *Intergovernmental Panel on Climate Change (IPCC)* put the matter starkly in its Assessment Report by concluding that "the warming of the climate system is unequivocal, as is now evident from observations of increasing global average air and ocean temperatures, widespread melting of snow and ice, and rising global average sea level" (Intergovernmental Panel and Climate Change 2007, p. 6). The IPCC reports with very high confidence—more than 90 % probability—that man-made carbon dioxide emissions are the ones driving climate change and the IPCC projects global warming at the rate of 0.2° per decade over the next 20 years under a variety of world economic scenarios. Table 6.1 sets out estimated global carbon emissions of different types of fossil fuel to illustrate the differences in their carbon content.

Table 6.1 Global carbon emissions of fossil fuels (tons per ton of oil equivalent—TOE)

Fossil type	Tons of carbon per TOE	CO_2
Coal	1.08	3.96
Fuel oil	0.84	3.07
Natural gas	0.64	2.34

Source: Khatib (2010, p. 42)

From Table 6.1 it is noted that coal has the highest carbon content, while natural gas is relatively the cleanest of the three fossil fuels. The relatively high level of fuel oil emissions has implications on future energy supply security for producers, as it will encourage national emission reduction policies and global agreements to limit these emissions through reducing the use of fossil fuels, despite its important use in all walks of life. At the same time, it will encourage the development of carbon-free renewable and similar forms of energy as illustrated in Fig. 6.1.

Fig. 6.1 Global energy consumption growth in 69 countries in 2015

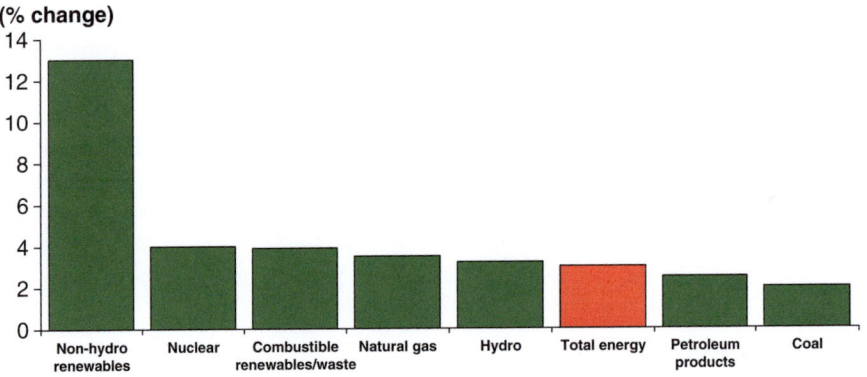

Source: The Economist Intelligence Unit, 2015

According to estimates by the Economist Intelligence Unit (EIU), demand for renewable energy is predicted to rise by around 13 % in 2015, as "dirty" coal is seen as more hazardous, forcing global governments to impose tighter environmental rules (EIU 2015). According to these forecasts, the growth in renewables will outpace that of petroleum too. At the same time, falling oil prices, as occurred during 2014/2015, might affect growth in the renewable market as suppliers might abandon investment in this sector in favor of cheaper oil, but this might depend on how long and sustained oil price falls continue.

In discussions about energy-related carbon dioxide emissions, there is often a distinction made between the OECD (Organization for Economic Cooperation and Development) countries and non-OECD developing countries. Currently the annual emissions from non-OECD countries exceed the total emissions of the OECD bloc as illustrated in Fig. 6.2.

The projected average annual increase in non-OECD emissions is at nearly five times the increase projected for the OECD countries by 2030, with non-OECD emissions projected at 26.8 billion metric tons, exceeding the projected OECD

Fig. 6.2 World energy-related carbon dioxide emissions, 2005–2030

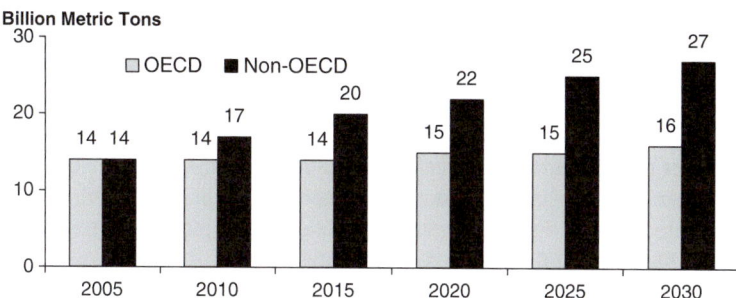

Source: Energy Information Administration (EIA), International Energy Annual 2005 (June–October 2007). Projections: EIA. World Energy Projections Plus (2008, p. 89)

emissions by 72 %. However this is on the assumption that existing emission laws and policies remain unchanged and *that no new greenhouse emission policies are introduced*. As discussed later, this is a pessimistic assumption as there is now a more consensual global approach on global emission policies, at least on *intentions*.

The relative contributions of different fossil fuels to total energy-related carbon dioxide emissions have changed over time. While in 1990 emissions from liquids and other petroleum were an estimated 42 % of world total, by 2030 this is projected to be at around 32 %, with coal emissions projected to increase to 44 % by 2030, illustrated in Fig. 6.3.

Fig. 6.3 World energy-related carbon dioxide emissions by fuel type, 1990–2030

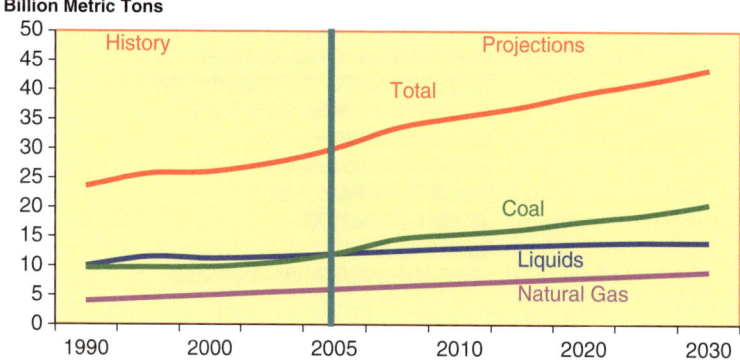

Source: Energy Information Administration (EIA), International Energy Annual 2005 (June–October 2007). Projections: EIA. World Energy Projections Plus (2008, p. 89)

From Fig. 6.3, it is projected that carbon emission from natural gas will stabilize at between 20 and 21 % by 2030. Coal, as noted earlier, is the most carbon intensive of the fossil fuels and, according to the EIA, is the fastest-growing energy source especially for China and India. According to the EIA, in 1990 China and India together accounted for 13 % of world carbon dioxide emissions which rose to 23 % by 2005 and are projected to reach 34 % of world emission, *with China alone responsible for 28 % of the world's total.*

Given such trends from the fastest-growing developing countries like China and India, what has been the pattern of energy-related carbon dioxide emissions between OECD and non-OECD countries and their forecasted trends? This analysis will also help to better understand the political pressures and alliances that are current among the nations most likely to be impacted by international emission reduction agreements. Figure 6.4 illustrates the growth rates in energy-related carbon emissions for major OECD and non-OECD economies.

Fig. 6.4 (a) Average annual growth in energy-related carbon dioxide emissions in the OECD economies, 2005–2030. (b) Average annual growth in energy-related carbon dioxide emissions in the non-OECD economies, 2005–2030

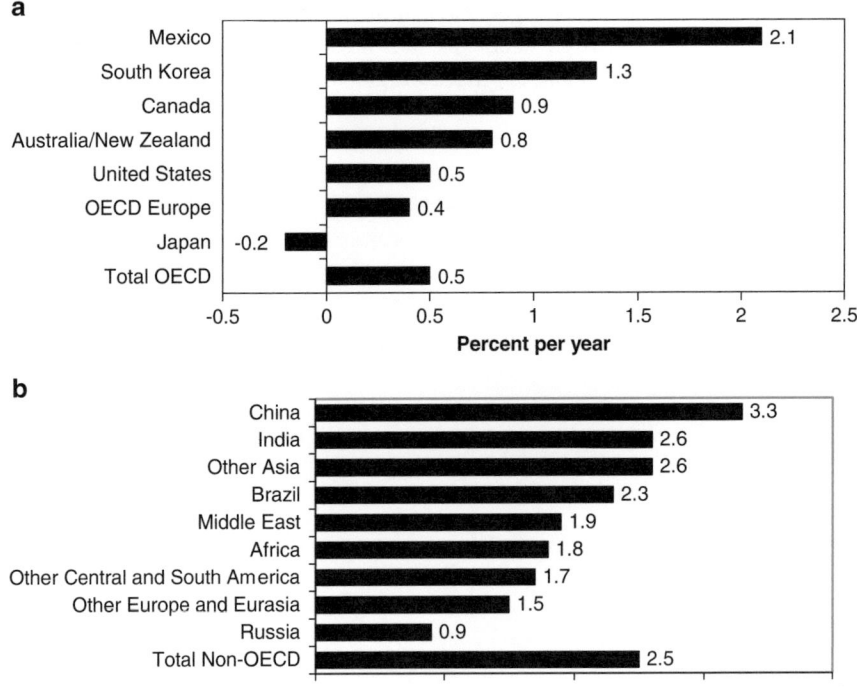

Source: Energy Information Administration (EIA), International Energy Annual 2005 (June–October 2007). Projections: EIA. World Energy Projections Plus (2008, p. 89)

The highest rate of increase in annual emissions of carbon dioxide among the OECD countries is projected for Mexico, an oil producer, while for all the other OECD countries, annual increases in carbon dioxide emissions are projected to be less than 1.5 %. South Korea is projected to be the second largest emitter after Mexico, given the continuing industrialization of the South Korean economy. Japan is projected to *decrease* its energy-related emissions by an average of 0.2 % per year from 2005 to 2030. The picture for the non-OECD countries illustrated in Fig. 6.4b is different. Only Russia is expected to have an average annual remittance of less than 1 % per annum from 2005 to 2030, with the non-OECD countries total carbon dioxide emissions projected to average at 2.5 % compared with 0.5 % for total OECD countries for the period 2005–2030. The highest growth rate among the non-OECD countries is for China, at around 3.3 % annual growth rate reflecting the country's heavy reliance on fossil fuels, especially coal, and highlighting the importance of China "buying in" to any lasting fossil reduction agreements.

For various reasons discussed later, this is not a given fact despite bilateral government agreements on emission reduction pledges, and a lot more ground has to be covered by leading emission countries before a meaningful global emission reduction can take place. This is not to say that there remain some "eco-skeptics" who point out that present knowledge on global emissions and future predictions lead to *decision-making under uncertainty* and that more global warming scientific evidence is needed (Michaels et al. 2004). Their arguments center around several criticisms of official data which claims that reliable temperature records exist which show how much the planet has warmed in the last century and that computer projections of future climate, while "not perfect," simulate the observed behavior of the past so well that they serve as a reliable guide for the future (Michaels et at. ibid). The key is to deal appropriately with what has been elegantly described as the *known-unknowns*. These known-unknowns are possible threefold:

- *First*, natural variability whereby the climate system is chaotic and small changes in one location at one point in time can lead to large differences at other locations at some future point in time (the so-called *butterfly wing flapping* effect)
- *Second*, changing boundary conditions, whereby climate is affected by many factors which are considered to be separate from the climate system, such as volcanic eruptions and solar output
- *Third*, our current scientific understanding of how the climate behaves and how it responds to changing boundary conditions such as a rapid increase in atmospheric concentrations of greenhouse gases (GHG) (Stainforth 2005)

While such uncertainties will always exist, some objective and more precise scientific assessment can be carried out using satellite observations of atmospheric variables which are highly technical and can provide observations which are undisputed, especially concerning Earth surface temperature measurements. According to the Intergovernmental Panel on Climate Change (IPCC), global warming has been rising, the global average surface temperature of the Earth has increased by approximately 0.8 °C since 1750, and the rate of warming has increased with a projected mean temperature over the twenty-first century of between 1.1 and 6.4 °C (IPCC 2007; Bolin 2008).

The Long and Winding Road from Rio to Lima (Through Kyoto): Political Pressure and Shifting Alliances

Climate change has gained significant political traction over the years and, if acted upon by signatory nations, would require a radical change in how the world produces and consumes its energy, with potential significant impact on economic growth and citizens' well-being. That being so, it is no wonder that global agreements are moving ahead in fits and starts as illustrated in the next table that sets out major global environment agreement milestones, their key achievements, and outstanding issues.

Table 6.2 indicates that *while* there is still a long way to go before translating the specific emission reduction pledges made since the process began in Rio in 1992, some progress is better than none in the face of mounting evidence of climate change. To date there have been 21 *conference* meetings, starting with the first in Berlin in 1995 and the last being in Lima in 2014. The major point of contention

Table 6.2 Key global environment agreements (1992–2014)

Year	Climate change agreement/name	Main objectives and outstanding issues
1992	Rio De Janeiro, Argentina	• First UN Conference on the Environment and Development—Earth Summit
		• *Outcome*: Framework Convention on Climate Change (*FCCC*) committing parties to reduce greenhouse emissions
1995	Berlin, Germany	• Parties to the UN FCCC outline specific targets on emissions
1997	Kyoto, Japan	• Kyoto Protocol adopted, December 11, 1997
		• *Protocol started 2008*
		• Agreed to the broad outlines of emission targets. Legally binding commitments to *Annex I* parties
		• Pitted Europe against the USA
		• US Congress did not ratify the treaty after President Bill Clinton signed it
		• The US Bush Administration explicitly rejected the Protocol in 2001
		• 2002: Russia and Canada ratify Kyoto Protocol, bringing treaty into effect on *February 16, 2005*, to contain emissions from greenhouse gases in ways that reflect national differences in emissions, wealth, and the capacity to make reductions
		• *2011 Canada became first signatory to withdraw*
		• December 13, 2012, Kyoto Protocol expired (but on December 8, 2012, at the end of the UN Climate Change Conference agreement was reached to *extend protocol to 2020 and set date of 2015 for the development of a successor document*). To be held in Paris in 2015
2009	Copenhagen, Denmark	• 18th Conference of the UN FCCC
		• Broke up without agreement on post-Kyoto Climate treaty due to conflict among major powers
		• Developing and developed countries would adopt parallel *nonbinding pledges* to reduce emissions
		• Prime objective preventing global temperature rising more than 2^C

(continued)

Table 6.2 (continued)

2010	Cancun, Mexico	• Adoption of specific pledges by countries for emission reductions
		• Called for a $100 billion per annum *Green Climate Fund* and a *Climate Technology Centre* and network, but the Green Climate Fund not agreed upon
		• Established a process of monitoring and verifications and that mitigation efforts undertaken with domestic resources would be monitored domestically, while those undertaken with international resources would be monitored internationally
		• Reconfirmation to keeping to the 2^c global temperature rise
2012	Doha, Qatar	• UN FCCC met and the European Union pledged to extend Kyoto binding on the 27 EU states up to 2020 pending an internal ratification procedure
		• No global consensus on emissions reached
		• Language on loss and damage formalized for the first time in conference documents
		• Little progress made to funding the *Green Climate Fund*
		• Final communiqué committed to finalizing post-Kyoto treaty to be held in Paris in 2015, with the new treaty to take effect in 2020
		• *As of May 2013, 191 countries have ratified Kyoto agreement*
2014	Lima, Peru	• UN Climate Change Conference
		• All countries asked to submit plans for curbing greenhouse emissions known as *Intended Nationally Determined Contributions* (INDCs) to the United Nations by *March 31, 2015,* as a core document for the Paris 2015 deal
		• No review to compare each nation's pledges after China refused
		• Donation to *Green Climate Fund* reached $10 billion
		• Cutting greenhouse gas emissions to *net zero by 2050*

Source: United Nations, IPCC (2007), Yergin (2011)

that remains at the heart of the negotiations in these *Framework Conventions on Climate Change* (*FCCC*) or conferences as they are called is the so-called "North-South" face-off between developed and developing nations. Almost all agree that some 75 % of total accumulated emissions of CO_2 between 1860 and 1990 had come from the industrialized nations, but they only had 20 % of world population. As carbon emission limits seemed to take on greater urgency, developing nations experiencing rapid economic growth like China, India, and Brazil became more outspoken in their opposition to limits being imposed on their use of hydrocarbons and the constraints this could impose on their planned economic growth (Yergin 2011, p. 480). As Table 6.2 illustrates, global powers and lesser powers tried to obtain the least resistant threshold acceptance levels to agreeing on specific emission targets, with the *Kyoto Protocol* starting in 1997 but finally signed in 2005 being the central plank of global consensus.

However, even the acceptance of the *Kyoto Protocol* was not without its obstacles as Canada withdrew from the *Protocol* in 2012. Various FCCC conferences after *Kyoto* such as *Copenhagen, Cancun, Doha,* and *Lima* have made incremental progress, especially in trying to assuage developing countries concerns, but realizing that, in the final analysis, global emission success could *only be achieved if the*

"Big Emitters" like the USA, China, India, and Russia agreed to control their own emissions. In all these negotiations, the oil producers and OPEC did not, until more recently, present a cohesive front bearing in mind that oil producers stood to lose more if the most stringent fossil fuel emission target cuts are imposed (Ghanem et al. 1999; Barnett et al. 2004).

According to observers of ongoing climate change talks, the *Kyoto Protocol* raised several major questions. The first is that it initially pitted the European Union (EU) against the USA leading to a standoff, with the Europeans wanting the USA to make deeper emission cuts which the USA refused. This was resolved by the personal intervention of US Vice President *Al Gore* (who later went on to be awarded the Nobel Peace Prize in 2007 along with the Intergovernmental Panel on Climate Change). The result was that the USA, Europe, and Japan all ended up with roughly the same binding targets of between 6 and 8 % lower by 2008–2012, compared with 1990 (Yergin 2011, p. 484). The second issue was the "North-South" divide, with the USA, in particular, not accepting a treaty that would put US industry at a competitive disadvantage versus countries such as China with its high greenhouse gas emissions. As a compromise, *Kyoto* agreed with the establishment of the *Clean Development Mechanism* under which companies from the developed countries would invest in "clean energy" projects in developing countries. Another major issue arising out of *Kyoto*, and still lingering to date, was how to implement gas emission reduction—the perennial *who pays what*, especially in a post-global financial crisis world and large national deficits, particularly in Europe. The EU at *Kyoto* wanted mandates and direct intervention, calling it *policies and measures*, while the USA was committed to a free-market trading system along the lines of "acid rain" for emission trading. In the final *Kyoto Protocol*, market emission trading was embedded. The implication of such an emission trading regime is obvious: advanced economies with a deficit in their emission reduction targets could meet their Kyoto commitments by buying allowances from developing countries with a "surplus," and, unless other commitments were made by the advanced economies to reduce the total surplus in allowances, such a trade would not actually result in emissions being *reduced*.

To accelerate all possible methods to reduce global emission, the *Kyoto Protocol* and others that followed introduced several schemes. These included:

- *Clean Development Mechanism* (*CDM*): This was expected to produce by 2012, the end of the Kyoto commitment period, a 1.5 billion tons of carbon dioxide equivalent (CO_2) in emission reductions with most of the reductions coming from renewable energy commercialization and energy efficiency and fuel switching to other emission friendly fuels. The largest CDM potential contributors are estimated for China (52 %), India (16 %), and Brazil (7 %). The CDM was designed at Kyoto in such a way that developing countries do not bear the full cost for limiting emission on the general assumption that it was a "win-win" situation whereby developing countries would face *quantitative* commitments in later commitment periods and that developed countries would meet their first round commitment of the *Kyoto Protocol*.

- *Financial commitments*: The *Kyoto Protocol* reaffirmed a key principle that the more developed nations have to pay significant amounts and supply technology to other countries for climate-related studies and emission reduction projects. This could be channeled through the *Adaption Fund* that was established at the *Kyoto Protocol* to finance specific adaptation projects in developing countries that signed up to the *Kyoto Protocol* as an inducement.

Does Kyoto Still Have Life?

The official meeting of all states party to the *Kyoto Protocol* is officially called the "Conference of the Parties" and is held every year as part of the Untied Nations Climate Change Conference which serves as the formal meeting of the UNFCCC. Non-Kyoto Protocol members can attend as observer nations. Table 6.2 has set out some of the major conferences like Berlin, Copenhagen, Doha, and Lima in 2014. On December 8, 2012, at the end of the 2012 United Nations Climate Change Conference, an agreement was reached to extend the *Kyoto Protocol* to 2020 as it officially expired in 2012. The successor *Kyoto* document is set to be signed in Paris during the 2015 meeting.

Until a more lasting agreeable successor to Kyoto is formulated in Paris in 2015, it is important to highlight some of the continuing issues hindering *Kyoto* compliance. One of these involved the surprise withdrawal of Canada from the *Kyoto Protocol* in 2012. In 2011 Canada, Japan, and Russia stated that they would not take on further Kyoto targets. While Japan and Russia continued as members of the Protocol with reservations, it was Canada that withdrew citing concerns that it was liable to "enormous financial penalties" of around $14 billion under the Kyoto treaty unless it withdrew (Guardian 2011). The facts were obvious: Canada had committed to cutting its greenhouse emissions to 6 % below 1990 levels by 2012, but in 2009 its emissions were 24.1 % *higher* than in 1990 and rising (UNFCC 2011). One of the major Canadian objections prior to their withdrawal centered on China and India's "unequal" emission control targets, which once again brought to the fore the concept of *climate justice* on how to strike a balance between lower emissions and the higher economic and social vulnerability of some key developing nations to world climate change controls. Differences on "climate justice" ensured that the *Doha Climate Summit* concluded without concrete agreement on emission reductions (O'Connor 2012).

The final communiqué of the Doha Summit committed to finalizing a post-Kyoto treaty governing greenhouse emissions by the Paris 2015 conference, with the new treaty to take effect in 2020. Underlying all the discussions on a post-Kyoto treaty has been the acrimonious rivalry between the USA and a developing country bloc led by China, whereby the USA has been determined to block progress on climate treaty emission agreements that uphold the *Kyoto Protocol's* classification of China, together with India and other developing nations, as "low income countries" and, as such, not subject to *binding* emission targets. The USA has objected to references in climate change negotiations to "common but differentiated responsibilities and respective capabilities, in light of different national circumstances" (United Nations 2014, p. 2).

Has the situation improved since Doha? By some accounts, the last UN climate conference in Lima in 2014 evidenced some modest agreement about the building blocks of a deal due to be signed in Paris in 2015, with all countries asked to submit plans for curbing greenhouse emissions known as *Intended Nationally Determined Contributions* or *INDCs* to the Untied Nations, but again there were few obligations to provide details to compare each nation's pledges, except to set a long-term goal of a cut in greenhouse gas emissions to "net zero by 2050" (Guardian 2014). OPEC's largest member Saudi Arabia indicated that it saw this zero-carbon target by 2050 as "not realistic" and its chief climate negotiator *Khalid Abu Leif* added that the "zero-emissions concept—or let's knock fossil fuels out of the picture—without clear technology diffusion and solid international cooperation programs, does not help the process." He continued by stating that the zero-carbon target was unrealistic "when two billion people do not have access to energy" (Saudi Gazette, Dec. 10, 2014). However, in a remarkable statement by OPEC's largest fossil producer, Saudi Arabia, its Oil Minister Ali Al Naimi stated in May 2015 that "Saudi Arabia recognizes that eventually, one of these days, we are not going to need fossil fuels—I don't know when, 2040, 2050 or thereafter. So we have embarked on a program to develop solar energy" and hopefully Saudi Arabia, instead of exporting fossil fuel, will one day be "exporting gigawatts of electric power" (Smith and Chmaytelli 2015). The minister reiterated the country's commitment to solar power during the June 2015 OPEC meeting (Coulter and Chilcote 2015).

US-China Breakthrough

A breakthrough on agreeing to climate change emission curbs that the USA in particular can live with would be a major "win-win" for all parties. In 2007 by some measure, China's carbon dioxide emissions had exceeded those of the USA but not on a per capita basis, which the Chinese repeatedly pointed out to its critics (Yergin 2011, p. 509).

The Chinese also emphasize that their country is still a relatively poor nation making the economic and social transformation and transition that the USA, Europe, and Japan made many decades before and that it should not be penalized from aspiring to other countries' standard of living. These arguments have been at the center of Chinese global climate negotiations (Lewis 2008), and the Chinese are well aware that internationally there would be no global climate change pact without them effectively participating. The Chinese development arguments are made even more stridently by India, and the country has also played a major role in trying to obtain advantageous conditions, exemptions, and other less onerous emission targets (Hinton 2010), earning itself the label of the *climate agnostic* when Indian Environment Minister *Jairam Ramesh* stated that the climate world was divided into three "climate atheists, climate agnostics, and climate evangelicals" and that India was in the middle category (Yergin 2011, p. 511). Putting bread and butter before its citizens was more important to India than climate change, a point that was underscored earlier by the comment made by the Saudi Chief Environment Negotiator Abu Leif concerning two billion people without energy.

However even the most intractable position sometimes changes, and often it is not through political pressure but through self-interest as Chinese attitude changes to climate issues illustrate. Severe droughts and floods in China (as well as other unusual patterns of climate changes globally during 2014 and 2015) have highlighted the risks of doing nothing to arrest climate change. The Chinese have become more concerned about the impact on their country's water supplies if global warming affects their water sources in the Himalayas and the Tibetan glacier plateaus that feed China's rivers. It is rare for senior Chinese officials to publicly criticize their government policies and it was somewhat of a surprise when *Zheng Guogang*, the head of China's meteorological administration, stated that warming temperature exposed his country to a "growing risk of climate change and climate disaster" that could have a "huge impact" of China, reducing crop yields and harming the environment (BBC 2015[h]). Mr. Zheng went on to warn that major infrastructure projects such as *Three Gorges Dam* would be affected. The country's leadership may have concluded that global climate change cooperation, especially with the USA, would improve their relations with the west and minimize trade tensions. On November 12, 2014, there was a historic moment when US President Obama and Chinese President *Xi Jinping* committed to target for cuts in their respective nations' carbon emission which, according to analysts, "has fundamentally shifted the global politics of climate change" (Davenport 2014). While some cautioned that the emission reduction targets put forward by the USA and China will not be enough to prevent an increase in global atmospheric temperature of 2 °C, others saw it a powerful signal to the world that the two biggest emission emitters are coming together by announcing concrete numbers and time lines. Figure 6.5 illustrates the pledges made by the USA and China.

Fig. 6.5 China and US climate goal pledges

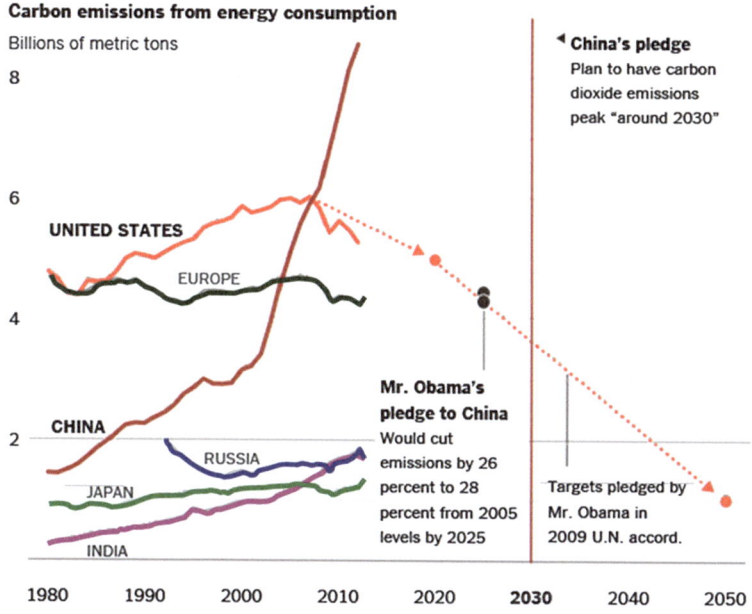

Sourse: EIA (2008), Davenport (2014)

Both countries made some bold forward-looking pledges to each other, with the USA pledging to cut emissions to 28 % from 2005 levels by 2025, while China's pledge was somewhat more vague, promising to have carbon dioxide emissions "peak around 2030," with some characterizing this as nothing more than "business as usual" by China. However, a vague pledge is better than no pledge, and this Sino-US agreement has put pressure on other countries to follow suit, including other major emitters like Australia, India, Russia, and OPEC oil producers Venezuela and Saudi Arabia as the Saudi Oil Minister's latest statement indicates. The Sino-US agreement could also provide a much needed shot in the arm to the planned 2015 Paris conference which is supposed to draft a new post *Kyoto Protocol*.

India's Conversion from "Climate Agnostic" to Possible "Climate Evangelical?"

The landmark election victory in May 2014 of Indian Prime Minister *Narendra Modi* has shaken traditional Indian politics, and it seems some long held positions on Indian climate change. While India's global emissions are far lower than either those of the USA and China, the need for sustained economic growth to lift millions of its citizens from a poverty trap has put considerable political pressure on the new Prime Minister who has been dubbed India's "Mrs. Thatcher" in trying to tackle deep-rooted economic interests and trade union power. In January 2015, India went the China route and inked another landmark agreement when President Obama visited India and announced a new US-India agreement on clean energy and climate change. In its press release, the US White House noted that both countries pledged to enhance cooperation in climate and clean energy goals. These involved, among others:

- Enhancing bilateral climate change cooperation to ensure a successful Paris agreement
- Cooperating on hydrofluorocarbons (HFCs), concerning the phase down of HFCs.
- Expanding partnership to advance clean energy research, a $125 million joint US-Indian government program, in solar energy, building energy efficiency and advanced biofuels
- Accelerating clean energy finance to create a market environment that will promote trade and investment in this sector
- Starting technical cooperation on heavy-duty vehicles and transportation fuels and how to reduce the environmental and emissions impact of this sector (White House, Jan. 25, 2015[a])

Analysts gave the US-Indian agreement a favorable review, pointing out that it will help turn India's bold renewable energy targets into reality and that rather than relying on one major plank of collaboration, the agreement is a comprehensive set of actions that, taken together, represent a substantial step in advancing low-carbon developments while also promoting economic growth and expanding energy access to the needy (Waskow and Bapna 2015). Building on India's 100 GW solar capacity goals, Prime Minister Modi announced the country's intention to increase the overall share of renewable energy in the nation's electricity supply. Roughly 300 million Indians—nearly 25 % of the population—lack access to electricity, and solar power which is already cheaper than diesel in some parts of the country and may soon be as cheap as conventional energy can put affordable and clean power within the reach of many (Waskow and Bapna 2015).

Both Chinese and Indian leaders are interested in pursuing economic growth, fueled by low-carbon energy, especially if that energy can be obtained more cheaply with foreign assistance—as the US-India agreement promised—as well as access to renewable energy technology which the USA also pledged. Through a combination of pragmatism and altruism, it seems that even the most "climate agnostic" can be converted. The new Indian Prime Minister put it very well when he made it clear that he sees it as incumbent on *all* countries to take action on climate change and that rather than be motivated by international pressure, what counts is "the pressure of what kind of legacy we want to leave for our future generations. Global warming is a pressure … We understand this pressure and we are responding to it" (Sunday Times, Jan. 25, 2015).

From Fringe to Mainstream: Enter the "Greens"

The "green" or environmentalist "eco-warriors" of a few decades ago have now joined mainstream, with political representatives in many European parliaments to ensure their voice on the environment is heard. Public campaigning, including some high-profile cases involving mass demonstrations and targeting of so-called "polluting" companies, has raised environmentalist campaigners' profile as well as their message. The result is that national and international companies have started to demonstrate their green credentials, focusing on climate change and adapting their businesses in an age of more rigorous emission regulation. The name of the game is now ethical investment and economic sustainability to protect a company's brand and to ensure long-term survival and wider stakeholder acceptance.

By the beginning of the twenty-first century, climate change was gaining the attention of most major international companies, including leading IOCs such as BP, Shell, and those involved in energy transmission and research such as General Electric. In 2007, nine leading industrial companies and utilities along with four environmental organizations formed the *US Climate Action Partnership or USCAP*, to promote climate legislation. By 2009 the membership had grown to 25 companies while the Global Climate Coalition which had opposed climate change regulation dissolved itself (Yergin 2011, p. 459). Advocates of ethical investment and economic sustainability has seen some institutions "divest" from fossil fuel investments and the movement seems to be gathering force. *Stanford University* of the USA saw some 300 professors, including *Nobel Laureates*, calling on the university to rid itself of all fossil fuel investments from its reputed $21.4 billion endowment fund (Goldenberg 2015). The petition called for disinvestment from all oil, coal, and gas companies, and the fossil fuel divestment campaign has not stopped with academic endowment funds but with other major investors. In 2014, the heirs to the Rockefeller oil fortune withdrew their $860 million philanthropic fund from investments in tar sands, coal, and oil, and campaigners claimed that they have persuaded other investors to withdraw a total of $50 billion from fossil fuel investments over the next 5 years. *Harvard University* however, with its $32 billion endowment fund, rejected a similar petition to remove fossil fuel from its investments citing that such actions would harm the economic sector and employment (Goldenberg 2015).

The sudden enthusiasm for "eco-friendly" business has raised questions on whether the conversion of big business is truly committed to climate change or simply riding out a public relations' wave. Just repainting a company brand from red to different shades of green or blue does not make for a true eco-conversion, but the art of management is to preempt change and move ahead of others. What was once an arena of fierce antagonism between multinationals, especially in the energy sector with activist such as *Greenpeace*, has been transformed to a certain extent to what amounts to a mutual love fest, with big business outdoing each other in terms of their environmental friendliness. Misgivings about corporate environmentalism exist, with some of the self-proclaimed "green" manufacturers finding themselves being investigated for false advertising and other offenses against consumers (Mattera 2007). By the early 2000s, some companies sought to depict themselves as being not merely in step with the environmental movement but at the forefront of a green transformation. This is best illustrated by British Petroleum publicizing its investments in renewable energy and saying that its initials—*BP*—really stood for *Beyond Petroleum*, despite the fact that BP's operations continued to be dominated by fossil fuels (Mattera 2007).

Such skepticism has led to a host of questions as to the motives for eco-friendly corporate conversion. These range from whether corporations are fooling themselves, in which case they will realize that environmental constraints are difficult to implement and costly, leading to reneging on their green promises, to interpretations that companies are taking voluntary steps that are indeed genuine, but inadequate for the problems at hand, and are only meant to prevent stricter and enforceable regulations. At least Canada realized the implication of their *Kyoto Protocol* commitments and decided to withdraw in 2011. Whatever the true motives, the fact remains that global environmental concerns are no longer in the back burner, but are at the center of policymaking and long-term business decision-making. The environmentalist pressure groups have now seen religious figures calling for fossil fuels to be "progressively replaced without delay" with no less a person than Pope Paul issuing such an encyclical, and urging the richer world to make changes in lifestyle and energy consumption to avert the unprecedented destruction of the ecosystem (BBC, 2015ᶜ). Many, including the UN's climate change Chief Chritiana Figueres, hoped that the Pope's message will influence the Paris climate change talks (BBC, ibid). To add to the pressure, Islamic environment and religious leaders have also called on rich countries and oil producing nations to end fossil fuel use by 2050 and urged politicians to agree to a new treaty limiting global warming to 2 degrees or 'preferably 1.5 degrees' (McGrath, 2015).

Global Climate Change: Who Pays?

A significant factor for the slow implementation of pledges made for climate change has been the issue of cost and how to measure these costs and to allocate them on an *equitable basis*. There are wide ranges of estimates both for the marginal cost levels required to achieve various target reduction levels and their corresponding impact on countries' gross domestic products (GDP). According to the IEA, if by 2030 the world GDP were 1 % lower as a result of emission mitigation efforts, it would mean an *annual cost of $1.5 trillion in constant 2000 dollars*. As the IEA succinctly puts it, "estimated future costs must be weighed against future benefits in the form of avoiding human-caused climate disruptions" (IEA 2011^a, p. 92). According to a study carried out by two of the UK's universities (*LSE and Leeds University*), it has been estimated that richer nations need to fund at least *$400 billion a year* into the developing world by 2050 to help cut greenhouse gases and fight climate change and quadrupling of aid pledged, with estimates as high as $2 trillion, may even be required (Morales 2015). These are truly astronomical figures and in the financially stressed Western economies, even the most ardent climate change campaigners will balk.

The problem of global warming cost estimates is adopting strategies based on *sequential decision-making* whereby decisions to combat global warming are made with incomplete information and that such decisions may turn out to have large potentially long-term impact, especially on economic development of future generations by diverting resources away from more pressing (present) socially and economically beneficial opportunities. Another aspect of evaluating the cost of global emission is the choice of the social discount rate used by governments to compare the economic effects of different policy decisions over time. The longer the time period, and the more uncertain the outcome, especially concerning predicting future GHG emissions, makes the choice of social discount rates more difficult for one country, let alone between different countries with different social discount rates. Given such uncertainties, emission "scenarios" are designed to project future GHG emissions, and the Intergovernmental Panel on Climate Change (IPCC) has worked on different scenario projects called *Special Reports on Emission Scenarios* or "SRES" under a wide range of possible future emission levels and widely differing assumptions of future social and economic changes, including population levels of 15 billion people by 2000 under one scenario and seven billion people under another scenario (Metz 2001). Such projection uncertainties have spawned a host of research literature on the economics of climate change, with one of the first proponents being Nordhaus who used a dynamic integrated climate model or *DICE*, to take into account the latest findings and some of the uncertainties about the major risks of climate change in his seminal article "*To slow or not to slow*" (Nordhaus 1991).

The first *DICE* models had in-built assumptions on growth, damages, and risks, which together, in the opinion of later researchers, resulted in gross underassessment of the overall scale of the risks from *unmanaged* climate change (Stern 2013). In later research, some of the earlier Nordhaus assumptions were relaxed to assess the consequences in terms of optimal emission reductions and carbon prices, atmospheric concentrations of carbon dioxide, and global mean temperature. The new version of the *DICE* model suggested that the risks for climate change were bigger than portrayed by previous economic models and strengthens the case for strong cuts in emission of GHG (Dietz and Stern 2014). The new research suggests a global carbon price should be from \$32 to \$103 per ton/CO_2 by 2015 and rise to between \$82 and \$260 per ton/CO_2 by 2035 (Dietz and Stern 2014, p. 22). These estimates, however imprecise, still do not answer the central question of *who pays for emissions*? The issue of "equity" would look at the problem from the perspective of who has contributed most to GHG, and because the developing countries would argue that the industrialized countries have contributed more than two-thirds of the current stock of human-induced GHGs in the atmosphere, they should bear the largest share of costs with this accumulated stock of emissions being described as an *environmental debt*. However, in terms of efficiency and looking to the future, this approach is not going to be supported by the advanced economies, and the question of historical responsibility then translates to ethics and moral obligations. To obtain a better understanding on who might pay what for GHGs in the future, one has to assess projections for carbon dioxide emissions on a per capita basis for the major OECD and non-OECD emitter economies. This is illustrated in the next set of Figs. 6.6 and 6.7.

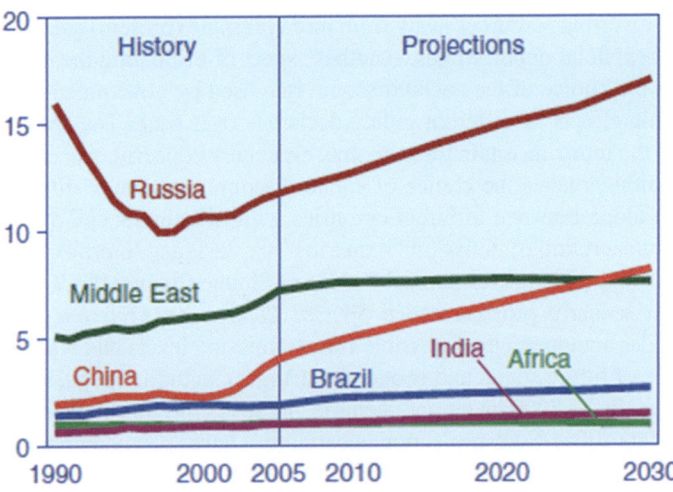

Fig. 6.6 Non-OECD carbon dioxide emissions per capita by country and region, 1990–2030

Source: *History*: Energy Information Administration (EIA), International Energy Annual 2005 (June–October 2007). 2030: *Projections*: EIA. World Energy Projections Plus (2008, p. 98)

Fig. 6.7 OECD carbon dioxide emissions per capita by country and region, 1990–2030

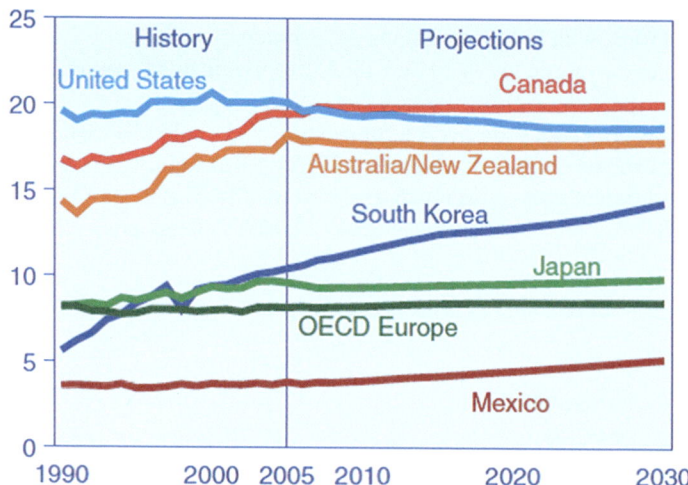

Source: *History*: Energy Information Administration (EIA), International Energy Annual 2005 (June–October 2007). 2030: *Projections*: EIA. World Energy Projections Plus (2008, p. 98)

From Figs. 6.6 and 6.7, we note that the OECD economies have higher levels of carbon dioxide emissions per capita due to their higher levels of income and fossil fuel use per capita, but that the projections forecast some improvement in country emissions. The US emissions per capita are projected to *fall* slightly from 20 metric tons in 2030, while Canada's emissions are projected to *rise* from 19 metric tons to 20 metric tons over the same period in the absence of binding constraints, once more underscoring Canada's decision to withdraw from the *Kyoto Protocol*. Among the OECD countries, Mexico, an oil producer, has the lowest level of per capita in 2005 and projected to rise to 5 metric tons by 2030.

By comparison with Mexico, in 2005 other oil producers in the Middle East had substantially higher CO_2 emissions per capita with Arab OPEC members, the UAE at 58 metric tons, Qatar at 48 metric tons, Kuwait at 32 metric tons, and Saudi Arabia at 17 metric tons. With their smaller populations, these countries are among the highest per capita emitters in the world (Khatib 2010, p. 65). Such per capita emission is not only attributed to economic growth and to high per capita electricity usage, increasing at a rate of 7–9 % annually, more than three times the world average, but also to wastage and overuse owing to generous subsidies as highlighted in earlier chapters. Such emission figures for the Gulf countries have prompted calls for promoting environmental integration and to make it a higher political priority for the countries of the Gulf Cooperation Council (Abdel-Rouf 2015).

Among the non-OECD countries, another major non-OECD oil producer Russia has the highest projected increase in carbon dioxide emissions per capita, rising from 12 metric tons per person in 2005 to 17 metric tons in 2030 due to a projected decline in Russia's population averaging 0.6 % per year from 2005 to 2030. However the population decline leads to a higher rate of increase in emissions per capita by

2030, but still lower than Arab OPEC members (excluding the non-oil-producing Middle East countries). The lowest levels of per capita emissions in the world are in India and Africa, with Indian emissions per capita projected to rise from 1.0 metric ton in 2005 to 1.5 metric ton in 2030, with Africa to remain at around 1.0 metric ton for the whole period.

The above projections are average projections and could change under different economic growth or energy price scenarios. Figure 6.8 examines carbon dioxide emissions under different economic growth scenarios for OECD and non-OECD countries, while Fig. 6.9 examines both blocks in three alternative energy price scenarios.

Fig. 6.8 Carbon dioxide emissions in three economic growth cases—2005 and 2030

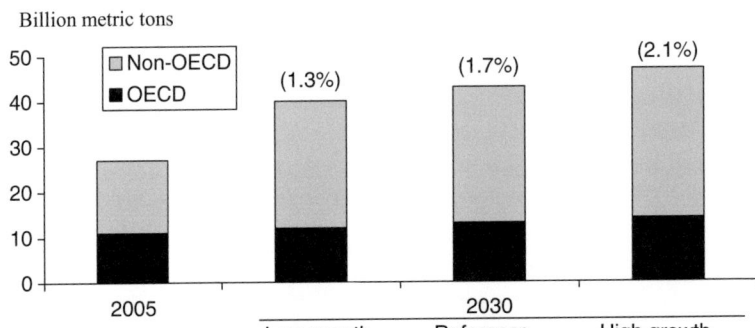

Source: 2005: Energy Information Administration (EIA), International Energy Annual 2005 (June–October 2007). 2030: Projections: EIA. World Energy Projections Plus (2008, p. 98)

In setting out the alternative growth scenarios in Fig. 6.8, the US Energy Information Administration notes that economic growth is the most significant factor underlying the projections for growth in energy-related CO_2 emissions in the immediate future, as the world continues to rely on fossil fuels for most of its energy use. As noted earlier, this is the argument that OPEC has been making—that oil is a vital component to the world's continuing economic growth and talk on "zero-carbon" emission for this vital energy source is unrealistic. In Fig. 6.8, in the high growth case, world carbon dioxide emissions are projected to increase at an average rate of 2.1 % annually compared with 1.7 % in the reference case and 1.3 % in the low growth case. However the projected rates of growth for the OECD and non-OECD countries differ in the three scenarios, with the OECD growth rates—from high, reference, to low growth rate scenarios—being 0.9 %, 0.5 %, and 0.2 %, respectively, annually from 2005 to 2030. The corresponding annual rates of growth for the non-OECD countries are 2.9 %, 2.5 %, and 2.1 %, respectively, for high, reference, and low growth scenarios. The total worldwide CO_2 emissions are projected by 2030 to be 38.4 billion metric tons for the low, 42.3 billion metric tons for the reference, and 46.6 billion metric tons for the high growth scenario.

Fig. 6.9 Carbon dioxide emissions in three alternative price cases—2005 and 2030

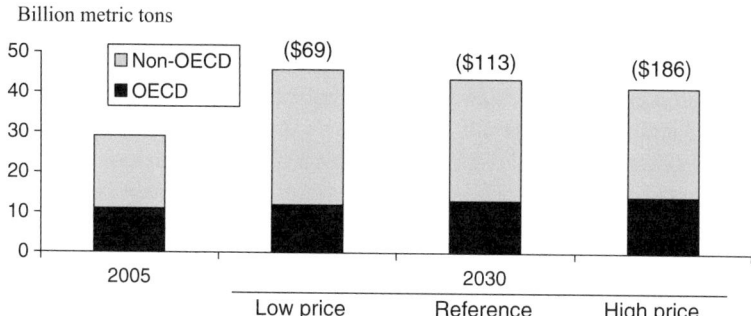

Source: 2005: Energy Information Administration (EIA), International Energy Annual 2005 (June–October 2007). 2030: Projections: EIA. World Energy Projections Plus (2008, p. 98)

Under different price scenarios, this time there is a projected *decrease* in carbon dioxide emissions when energy prices are high, amounting to a forecasted 40.1 billion metric tons in global emissions in the high-price scenario, compared with the 46.6 billion metric tons under the high economic growth scenario in Fig. 6.8 or nearly 14 % difference in emission rates. Conversely, total carbon dioxide emissions are projected to be higher in 2030 at the low price level of $69, reaching 43.4 billion metric tons, compared with 38.4 billion metric tons under the low economic growth rate scenario.

According to the EIA's 2030 forecasts, in the various price case scenarios, world oil and natural gas prices are affected more strongly than coal. As a result, in the higher price scenarios, both oil and gas lose global market share to coal whose share is forecasted to reach 30 % in the high-price scenario and 27 % in the low-price scenario. At the forecasted $186 per barrel of oil in 2030, coal becomes more attractive over liquid fuel, but as noted earlier in the chapter, coal is also the most pollutant energy. At the lower forecasted $69 per barrel low-price scenario in 2030, there is little economic incentive for nations to turn to other forms of energy like coal. It will be interesting to see if such a scenario continues to hold given the fall in oil prices during 2014 and whether coal-consuming nations like India and China will ease on using this form of energy at the expense of cheaper oil imports.

Paris 2015: A Half Glass Full Better Than None

As noted earlier, the *Paris* climate change conference is due to be held in December 2015, when the parties are expected to sign a universal agreement to replace *Kyoto* that would take effect from 2020. While hopes are high that momentum would continue on climate change action, expectations for a truly meaningful agreement to emerge are somewhat low. The key reason is that given the costs of climate control, there is limited prospect of an agreement where mitigation commitments will be binding, similar to the ones introduced at *Kyoto*, wherein a failure to meet the original mitigation targets was to be reflected in tighter targets in later commitment periods and was one of the major reasons for Canada to withdraw from the *Kyoto Protocol*.

Another reason why the Paris conference might still be lacking teeth is the unresolved issue of *enforcement*. Although governments may be willing to make ambitious commitments, like the Sino-US agreements, it is unlikely that targets will be met if their attainment turns out to be either more difficult or more costly than expected. As such, a consensus is emerging that it is far better to have almost everyone signing up to the new *Paris* agreement, even if the agreement is still vague on enforcement which is *the* key parameter. This is different from the Kyoto philosophy *where it was more important to obtain binding agreements that hopefully would include everyone*. This did not work and it was left to later bilateral deals between the world's largest emitters—the USA, India, and China—to try and fill the Kyoto gaps, although still left agreements somewhat vague on enforcement. The difference between *Kyoto* and *Paris* is that while the first espoused a *top-down* global agreement on mitigation, the second seems to be adopting a *bottom-up* approach that emphasizes national commitments from most countries, along with international agreements on transparency and accountability, and a *gradual* tightening of these commitments (Robinson 2015).

The reason for this gradualist approach is based on political reality: leaders of the negotiating countries have domestic political agendas that heavily influence their negotiating positions and the leeway that they have in forming and keeping likeminded country coalitions. It is not surprising that consensus is often difficult to achieve. Saudi Arabia, for example, has now become more active in taking the initiative in climate change negotiations on behalf of many other Arab oil producers and some developing countries, and this has only come about by expanding the remit and technical skills and abilities of the national negotiating team, whereby according to one of the climate change advisors, "we now even examine very carefully the footnotes to any proposed agreements" (Al Harthi 2015). When agreements are reached, they usually reflect the interests of the most powerful countries with the highest levels of emissions such as the USA, China, India, Brazil, and Saudi Arabia, OPEC's largest fossil producers. What will most likely emerge from *Paris* is that each country is expected to make *nationally determined contributions, or INDCs* (*Intended Nationally Determined Contributions*), which are probably easier to agree on, as there are no legal or financial penalties for noncompliance, unlike *Kyoto* which caused Canada to withdraw. In the INDC scenario, no country will be accused of being either an obstacle or of being responsible for a failure in *Paris*. To ensure that *Paris* is not just talk but there are some tangible inducements for developing countries to come on board, 27 countries including seven of the largest developing countries have agreed to contribute more than $10 billion to capitalize a new financial instrument for developing countries, known as the *GCF*, which was pointed out earlier as being one of the concerns of OPEC.

Contributions to the INDCs will most likely be built on the basis for all countries which implies that there will be less of a stark distinction between *Annex I* and *non-Annex* I countries of the *Kyoto Protocol*, which put a more onerous emission abatement regime on (more advanced) *Annex I* countries. In *Paris*, there will probably be more consensus on the need to retain the principle of "common but differentiated responsibilities and respective capabilities" which OPEC has been strongly advocating as one of their pillars for climate change agreements. According to some analysts, *Paris* could come up with some tentative agreements on a way forward for implementing some practical measures to combat emission, based on principles of "equity and responsibility." Some of these measures include incentives in the reduction of production of fossil fuels, such as falling fossil fuel prices for producers, offering opportunity for governments to eliminate or sharply reduce fossil fuel subsidies; introduction and increase of carbon taxes; direct compensation to the owners of fossil fuel resources who leave fossil fuels in the ground; and tangible ways of combating climate change such as improved energy efficiency and financial incentives and R&D support to carry this out. The onus for a more consensual climate change meeting in *Paris* is present.

OPEC and the Environmental Bandwagon

The global environmental bandwagon is moving ahead, maybe at a snail's pace in the view of some, but it is moving along with compromises and long-term goals now set by those that were implacable opponents of the *Kyoto Protocol* in the first place. What has been OPEC's collective position on the matter? Of more importance to the organization, what is the position of OPEC's largest members? From the analysis of publicly available statements by senior OPEC officials and country representatives to climate change conferences, it would seem that while OPEC is more than willing as a group to play its responsible share in curbing global emissions, there is a feeling that the organization is being unfairly targeted and stands to lose from more stringent and compulsory emission abatement policies while wearily accepting that the ant-fossil bandwagon was moving ahead as Minster Ali Al Naimi's earlier comments suggest.

According to some, the initial OPEC position to climate change was constant with two key issues raised: the rejection of new developing country commitments and what later became known as the *impact of response measures*. In the first, the then OPEC Secretary-General *Alvaro Silva-Calderon* insisted on the principle of *common but differentiated responsibilities* with the understanding that industrialized countries bear the lion's share of response measures due to past emissions. The second, *impact of response measures* has been to highlight the adverse social and economic impacts of measures taken to reduce greenhouse gas emissions on countries like OPEC who are dependent on one source of revenue (Muller 2006). According to OPEC's Director of Research, *Shukri Ghanem*, the "*Kyoto Protocol*, if fully implemented would lead to a dramatic loss of revenue for oil-exporting developing countries including OPEC's own members and the financial impact on our countries has been estimated at tens of billions of US dollars per year, according to OPEC's calculations" (Ghanem et al. 1999). Other researchers, while not so pessimistic about forecasted revenue losses, have indicated that based on global

Table 6.3 Distribution of losses among OPEC countries on the basis of OPEC's World Energy Model (OWEM)

	% of OPEC revenue 1999	Losses in 2010 (billion $)	Losses in 2010 as a % of 1999 GDP	Ranking in terms of losses as % of GDP
Saudi Arabia	28	4	2	5
Iran	11	1.5	0.4	9
Venezuela	10	1.4	0.7	7
Nigeria	9	1.3	1.2	6
Iraq	9	1.3	2.2	4
UAE	9	1.3	3.1	2
Kuwait	7	1	2.2	4
Libya	6	0.9	2.3	3
Algeria	5	0.7	0.5	8
Indonesia	3	0.4	0.07	10
Qatar	3	0.4	3.3	1

Source: Barnett et al. (2004, p. 2085)

energy-economy models of the impact of the *Kyoto Protocol* on energy exporters, the members of OPEC believe that the Protocol's implementation will slow growth in their revenues from oil exports (Barnett et al. 2004). The models suggest that policies and measures to implement *Kyoto* will increase oil prices to consumers and reduce demand in developed countries, thereby driving down global oil demand and prices received by producers. The models used, with refinements, were OPEC's World Energy Model (OWEM). These showed that regardless of how the *Kyoto Protocol* is implemented, OPEC's ability to influence the price of oil will not be diminished, something which the post 2014 oil price fall and OPEC market-share loss indicated was a major flaw in the OWEM assumptions. Table 6.3 summarizes the distribution of losses among OPEC members in 1999—excluding Ecuador which joined OPEC in 1973, left in 1992, and rejoined in 2007.

Table 6.3 suggests that declines in oil revenues will have less of an impact on some OPEC members than others. Countries like Indonesia (in 1999 a member of OPEC but left in 2008), Iran, and Nigeria with their larger share of agriculture in the GDP will feel more the impact of climate change in this sector of the economy than less diversified oil economies like Saudi Arabia, Kuwait, the UAE, and Qatar. At the same time, the less diversified oil economies of OPEC hold substantial non-oil investments and assets overseas, and some of these investments could become increasingly risky in a world where the climate is changing (Barnett et al. 2004, p. 2085).

The possible losses that OPEC countries may incur have triggered a debate on a so-called green paradox (Sinn 2008), whereby this paradox is based on the assumption that suppliers of oil feel threatened by a decline of future prices due to a gradual reduction of oil consumption in emission-abating countries. If this reduction reduces the discounted future value of oil prices more than at present, then oil-producing countries will be better off expanding production in the short run. This will then increase oil consumption and perversely *accelerate* global warming.

Some analysts argue that the 2015 surge in Saudi production to over ten million bpd is in fact on attempt by Saudi Arabia to push oil prices *lower* on fears of a possible flagging in long-term oil demand and that climate change and high crude prices will boost energy efficiency, encourage renewable energy, and accelerate a switch to alternative fuels such as gas, especially in the emerging markets that they count on for growth (Waldman 2015). This is an interesting proposition, but is more likely to be a side effect of Saudi oil policy, rather than a conscious decision to delay *climate* change energy switches.

According to Sinn (2008), even in cases where oil producers *do not* increase production, there may be another perverse effect in the form of *carbon leakage*, whereby a CO_2 reduction policy in home reduction will raise domestic energy costs which will bring a comparative advantage to firms in those non-abating countries. This may induce firms to migrate from the high-cost home country to non-abating countries, creating an *environmental capital flight*. Both carbon leakage and environmental capital flight will lead to an expansion of production in non-abating countries and help offset some of the emission abatement in the home country.

How realistic is the basic assumption underlying the "green paradox," whereby oil producers fear a decline in future prices due to a gradual reduction of oil consumption in energy emission abating countries? There is some support for this assumption derived from *decreasing oil intensify of production*, i.e., the total primary use of oil per unit of output due to the more efficient use of oil, increasing utilization of alternative and renewable energy sources, and a shift in the composition of national output toward less-oil-intensive service-oriented sectors. According to the OECD, oil intensity of production has steadily declined in OECD countries by slightly less than 50 % over the period 1970–2003, while in developing countries it increased slightly less than 30 % until 2000, when it started to marginally decline (OECD 2004).

Figure 6.10 illustrates two different concepts of energy and carbon intensity and how shifts in the composition of national output can influence these two measures.

Fig. 6.10 Energy vs. carbon intensity

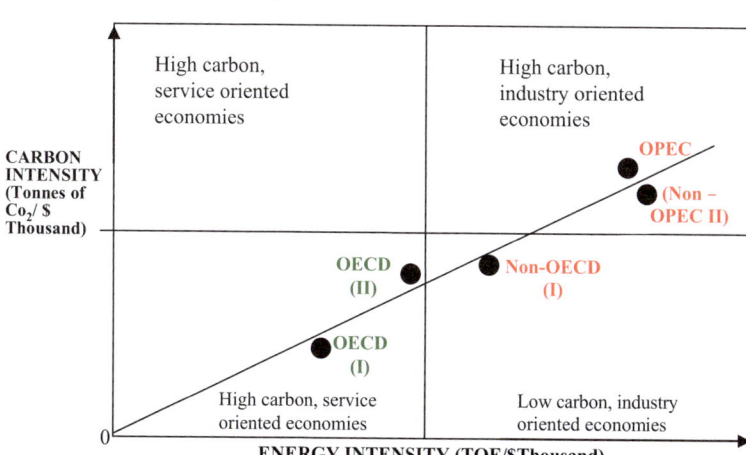

In the above figure, *energy intensity* measures the energy needed using tonne of oil equivalent (TOE) to produce one unit of economic output, while *carbon intensity* measures the omissions generated in tonnes of CO_2 to produce the same unit. Both metrics are widely used by policymakers when setting economy-wide environmental targets. Both metrics are closely linked because most carbon emissions occur as a result of energy use. According to research carried out, while some countries are reluctant to implement strong CO_2 mitigation policies, all countries are in favor of enhancing their energy efficiency to boost economic productivity (KAPSARC 2013, p. 4). This is because of the benefits such targets bring them in terms of moderating energy consumption and reducing waste. This was highlighted earlier with the contrast between Arab Gulf OPEC members' high per capita energy consumption and OECD countries. As such, these are worthwhile goals irrespective of the issue of climate change. In Fig. 6.10 most OECD countries like the UK, Germany, and Japan are in the low-carbon, service-oriented economies (*OECD I*), while the USA is in the higher carbon emission range (*OECD II*). By comparison, non-OECD and non-oil-producing countries like India are to be found in the low-carbon, industry-oriented economy (*non-OECD I*), with non-OECD OPEC and non-OPEC oil producers like Russia in the high-carbon, industry-oriented economy universe.

OPEC's Policy Challenges: Choices, Choices

OPEC is faced by a bewildering variety of policy challenges on global warming, with different choices and outcomes. Figure 6.11 illustrates some of the key drivers and challenges faced by OPEC's policymakers.

Fig. 6.11 OPEC's policy minefield choices

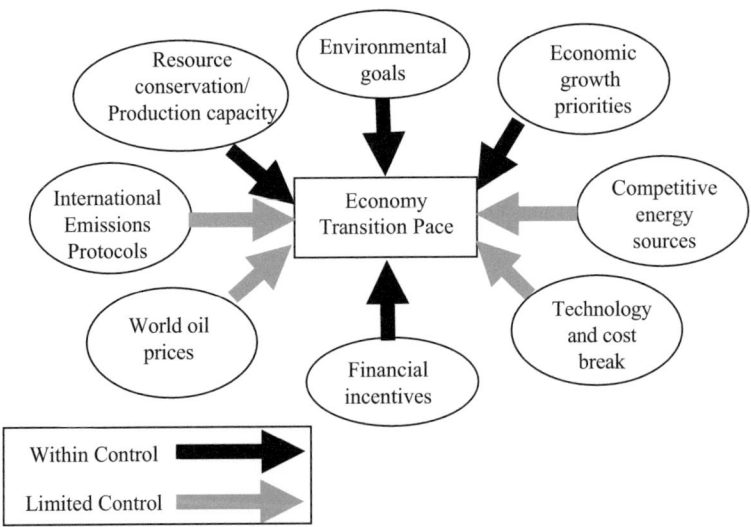

For OPEC members, understanding the various trade-offs among the above objectives (those within their control) is helpful in evaluating the effectiveness of their policies. The problem for OPEC members is that they are faced by many factors outside their control or with limited control. OPEC has seen world prices fall sharply during 2015 and seemed to have taken the role of a bystander, unlike previous periods when it was accustomed to be in the driving seat. However, in other "limited control" areas, OPEC can *influence* the outcome, like taking the lead in new technology breakthrough to reduce carbon emission and become a more active participant in putting forward novel solutions in international emission protocol meetings. OPEC members with financial resources can also champion their own non-fossil fuel alternative energy mix, such as solar power to ensure resource conservation of their prime oil asset as Saudi Arabia seems intent on doing in the case of solar energy.

It is important to assess in closer detail the various initiatives that OPEC's largest producer Saudi Arabia has carried out in the various fields of emission control, as Saudi leadership in this respect could influence the position of many of the smaller OPEC producers to enable a more unified position be adopted at future climate change conferences. According to Deputy Petroleum Minister *Abdulaziz bin Salman*, Saudi Arabia has undertaken the following initiatives:

- Continuing to make large investments to develop its gas network and reduce flaring
- Using treated sewage effluent for irrigation and industrial applications for cooling and capturing methane to use it in future applications
- Pilot projects to convert waste to energy and capturing methane to be used for electricity generation
- Preparing Saudi Arabia's intended National Determined Contribution to be ready for submission in Paris with four main actions planned: first, by increasing energy efficiency; second, utilizing renewable energies such as solar and wind; third, developing carbon capture utilization and storage and carbon dioxide enhanced oil recovery; and fourth, switching from liquid to gas

The Deputy Minister went on to advise that approximately 150 professionals from 30 organizations are actively engaged in the design and implementation of the Saudi Energy Efficiency Program's 71 initiatives "with a potential savings of 1.5 million barrels of oil equivalent per day by 2030" (Abdulaziz bin Salman 2015). According to the Deputy Minister, the Kingdom joined the Carbon Sequestration Leadership Forum in 2005 and Riyadh will be hosting the Forum's ministerial meeting in November 2015. Furthermore the Kingdom has made many advances in this area, with carbon dioxide storage in saline water aquifers and carbon dioxide enhanced oil recovery in Saudi Aramco's *Uthmaniyah* oil fields from the carbon dioxide captured at *Hawiyah* NGL Plant. These technology experiences will be invaluable to other OPEC members.

How have the various OPEC officials addressed these and other policy challenges facing the organization on environmental issues? Table 6.4 sets out a brief survey of public comments made by OPEC officials over time and the policy outcomes that the organization hopes for.

Table 6.4 OPEC's environment cooperation options and concerns

OPEC source/location	Issue and proposed solution
• Mohammed Barkindo, First International Conference on the Clean Development Mechanism, Riyadh, Sept. 2006	• OPEC members proposed *Compensation Fund* in run up to *Kyoto* to counter adverse effects of implementation of response measures
	• Financial penalty would be levied proportionate is noncompliance and paid into a *Clean Development Fund* to be used to support sustainable development projects in developing countries
	• Accessing investment capacity building and technology transfer to help stabilize emission levels. A necessary policy ingredient
	• None of the above has gone to plan (*as of 2006*)
	• *Solutions*
	– Technological options that allow for continued use of fossil fuel such as *carbon capture and storage* (CCS) in conjunction with CO_2 enhanced oil recovery offer not only storing CO_2 but increasing oil reserves in mature fields
	– Active participation in the IEA's Greenhouse Gas R&D program
	• To date European Union has been focused on emission trading to the exclusion of others, with little mention of Clean Development Mechanism (CDM)
• Mohammad Al Sabban, Saudi Arabian Envoy to Climate Negotiations, Energy Dialogue Conference Riyadh, Nov. 21, 2011	• Saudi Arabia and OPEC partners asked to bear too much of the burden of cutting greenhouse emissions because their economies depend on oil and natural gas revenues
	• Saudi Arabia has not asked for compensation for the loss of income from oil sales as consumers look to obtain energy from cleaner fuels. The Kingdom wants technological assistance from developed countries
	• Saudi Arabia thinks that a second commitment period for the *Kyoto Protocol* is a must, and without having *unconditional* emission reduction numbers from developed countries for the period beyond 2012, it will be impossible to have any agreement in Durban
	• Saudi Arabia and other developing countries will not agree to negotiate the United Nations Convention on Climate Change
	• The Gulf States want carbon storage and capture to be included in the CDM

(continued)

Table 6.4 (continued)

OPEC source/location	Issue and proposed solution
• OPEC Secretariat Special Seminar Meeting, Vienna, March 18–19, 2009	• Special OPEC seminar bringing together key OPEC officials, IOC representatives, and government energy-related bodies (*Came in the wake of the 2008 Global Financial Crisis and collapse of oil prices to $40 pb levels in 2009 from June 2008 $139 peak*)
	• Oil producers urged developed countries to take the lead in greenhouse gas mitigation efforts, given their *historical responsibility* and their technological and financial capabilities
	• *Gholam-Hossein Nozari* (*Iran Minister of Petroleum*): consumer energy policies, including unfair tax systems, hindered OPEC efforts in meeting its commitments
	• *Ali Al Naimi* (*Saudi Oil Minister*): the recent extraordinary economic and financial events had created the global crisis and caused a dramatic price slide for oil and gas. Compounding this outlook are calls to lesson or end dependence on oil, particularly from certain regions
	Don't pin too many hopes on alternative energy source since they might not meet their high expectations and instead divert investment from proven hydrocarbons
	• *Dr. Shukri Ghanem* (*Chairman, Management Committee, Libya National Oil Corporation*): increased energy efficiency, clean fuel technology, carbon capture, and storage and renewable energy are all important subjects in helping reduce damage to the environment
	• *Suleiman Al Herbish* (*Director General, OPEC Fund for International Development*): electricity is not always the most appropriate energy source for all needs. Petroleum products are needed for transportation, cooking, and heating as well as to produce electricity. The biggest environmental tragedy facing the globe is human poverty

Source: OPEC (2009[a]), Daya (2011), OPEC (2006)

The above list of comments by senior OPEC officials indicates a hint of frustration and exasperation about the lack of appreciation of the value of fossil fuel to the well-being of the world. It is also interesting to note that the comments of the OPEC officials made in 2009 in the aftermath of the global financial crisis and the sharp fall in oil prices after record highs echo the OPEC official's meeting in late 2014 and 2015 when oil prices also fell sharply bringing a sense of *deja vu*. All OPEC officials seem to agree that there is a need to cooperate to mitigate the effects of global emission and that some other solutions such as *carbon capture and storage* (*CCS*) could mitigate, to an extent, the effects of harmful emissions, but the feeling of "injustice" and lack of equitable sharing of costs based on historical emission responsibility is also a common thread from the OPEC statements. The payment of compensation to OPEC members for potential future losses of oil revenues has also been highlighted, but as some have commented, assessing such an impact requires a distinction between the impact of other *unrelated* "policies and measures" to employment (PAMs), from those taken *pursuant* to the protocol to enable a disaggregation of the effect of climate change PAMs between the two (Barnett et al. 2004).

The issue of taxes on carbon emission is also another area of concern to OPEC members, as this will raise the price of oil to consumers and reduce demand further and reduce OPEC's market share in the future. Some have pointed out that any form of a "carbon tax" may in fact increase the "rent" that governments in the energy-importing countries have in the oil market and contribute to a further transfer of wealth from all producers to consumers (Mabey et al. 1997, p. 274). OPEC itself has repeatedly pointed out this "unfair" transfer of wealth from producer nations to consumer nations by highlighting that the *Group of Seven* (*G7*) developed countries (the USA, Canada, Japan, Germany, Italy, Britain, and France) earn around 70 % *more income* from oil than OPEC members earn from petroleum exports (OPEC 2001).

This is illustrated in Fig. 6.12 that sets out the breakdown in dollar terms of a composite barrel of oil for the period 2011–2013 for key industrial countries.

Fig. 6.12 Composite barrel and its components ($/b) (2011–2013)

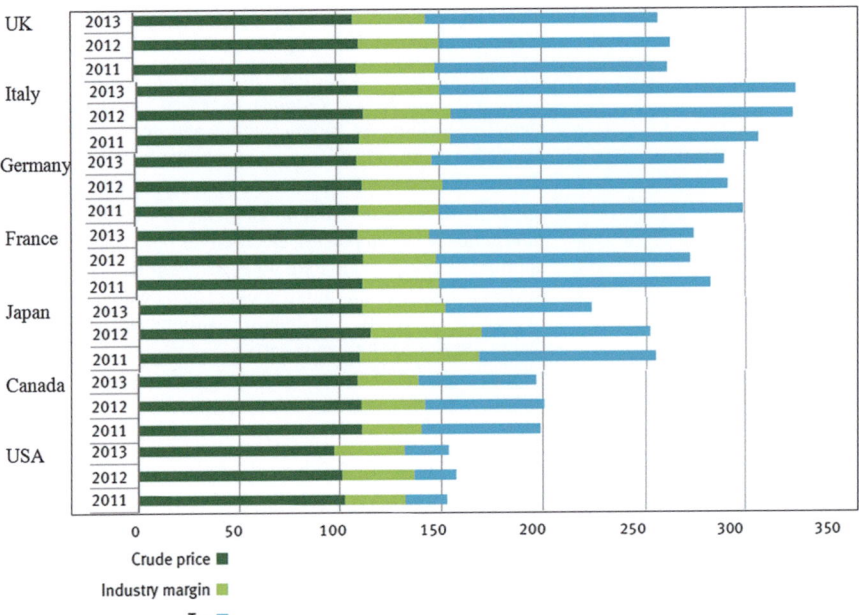

Source: OPEC Annual Statistical Bulletin (2014, p. 91)

The above figure illustrates the sharp tax margins per barrel for some countries like the UK, Italy, Germany, and France with taxes levied being almost as much as the price of the crude oil, while taxation levels are smaller in the USA and Canada. In some instances, for countries like Italy, the combined industry margin and taxation are more than double the price of crude oil per barrel.

OPEC has made strenuous efforts to ensure that oil producers' concerns at such an inequity in wealth generation are acknowledged at the various global environment meetings. In 1997, OPEC's statement at the COP-3 (Kyoto) stated that "the principle of compensation *must* be built into the protocol" that was to emerge from Kyoto (OPEC 1997, p. 4), while at the COP-4 (Buenos Aires), the OPEC Secretary-General *Dr Lukman* stated the organization's position on compensation and mutual progress as follows: "without a favorable disposition towards the compensation issue among the parties to the conventional, how can fossil fuel producers be expected to give their wholehearted blessing to measures that could wreck havoc on their economies?" (OPEC 1998, p. 2).

The oil producers are aware that appealing to equitable solutions for compensation might or might not materialize and that in the meantime, those OPEC members with a narrower economic base have to restructure their economies to be able to meet their population's future expectations. As Saudi Arabia's chief climate negotiator Khalid Abu Leif put it, "we know we are in a race with time. Climate change and economic diversification for us is hand in hand" (Arab News, Dec. 10, 2014[d]).

The global climate change bandwagon is not waiting for anyone and OPEC has to readjust, with the organization itself adopting technology that minimizes carbon emission and convincing others that CCS is something worthwhile for all parties, both producers and consumers, to consider as one possible solution for CO_2 emission. For this to happen on a global scale, and to be able to support the smaller OPEC and non-OPEC producers, a portion of current OECD oil taxes could be set aside to assist in CCS projects.

Chapter 7
Charting a New OPEC Role: Avoiding the Coming Perfect Storm

> *Cheer up. The worst is yet to come.*
>
> Philander Johnson

Oil: A Blessing or a Curse?

There is a finite amount of oil in the world, and while new technology will push back the boundaries of "peak oil," current oil endowments depend on the rates at which oil is being produced and how it is being used for the good of nations. The modern oil era of the twentieth century has brought about many benefits to countries and citizens' way of life and home comforts. High oil prices have brought about shifts in financial assets from consumer to producer nations. At the same time, high prices have promoted energy efficiency measures and the introduction of substitutes leading to oil supply disruptions and oil price volatility. These have been compounded by other factors that have become important for the future use of oil like environmental concerns.

For many oil producers, especially the larger OPEC economies like Saudi Arabia, Venezuela, Iraq, Iran, and Nigeria, the transfer of financial assets from oil-consuming nations has transformed them into so-called petro-states where the main concern of these countries is how to maximize their resource "rent-seeking" opportunities from oil sales (Karl 1997). This has often come at the expense of domestic innovation, hard work, risk taking, and entrepreneurship and developing private sector-driven competition-oriented economies. Instead of oil being a blessing to foster such economic and social traits, it has become, for many of the petro-states, a curse, with the state sector dominating all major aspects of economic and social life, leading to bloated bureaucracy, inefficient subsidies, micromanagement, and corruption. In such countries, political activity revolves around the struggle to distribute wealth, rather than the creation of a sustainable source of private wealth generation and productivity.

The notion of a "resource curse" is well documented in literature and is often referred to as the *Dutch disease*. This term describes the economic consequences of dependency on a major natural resource, such as natural gas exports which affected the Netherlands in the 1960s. As new gas wealth flowed to the country from gas exports, the rest of the Dutch economy suffered, with the currency overvalued and non-energy exports becoming more expensive and domestic businesses less competitive (Corden 1984). To partially overcome for such a disease, sovereign wealth funds were created to absorb large inflows of revenues and prevent it from flooding into economies and insulating domestic businesses. However, the true oil curse lies in the near-dependency of petro-states on oil revenue and their seemingly incurable fiscal rigidity in raising other forms of revenues, especially during periods of falling oil prices, and yet raised social expectations and political expediency leads to ever-increasing government expenditures, a so-called "reversed *Midas* touch" (Yergin 2011, p. 109). Some oil producers, even those with substantial foreign reserves like Kuwait, are considering diversifying their sources of income by imposing a corporate tax on local companies after discussions with the International Monetary Fund (Reuters 2015[a]). Introducing a new corporate tax would be a major and politically sensitive policy shift not only for Kuwait, but for the Gulf Arab oil producers, but with Kuwait's budget surplus shrinking by around 26 % for the 9 months through December 2014, some predict that Kuwait could run a budget deficit in 2015 if oil prices stay below $60 a barrel level for 2015 (Reuters 2015[a]). Saudi Arabia, OPEC's largest producer, has forecasted a budget deficit of SR 145 billion ($38.7 billion) for 2015, with an estimated 83 % of budgeted revenues of SR 715 billion ($190 billion) coming from oil (Jadwa Investment 2015[b]).

Tinkering with corporate taxes and limited raises in the price of subsidized fuel will not bring in significant revenues to such petro-states, compared with meaningful economic diversification.

OPEC: Can It Ensure a "Fair" Market Price?

Since the early 1970s, energy security has been interpreted, if somewhat narrowly, as being synonymous with adequate and stable availability of oil at an "acceptable" price. As noted in the volume, changes in oil technology and emergence of other energy players have altered this narrow view. New suppliers have increased, especially nonconventional, as have proven reserves and exploitable stocks in offshore fields. Prices, while still volatile at times, have become flexible and more dictated by market forces rather than by cartel pricing or quota production agreements. The previous narrow definition of energy security can now be viewed in a wider spectrum to include energy and environmental management, given the increased emphasis on global emission control.

OPEC today operates in a multi-polar energy world, trying to find its way through a maze of complex supplier and consumer relationships with new "petro-superpowers" like the USA and Russia and consumer powers like China and India, leading to a potential energy conflict if a stable and mutually beneficial energy market does not evolve to meet conflicting resource objectives (Klare 2008). Finite resources, escalating global population and energy demand, and the location of energy resources in regions torn by strife and ethnic and political unrest all combine as preconditions for energy conflict. As of June 2015, 3 of OPEC's 12 members (Libya, Nigeria, Iraq) were involved in conflicts of one sort or another, while 2 others (Iran, Venezuela) were facing either internal or external political pressures. According to analysis of resource conflict (Klare 2002), the major reasons for such conflict can arise due to the following underlying resource characteristics:

- If resources are *scarce*
- If resources are deemed *valuable*
- If resources are *lootable*
- If resources are *strategic*
- If resources evidence *unstable prices*

Table 7.1 summarizes some of the major resource conflicts that took place as well as current oil resource disputes between bordering nations.

Table 7.1 Oil resources conflicts underlying factors

Conflict area	Potential underlying factors
Iran-Iraq (1980–1988)	Iran-Iraq war : dominance by either party
	• *Strategic*
Iraq-Kuwait (1990–1991)	Iraq occupation of Kuwait
	• *Lootable*: accusations by Iraq of Kuwait "looting" Iraqi oil through lateral drilling
	• *Unstable prices*
Iraq-USA (2003–2011)	US occupation of Iraq
	• *Strategic*
	• *Valuable*
	• *Lootable*
Libya (2011–)	Overthrow of Libyan leadership
	• *Strategic*
	• *Valuable*
Sudan (1983–2005)	Civil War between North and South Sudan
	• *Valuable*
	• *Scarce*
South China Seas (1971–)	Involves disputes on territorial issues between Brunei, China, Taiwan, Malaysia, Vietnam, and the Philippines
	• *Strategic*
	• *Valuable*
	• *Scarce*
Caspian Sea (1980–)	Involves disputes on territorial issues between Russia, Iran, Azerbaijan, Kazakhstan, and Turkmenistan
	• *Scarce*
	• *Valuable*

While some of the above resource conflicts have been partially resolved, such as the Sudan civil war which tried to demarcate oil resources between the Sudanese warring parties, lingering tensions in the South China Seas remain a major concern, as this involves the new energy consumer "petro-superpower" China, with implications for long-term energy supplies to China from Arab OPEC members should Chinese dependency on these energy supplies decrease in the future. The escalating war of words between the USA and China on Chinese activities in the South China Seas only adds to global concerns.

OPEC will continue to face long-term structural changes and uncertainties especially on the global stage given the profound shifts in energy demand and supply patterns from non-OPEC and nonconventional producers should the US "shale revolution" spread successfully to other parts of the world, like China and Russia. According to OPEC, estimates for non-OPEC supply have continued to grow, averaging 56.23 million bpd in 2014, an increase of 1.99 million bpd over 2013 as illustrated in Fig. 7.1, with the major year-on-year growth originating from OECD North American countries like the USA and Canada as well as Brazil.

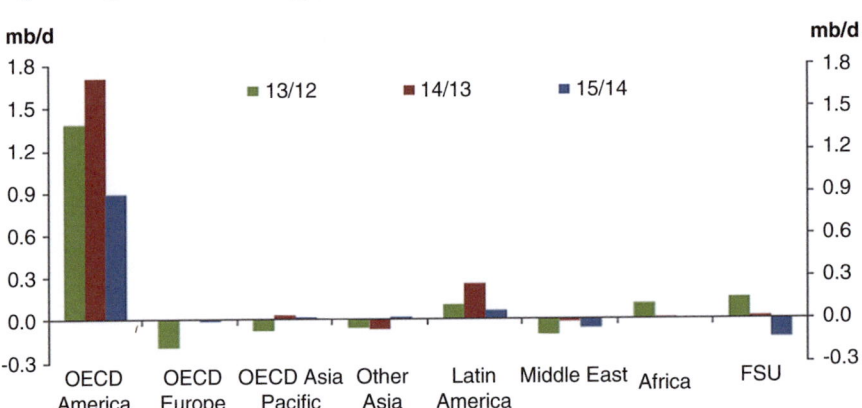

Fig. 7.1 Regional non-OPEC supply growth (year-on-year)

Source: OPEC monthly oil market report, Feb. (2015[b], p. 42)

Of more interest is that despite reduced oil prices from 4Q 2014, OPEC was still forecasting growth in US production in all categories of energy—both conventional and nonconventional as illustrated in Table 7.2.

According to OPEC, the major US shale players are expected to add output growth in 2015 with the caveat that "as drilling subsides due to high costs and low prices, production can be expected to fall, possibly late in 2015" (OPEC 2015[c], p. 56). Given these OPEC predictions, the organization has published its 2015 forecast for world oil demand and supply and OPEC's estimated crude oil production. This is set out in Table 7.3 and Fig. 7.2 which illustrate the call on OPEC in 2015.

In the above OPEC 2015 scenarios, the organization is forecasting a rebound in OPEC's production from 2Q 2015 with demand for OPEC crude reaching 29.2 million bpd for year-end 2015, compared with 29.09 million bpd for 2014. According to OPEC, its members produced 30.03 million bpd in 2014, indicating an overall surplus of around 940,000 barrels per day for 2014, highlighting the intense speculation prior to the November 2014 OPEC conference meeting on whether OPEC's leading

Table 7.2 US production in 2014 and 2015

US growth by source	2014 Estimated mbpd	2015 Forecast mbpd	Growth y-o-y *tbpd*
Eagle Ford shale	1.3	1.5	150
Bakken shale	1.2	1.3	140
Permian Midland	0.4	0.5	110
Permian Delaware	0.3	0.4	100
Niobrara	0.2	0.3	50
Others	0.4	0.4	30
Oil from unconventional gas	0.1	0.1	10
Total US tight oil	**3.9**	**4.5**	**590**
Unconventional NGLs	**2.0**	**2.1**	**50**
US Gulf of Mexico	**1.4**	**1.6**	**180**
Total US output	*7.3*	*8.1*	*820*

Source: OPEC monthly oil market report, Feb. 2015[b], p. 56

Table 7.3 Summarized supply-demand balance for 2015, mbpd

	2014	1Q15	2Q15	3Q15	4Q15	2015
(a) World oil demand	91.15	91.36	91.18	92.96	93.76	92.32
Non-OPEC supply	56.23	57.55	57.08	56.79	56.94	57.09
OPEC NGLs and nonconventional	5.83	5.89	5.98	6.08	6.18	6.03
(b) Total supply excluding OPEC crude	62.06	63.44	63.05	62.67	63.12	63.12
Difference (a−b)	*29.09*	*27.93*	*28.13*	*30.10*	*30.64*	*29.21*
OPEC crude oil production	30.03					
Balance	0.94					

Source: OPEC (2015[c], p. 87)
Totals may not add up due to independent rounding

Fig. 7.2 Balance of supply and demand and call on OPEC

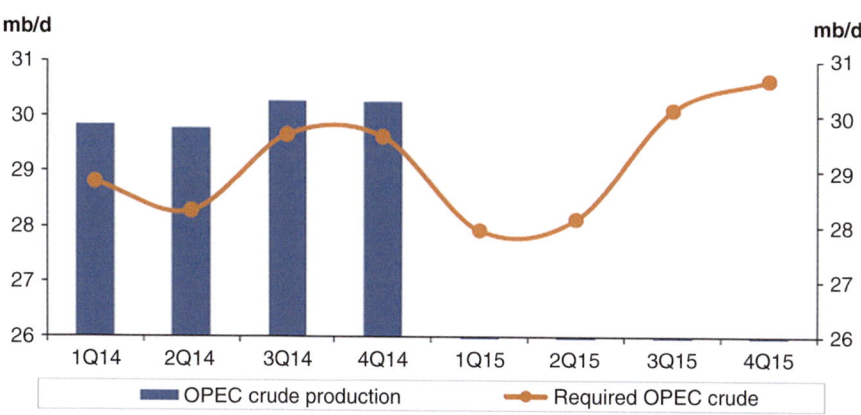

Source: OPEC (2015[c], p. 88)

producers, notably Saudi Arabia, Kuwait, and the UAE, would cut back on production to remove this "excess" of around 1 million bpd from the market to balance supply and demand and raise prices. As pointed out in the volume, Saudi Arabia and its OPEC Gulf allies did not cut production in fear of losing market share. However, when analyzing the production of individual OPEC members over the period December 2014/January 2015 when the OPEC decision was taken not to cut production, it is somewhat odd to note from both OPEC's own direct members' communication and independent secondary sources that Saudi Arabia, Kuwait, and the UAE actually produced *more* in that period while other OPEC members had lower production. This is illustrated in Table 7.4 which is based on non-OPEC sources.

According to OPEC's own *direct communication* figures from members, OPEC production for December 2014 totaled 31.054 million bpd, compared with 31.153 million bpd from non-OPEC sources, or around 300,000 barrels per day difference. This could be due to the fact that both Kuwait and Saudi production fell by around this amount when the joint Kuwait-Saudi *Khafji* production ceased due to "environmental reasons" in November 2014. In April 2015, a further 200,000 bpd was removed from the market when another Saudi-Kuwait oil field *Wafra* was shut down, ostensibly due to technical reasons but in reality due to disagreement between the two countries over the role of Saudi Chevron's contractor role in the neutral zone. More noticeable though from Table 7.4 are production cuts for Iraq, Libya, and, to a lesser extent, for Iran. The sharp fall in Libyan production from around 1 million bpd in 2013, to around 340,000 bpd in 2014, was due to the unstable political condition in that countries with different factions control Libya's oil production and exporting facilities, while Iraq's reduction was also associated with military conflict in oil-producing regions such as *Kirkuk*. How OPEC will accommodate the

Table 7.4 OPEC crude oil production based on secondary sources, tbpd

	2013	2014	2Q14	3Q14	4Q14	Nov. 14	Dec. 14	Jan. 15	Jan./Dec.
Algeria	1159	1151	1158	1167	1152	1158	1143	1130	*−12.6*
Angola	1738	1654	1646	1690	1678	1656	1655	1777	*122.1*
Ecuador	516	542	541	543	545	543	554	553	*−0.5*
Iran, I.R.	2673	2766	2768	2759	2765	2765	2779	2754	*−25.1*
Iraq	3037	3264	3266	3150	3422	3337	3632	3353	*−279.1*
Kuwait	2822	2774	2786	2794	2719	2699	2699	2777	*77.6*
Libya	928	473	222	614	678	673	475	343	*−131.7*
Nigeria	1912	1910	1895	1949	1898	1919	1896	1940	*44.4*
Qatar	732	724	729	733	700	698	687	680	*−6.5*
Saudi Arabia	9586	9683	9675	9747	9608	9584	9590	9683	*93.2*
UAE	2741	2760	2749	2791	2753	2741	2777	2841	*63.9*
Venezuela	2356	2333	2337	2329	2324	2323	2320	2321	*1.0*
Total OPEC	30,198	30,032	29,772	30,266	30,243	30,097	30,206	30,153	*−53.2*
OPEC excl. Iraq	27,161	26,768	26,506	27,116	26,821	26,760	26,574	26,800	*225.9*

Source: OPEC (2015[c], p. 61)
Totals may not add up due to independent rounding

return of "full" production by Iraq, Iran, and Libya will be discussed later in the chapter as this will further complicate the OPEC market surplus noted earlier in Table 7.3, *without* taking into consideration any further surge in both US and other non-OPEC supplies. There is some evidence that oil prices of around $55–$60 a barrel have prompted some IOCs to reconsider their energy projects due to higher drilling costs and remote locations of fields, such as *Chevron's* decision to pull out of a national gas project in Australia in the wake of similar moves by *Conoco Phillips, Hess Corp.,* and *Statoil ASA* (Paton 2015). Australia has the seventh biggest potential shale gas and the sixth biggest shale oil resources in the world, but as analyzed earlier in the volume, not many nonconventional resource countries can replicate the success of this US energy sector. With all these unknown factors, can OPEC obtain a so-called fair price and even see a return to higher oil prices?

Predicting Future Oil Price Curves

Even the most knowledgeable OPEC officials seem perplexed about the future direction of oil prices with some stating that oil prices will not rebound to $100 a barrel level, as these levels would draw more shale and output from higher-cost non-OPEC producers, according to Saudi Arabia's OPEC Governor Mohamed Al-Madi (2015[b]). The Saudi official went on to state that OPEC is not against shale oil and that it was welcome, but that it was "not fair for high-cost producers to push low-cost producers out of the market" and that the world needed *$40 trillion* of new oil investments in the next two decades to meet growing global demand. This is forecasted to grow by 1 million barrels a day every year for the next 15 years to about 111 million barrels a day according to Nasser Al Dossary, the Kingdom's OPEC national representative (Mahdi 2015[b]). Again, the Saudi OPEC representatives state that it is in the organization's interest to achieve "balance" in the market and that prices should be decided by the market and that the market was subject to supply and demand. Saudi Arabia and other major OPEC producers have realized that they cannot create a glut or shortage in oil simply on their own, and the rise of North American production, despite erratic oil prices, demonstrates other forces at work besides market supply and demand. This is illustrated in Table 7.5 which compares average US and Saudi production over the period 2008–2014.

From Table 7.5, the inexorable rise in US production over the period 2008–2015 illustrates the dynamics of the shale oil revolution in that country, while Saudi oil production was more erratic and was within a whisker of 10 million bpd in March 2015, almost the level reached in 2013. By the end of March 2015, Saudi production reached 10.3 million bpd. The surge in Saudi oil production, as well as those from the UAE and Kuwait, could lead OPEC to face difficult internal dynamics from the fact that Iraq, Iran, and Venezuela are major potential sources of capacity and growth, albeit at different paces in the short and long term. Normalization of relations with the West following a final successful nuclear deal will lead to Iranian production gains as well as a return of some sort of stability in Iraq and Libya will all lead to a challenge of Saudi Arabia's leadership. The Kingdom is aware of these

Table 7.5 American and Saudi oil production and oil prices

	US production (mbd)	Saudi production (mbd)	Brent price (year average or end of month)
2008	5.0	9.26	$96.94
2009	5.35	8.25	$61.74
2010	5.48	8.9	$79.61
2011	5.65	9.46	$111.26
2012	6.5	9.83	$111.63
2013	7.45	10.05	$108.56
Jan. 14	7.96	9.94	$108.16
Apr. 14	8.22	9.7	$108.63
Jul. 14	8.69	9.84	$104.94
Aug. 14	8.74	9.74	$101.12
Sept. 14	8.9	9.64	$94.67
Oct. 14	9.05	9.74	$84.17
Nov. 14	9.02	9.64	$71.89
Dec. 14	9.1	9.7	$55.27
Jan. 15	9.3	9.8	$54.30
Mar. 15	9.42	9.97	$57.10

Sources: EIA, Bloomberg, OPEC

potential production challenges and Saudi Aramco has stated that it is developing technology that could raise the proportion of oil extracted from its fields to "twice industry average," with a goal of 70 % field recovery according to Aramco's Chief Technology Officer *Ahmad Al Khowaiter* (DiPaola 2015^c). Saudi Aramco seems to be hedging its bet though on future developments in the energy sector. According to Khowaiter, the company "wants to emerge as a leading global integrated refining and chemicals company by the end of this decade" (DiPaola, ibid). This might not seem far-fetched, as Saudi Arabia has recently invested heavily in export-oriented petrochemical refineries and the country may now be perceived to be more as a "refiner" energy producer rather than as a crude exporter. In the long term its perceived "hegemonistic" influence on crude oil prices is diminished.

Can OPEC feel confident that it can predict, with a great deal of certainty, where future oil prices are heading? Saudi Oil Minister Ali Al Naimi is reported to have responded to the question of where oil prices are heading by replying that "if he knew the answer, he would be in Las Vegas." None of us know the future, said the minister, who went on to place the sharp fall in prices in 2014 to expectations and speculations and that it was "nothing about fundamentals" (Critchlow 2015). Whether the Saudi Oil Minister's comments were made in jest or not, it fundamentally illustrates the uncertainty that OPEC is facing in trying to make longer-term price predictions, as "getting it right" or "wrong" could have serious implications to the different members of the organization, and the lower the degree of forecast error, the easier it will be to manage their economies in face of growing energy competition.

As discussed earlier in the volume when analyzing different OPEC members fiscal breakeven and financial stress levels, the current straitened financial circumstances of many of the oil exporters could lead to a reduction in tensions among them and could prompt them to explore, officially and unofficially, ways to cooperate both among themselves and with non-OPEC members in order to prop up oil prices. Seemingly implacable position, such as unwilling to consider cuts, might lead to discussions on possible mutual steps of reduction by all parties, which by necessity have to be verified if such agreements are to last. The history of OPEC quota busting and noncompliance, explored earlier in the book, might not be a good omen to ensure that such commitments are adhered to. This has not stopped some members of OPEC in trying to obtain a degree of consensus on what to do next, especially concerning a common stand on production cuts. This was exemplified when the Algerian Energy Minister Youcef Yousfi opened up discussions with counterparts from fellow OPEC members Angola and Nigeria, as well as other non-OPEC members to sound out "possible responses to continued decline in oil prices," given that these three OPEC members, especially Nigeria, were part of the group that advocated a reduction in production at the November 2014 OPEC meeting, but grudgingly went along with the Saudi decision not to cut production (Alike and Okafor 2015). In May 2015, the Algerian Energy Minister Yousfi was replaced by Salah Khebri in a reshuffle of Algeria's Cabinet indicating that the country was facing fiscal pressure (Slimani 2015).

Despite statements that oil prices especially during periods of volatility have nothing to do with "fundamentals" and are driven by "speculators," the art of forecasting future oil prices has to take "fundamentals" into consideration, especially future supply and production constraints. One major constraint is adding new capacity which involves investment by both OPEC and non-OPEC producers. As noted earlier, it has been estimated that the industry as a whole requires significant investment over the next two decades of around $40 trillion. The question then arises: is OPEC willing to invest significant amounts if the oil price at which global supply and demand balance is widely uncertain? If such prices remain uncertain for the foreseeable future, a further question then arises: *should OPEC be willing to invest, and is it able to invest such amounts?* Availability of financial reserves and access to domestic and international financing are important elements in energy investment decisions and funding of new oil capacity depends on the extent oil prices recover toward countries' fiscal break-even prices. Figure 7.3 that follows illustrates that many OPEC members will hardly be in a position to finance significant upstream investments with their estimated financial reserves.

Fig. 7.3 Net financial reserves/debt of OPEC countries, $ bln, 2014

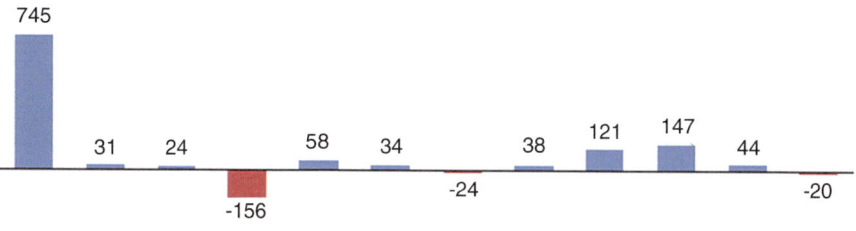

Source: Rosneft (2015, p. 160), IHS, BP Statistical Review of World Energy (2014), IEA, Wood Mackenzie, IMF

From the above figure, it is noticeable that most OPEC countries, with the exception of a few Arab OPEC members like Saudi Arabia, UAE, Kuwait, Libya, Algeria, and Qatar, lack the substantial financial reserves required to grow production and meet future "calls on OPEC" when global demand rises. To assess the range of potential interactions of future demand, supply, and prices, the International Energy Agency (IEA) considered three price cases by 2040 (IEA 2014). These were a *reference*, central case where prices trend toward $141 per barrel in 2040, a *low* oil price case of $75 a barrel, and a *high* oil price case of $204 a barrel. According to the IEA, at these forecasted prices, the likely "call on OPEC" oil would be expected to be between 43.7 and 65.3 million bpd in 2040. This compares to current OPEC production levels of around 30 million bpd as noted earlier in Table 7.3.

However, oil prices are not the only factor in determining the investment capacity of OPEC countries, as our previous analysis of unsustainable domestic consumption patterns and energy subsidies illustrated, adding to fiscal pressures on many countries. While energy efficiency initiatives and pronouncements have been mooted, energy pricing and subsidy reforms remain a real policy and political conundrum for many of the OPEC producers, especially those involving a high degree of "rentier-economy" social contract characteristic. Given OPEC members' economic and financial diversity, what are the organization's options in determining future oil prices?

Future Oil Price Curves: An Alphabet Soup of *V, L, W, U,* and *M*

A key problem for OPEC is rooted in an ambition to set *both* price and production levels at the same time, but as discussed, this ambition is now more of a hope than a reality and can only be achieved if current marginal non-OPEC producers exit the market. What has evolved over the past decade, with the rise of non-OPEC production, especially nonconventional oil, is that OPEC does not now control the majority of production. As such, the desire to set prices will work only if OPEC sets a price *and* defends it by increasing production if the price is above a "target" and by reducing production if the price is below the target. As noted in the volume, the internal allocation of OPEC production and the subsequent adherence to an agreed set of production level or quotas created problems due to non-adherence to what was agreed upon by OPEC members, as well as to substantial differences of interest concerning acceptable or fair prices, and how great the supply restrictions of each individual OPEC producer should be. In order to be able to influence oil prices in an effective manner, the organization must be able to do two things. First, determine a "fair price" for the group as a whole. Second, determine a "fair production" level for the group as a whole and the allocation of output among members. However, in reality, OPEC today is not the same organization as in its heyday when it was *the* marginal producer of oil, had global capacity, and a market clout to influence prices. Figure 7.4 illustrates the gradual erosion in OPEC's global market share and the different economic and geopolitical circumstances facing some key OPEC members.

Fig. 7.4 OPEC's declining share in global oil production and key members' geopolitical issues

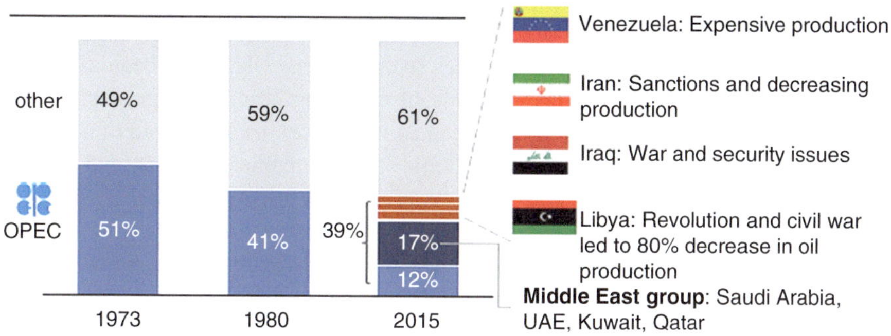

[Chart showing OPEC share: 1973: OPEC 51%, other 49%; 1980: OPEC 41%, other 59%; 2015: OPEC 39% (split 17% + 12%), other 61%]

- Venezuela: Expensive production
- Iran: Sanctions and decreasing production
- Iraq: War and security issues
- Libya: Revolution and civil war led to 80% decrease in oil production
- **Middle East group**: Saudi Arabia, UAE, Kuwait, Qatar

Source: Rosneft (2015. p. 16)

The above figure illustrates the decline in OPEC's share in global oil production to around 40 % levels. A key observation is that while OPEC's share has remained somewhat stable since 1980, OPEC is no longer a united centralized organization due to a range of economic and regional geopolitical issues facing some key members, especially Iran and Iraq which are addressed later. These issues ensure that the interests of OPEC members *are not aligned* and that, as Fig. 7.4 indicates, only a group of Middle East countries which have substantial financial resources and spare capacity to increase or decrease production (Saudi Arabia, UAE, Kuwait, and Qatar) can execute independent oil policies. Whether they will do so to the detriment of the other members and OPEC as an organization is still an open question.

Even assuming that the interests of all OPEC members are perfectly aligned, oil pricing information is not very reliable and requires some major refinements in order to present a more accurate picture of *actual* oil demand and supply conditions. One serious flaw, especially concerning supply data and reserves, is that oil market fundamental data which influences price are not conformed by *independent audits*. This is especially true for many of the stated oil reserves of Middle East producers and even for some North American shale oil reserves. The result is that forecasts of oil market equilibriums are often opaque, are changeable, and *exacerbate price movements*. This happens not only for national oil companies, but by global agencies which traders and investment analysts use to forecast long-term trends such as the *IEA*. This agency, for example, provided three different forecasts for 2014 world oil demand: 92.5 million bpd for January 2014, 92.7 million bpd for July 2014, and 92.4 million bpd for October 2014. While the IEA differences in projections are 100,000–200,000 bpd, or around 0.1/0.2 % of 92 million bpd, yet such is the state of market nervousness about the future direction of oil prices that even seemingly insignificant demand variation estimates can magnify market price volatility.

However, of more concern are reports of oil price manipulation, highlighted by the European Commission's unannounced inspections of possible manipulation of energy price benchmarks by several companies in 2013 (European Commission 2013). According to Bloomberg, 2 weeks after *Royal Dutch Shell* and *Platts* changed the way more than half of the world's crude was valued in 2013, the two companies along with *BP PLC* and *Stat-Oil ASA* were probed by the European Commission's Antitrust regulators about potential manipulation of oil prices (Kwiatkowski and Zhu 2013). The investigations sent a ripple effect of concern given the high degree of dependence on oil prices from agencies such as *Platts* which has been assessing the price of oil since 1923. The influence of reporting companies grew since the mid-1980s when the energy industry began using *market prices* instead of a system where they had been set by international oil companies (IOCs) and OPEC (Kuwiatkowski and Zhu, ibid). Such "market-based" price assessments made by reporting companies underpin long-term contracts, short-term spot transactions, as well as future settlements and derivatives. Among pricing companies, *Platts* assessment represents as much as 95 % of crude trades and 90 % of oil products and over-the-counter (OTC) derivative deals, according to *Total SA*, Europe's third biggest oil company. *Platts* determines daily prices in the OTC energy markets in one of two ways. The first is that transactions made online during prescribed times known as *Market-on-Close* (or "MOC" window) is used only for selected products that have industry agreed specifications. The second method involves employees of assessing companies discussing prices with market participants and traders each day and then determining the daily settlement price.

These methods are in the so-called physical markets and are different from the daily prices seen for *exchange traded futures*. High volume contracts such as *North Sea Dated Brent* are well supplied with data from traders, but other markets rely on reporters calling companies to assess price levels and depend on their sources being honest when they report their views. According to *Bassam Fattouh*, this leads to situations where market participants are under no legal or regulatory obligation to report their deals to price-reporting agencies or any other body and that whether participants decide to share information "depends on their willingness, their reporting policies and their interest in doing so" (Fattouh 2011[a]). Such a voluntary-based reporting regime has been a cause of concern for *the International Organization of Security Commission (IOSCO)* who concluded in October 2012 that price assessments could be "vulnerable to manipulation" because traders participate voluntarily, meaning that they may selectively submit only trades that benefit their positions (Kwiatkowski and Zhu 2013). Potential abuse of such voluntary, unregulated global pricing benchmarking is not only confined to the oil sector, as financial regulators' investigations into bank manipulation of the London Interbank Offered Rate or *LIBOR* and *ISDAFIX*, the benchmark for the $279 trillion swap market also demonstrated following the 2008/2009 financial crisis. It might be argued that such type of negative media coverage and reputational risk damage could ensure more self-regulation by companies involved in global price benchmarking, but conversely this could cause some major trading companies to withdraw from the price-reporting process which might lead to even more reduction in transparency from those remaining and compromising the overall effectiveness of pricing company services.

Others like Fattouh have argued that it is difficult to completely separate the linkage between financial layers highly interlinked with benchmarks through processes of arbitrage and development of products that link the different layers together and that information derived from financial layers "plays an important role in identifying the price level of the (physical) benchmark" (Fattouh 2011[a]) with different price benchmarks such as *Brent* and *Dubai,* for example, affecting each other, as illustrated in the figure that follows.

Fig. 7.5 The interlinkages between financial and physical layers

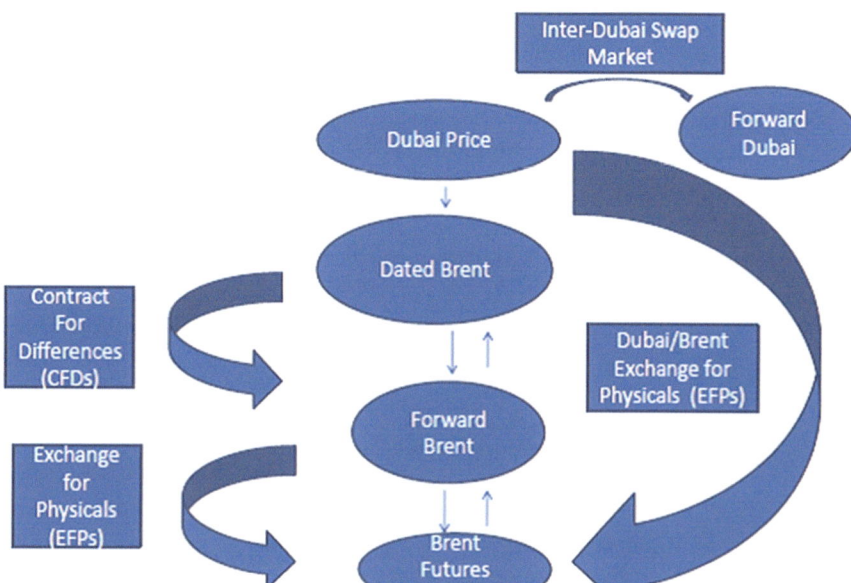

Source: Fattouh, Bassam "Inter-linkages and Regulation of oil Derivatives." Oxford Institute for Energy Studies, Jan. (2011[a])

From Fig. 7.5 it is noted how the price of *Dated Brent* is assessed using information from many layers such as forward markets, contracts for differences (CFDs), exchange for physicals (EFPs), and futures markets. The price of *Dubai* is often derived using information from the OTC *Dubai/Brent* swaps market and the *inter-Dubai* swap market. While the above may indicate that a seamless transfer of price information flows across the different layers, in practice the degree of transparency in financial layers is important as well as the skill of reporters, the choice of methodology and internal regulations and compliance procedures adopted. Concerns about these issues could very well move the oil price away from "true" underlying fundamentals. In conclusion, OPEC seems to have limited tools to influence the market price, specifically:

- OPEC does not set the oil price but has now publicly stated that so-called "market" forces do it.
- OPEC members do not participate in the futures or the OTC markets.
- OPEC members do not participate in the Platts "market-on-close," or MOC window.

As discussed later, this provides OPEC with an opportunity to redress the situation from being a bystander to become a more effective participant contributor to market price benchmarks. This is opposite to the current situation whereby the organization participation in influencing oil prices is either through production cuts or sending *signals* to the market through the media and in the longer term threatens to limit investment in new productive capacity affecting future supplies and prices. By all accounts, none of the above "strategies," if these can be categorized as such, have failed to reverse the sharp fall in oil prices witnessed in 2014.

The absence of a meaningful feedback pricing mechanism from OPEC means that markets would have a higher tendency to overshoot or undershoot and for prices to become more volatile than hoped for to assist OPEC policymakers in their fiscal and investment planning. Table 7.6 illustrates the effect on oil market prices from perceived key economic contributing factors.

Table 7.6 Market oil prices 1980–2014 and key contributing factors

Factor	1985	2008	2014
Oil Price (at 2013 $/bbl)	104 → 31 (−71%)	105 → 67 (−36%)	115 → 47 (−57%)
Spare capacity	24% of world demand	6% of world demand	5% of world demand
Marginal Upstream Operating costs	$10 - $20 /bbl	$20 - $50 /bbl	$30 - $50 /bbl
Marginal Upstream Full cycle costs	$30 - $40 /bbl	$35 - $60 /bbl	$60 - $100 /bbl
Demand growth	1979-85: −7% / 1985-89: +11%	2000-07: +13% / 2007-09: −2%	2009-13: +7% / 2014-20: +10%
Marginal oil type	Traditional fields	Heavy oil, off shore	Shale oil

Source: Rosneft (2015, p. 4), IEA (2014[d])

Table 7.6 illustrates that oil prices have been volatile during the years 1985, 2008, and 2014 but with volatility the highest in 1985 with around a 70 % drop in that year compared with 57 % levels in 2014. However, global spare capacity was five times *less* in 2014 compared with 1985, with excess production of less than 2 million bpd, mostly from Saudi Arabia. Over the same period, incremental marginal

Future Oil Price Curves: An Alphabet Soup of V, L, W, U, and M

oil production shifted from traditional fields to heavy/offshore and then to shale oil, with marginal upstream operating and full cycle costs also rising to reflect a shift from relatively lower cost traditional fields. However, unlike the previous 1980s and 2008 oil price crises, the fall in oil prices in 2014 was not accompanied by a fall in demand growth, with forecasted demand for 2014–2020 indicating a growth in global demand, unlike in the previous period.

Table 7.6 also highlights various shapes of oil price curves. Analyzing these oil price curves could give some indication on the likely shape of oil price recovery or not. Oil price trends come in a variety of shapes, almost resembling letters in an "alphabet soup" as discussed in the table below.

Table 7.7 Predicting oil price recovery shapes: a choice of "Alphabet Soup"—U, L, V, W, or M?

Price recovery shape	Key characteristics and enablers
L	• Oil prices fell sharply in 2014 and oil prices have taken this shape from $115 to $45–55 levels
	• Prices are at levels that would result in disinvestment from oil and depletion rates would rise; supply growth would slow and fall sharply under a supply shortage
V	• OPEC's ideal position, supported by IOCs
	• Envisions a sharp rally in price as marginal high-cost producers exit and cuts in investment capacity combine to push up prices
	• Unlikely as marginal producers will quickly re-enter markets
U	• More plausible recovery path than a sharp "V" rally, with prices falling further from $55 levels to around $40–45 and staying there for a while before recovering
	• Underestimates demand responses and overestimates the robustness of supply to sustain production below operating costs for high-cost nonconventional producers
W	• Envisages a rally in prices due to a "call on shale" replacing a "call on OPEC" as a new barometer
	• Low prices squeezing shale output growth and a price recovery result in a robust US supply response, creating another price dip and then a recovery to a new equilibrium
M	• High oil prices have spurred nonconventional investment and enhanced geology/seismic experience, with reduced costs leading to resilience in face of falling prices
	• Sharp rise in prices by 2017 leads to enhanced stimulus for US shale, overwhelming demand growth and bringing prices down
	• Entry of Iran and Iraq production ensures further downward price pressure

From Table 7.7 it becomes evident that no matter what is the predicted shape of future oil prices, unconventional oil, specifically US shale, is a key issue. Unconventional oil has been the most disruptive geopolitical energy factor in the markets such the 1970s and the danger for OPEC is that US shale, and elsewhere as the shale revolution eventually spreads worldwide, will not just put a cap on prices but could drive prices down again after a price recovery, with *M*-shaped recoveries, but each time at *lower* prices. Additional supplies coming from Iran and Iraq will only add to this downward *M*-shaped oil price curve.

OPEC's Quandary: Dealing with the Elephants in the Room: Iraq and Iran

While OPEC is grappling with a new oil world order, with a "call on shale" possibly replacing a "call on OPEC" as a new market balancing mechanism, the organization could also be facing other potential problems: namely, increased supplies from both Iraq and Iran. Other future large potential supplies could also come from Brazil and Venezuela. *All these additional supplies would take place in a word with shale.* How OPEC adjusts to these incremental supplies especially from fellow OPEC members Iraq and Iran will be a litmus test in whether the organization can adapt and survive new challenges.

Given that OPEC's Middle East members hold the key to future supply growth, it is important to assess individual country supply growth potential as well as their likely reaction to increased Iranian and Iraqi oil supplies. This is illustrated in Table 7.8.

Table 7.8 OPEC's Middle East players: supply growth potential and possible country reaction to Iran/Iraq increased production

Growth prospects	Country	Patterns of growth	Possible reaction to Iran/Iraq
• Limited growth	• Qatar	• Characterized by marginal increases	• Neutral
	• Abu Dhabi	• Predictable patterns of growth	• *Oppose*
	• Kuwait		• *Oppose*
	• Algeria		• Neutral
• Medium growth	• Libya	• Unpredictable patterns of supply growth, geopolitical issues	• Neutral
	• Iran	• Growth dependent on large investment flows/stock upgrade	
• High growth	• Saudi Arabia	• Largest potential. Low cost suppliers	• *Oppose*
	• Iraq	• New game changers	

From the above table, the key game-changing countries are Saudi Arabia and Iraq: the first by reason of its largest OPEC production role, significant financial, and reputed spare capacity, while the latter has large reserves and the capacity to build up its production to more than pre Iraq-Iran war, the 2003 US occupation, and United Nations sanction eras. The potential reaction of fellow OPEC members to increased oil production by Iraq and Iran is either one of neutrality or opposition, with Qatar, Algeria, and Libya most probably adopting a neutral stance, while Saudi Arabia, Abu Dhabi, and Kuwait being unofficially opposed to any significant increase. The neutral stance by the first group is either due to geopolitical factors (shared gas fields by Qatar with Iran) or focus on a different market share (Algerian and Libyan oil and gas exports to European markets, as opposed to Iraq/Iran exports to predominantly Asian markets). The opposition comes from countries who will feel threatened by erosion of their current market share in Asia. According to analysts, a key factor in the oil market that has underpinned the bearish sentiments on prices is that, despite ongoing political and military turmoil, the potential for Iraq to deliver a large increase in output is there. According to Fattouh, the Iraqi government has plans to keep adding 500,000 bpd each year over the coming years and that while these plans remain constrained by infrastructure limitations in the south of the country, the Iraqi government's original ambitious plans to reach 12 million bpd have been revised down to "9 million b/d of production capacity by 2020, to be maintained for 20 years" (Fattouh 2014[a], p. 52); however, the Iraqi government has announced a new more realistic target of 6 million bpd by 2020.

To allay Saudi Arabia's fears, the Iraqis have been quoted that they stood by Saudi Arabia's decision to maintain production during the November 2014 OPEC meeting, because "Saudi Arabia, along with the other Gulf countries, accounts for half of OPEC's production," according to Iraqi Oil Minister Adel Abdul Mahdi (2014). The Iraqi oil minister also felt that it was time to reopen talks with Saudi Arabia on the Iraqi-Saudi oil pipeline shut down since Iraq's invasion of Kuwait in 1990 (Mahdi 2014). Despite such sentiments, both Iraq and Iran have seen their production capacity overtaken by Saudi Arabia, from around same production levels in 1960, as illustrated in Table 7.9.

Table 7.9 Saudi Arabia, Iraq, and Iran daily oil production 1960–2013 ('000 b/d)

	1960	1970	1980	1990	2000	2013	Population 2013 (million)
Iraq	972.2	1548	2646	2113	2810	2979	35.1
(% of OPEC)	(11.7)	(6.8)	(10.4)	(9.8)	(10.1)	(9.4)	(8)
Iran	1067	3829	1467	3153	3661	3575	77.1
(% of OPEC)	(12.8)	(16.9)	(5.8)	(14.6)	(13.2)	(11.3)	(17.5)
Saudi Arabia	1313	3799	9900	6412	8094	9637	29.9[a]
(% of OPEC)	(15.8)	(16.8)	(39.2)	(29.8)	(29.3)	(30.5)	(6.8)
TOTAL OPEC	8273	22,534	25,280	21,481	27,601	31,604	438.4

Source: OPEC Annual Statistical Bulletin (2014, p. 8, 28), SAMA (2013, p. 38)
[a]Includes approximately nine million foreigners

The gap between Saudi Arabia and Iraq and Iran widened considerably by 1980, when Saudi Arabia's production reached nearly 40 % of total OPEC share and has remained at around the 30 % level since then. By comparison, Iran with nearly 18 % of OPEC's total population produces around 11 % of OPEC's total output, and Iraq with around 8 % of OPEC's population produces around 10 %. By comparison, Saudi Arabia accounts for around 7 % of OPEC's population (or around 5 %, if nine million foreigner residents in Saudi Arabia are excluded).

Such unequal production to population ratios and economic needs will put OPEC members at odds during periods of low prices. Increased production from countries such as Iraq and Iran will be justified on the basis that they are merely "catching up" and making for lost production and OPEC market share will not be easy to dismiss. This is illustrated in the next figure which sets out the historical pattern of production for both countries.

Figure 7.6 illustrates the most recent production peaks reached for Iraq (around 3.2 million bpd) and Iran (4 million bpd). In 2014, Iraq pumped 3.3 million bpd, the most since 1979 according to data compiled by Bloomberg as well as registering the highest level of crude exports in three decades in March 2015 of 2.98 million bpd (Al Ansary and Ajrash 2015). At the same time, in order to meet its planned ambitious production capacity goals, Iraq is installing two new export facilities offshore in the Gulf and expanding storage capacity to 15 million barrels a day by the end of 2015 to overcome bottlenecks that are curbing output (Al Ansary and Ajrash 2015). At present, Iraq's exports are constrained by the use of the southern port of *Basrah* with a capacity of around 1.2 million bpd and the use of the two oil pipelines to the Turkish city of *Ceyhan* from *Kirkuk*, with a capacity of around 1.6 million bpd, with the remaining output refined domestically. In May 2014, the first 150,000 bpd oil shipment took place from the Iraqi Autonomous Kurdistan region to the *Kirkuk-Ceyhan* pipeline, with plans to double capacity (Meric and Hacaoglu 2014). Given such constraints, how realistic is the Iraqi plan to increase production capacity as illustrated in Fig. 7.7?

Fig. 7.6 Historical crude oil production: Iraq and Iran, 1980–2013 ('000 barrels/day)

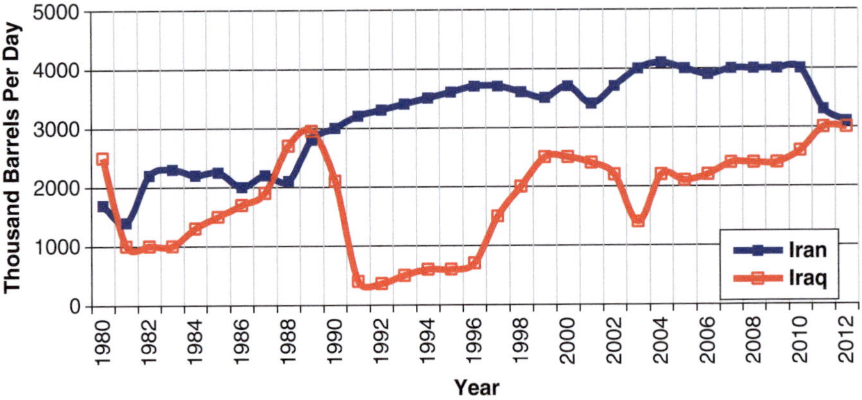

Source: United States EIA, OPEC Annual Statistical Bulletin (2014), BP Statistical Review of World Energy (2014)

Fig. 7.7 Iraqi planned production: official targets (million barrels per day)

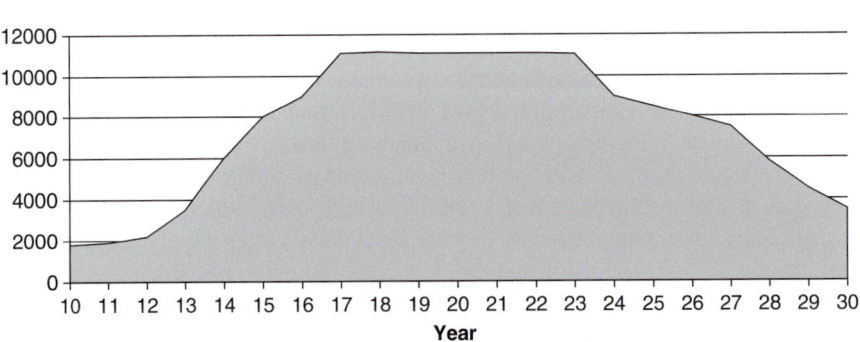

Source: Fattouh (2011ᵇ) "Oil market and OPEC behavior: Looking ahead"

Should OPEC, and specifically Saudi Arabia, feel threatened by these Iraqi planned production targets or are they mere mountains too high to climb? With their combined reputed reserves of 301 billion barrels of Iraqi and Iranian oil (Iraq, 144 billion; Iran, 157 billion), representing just under 25 % of total OPEC reserves of 1206 billion barrels in 2013, should Saudi Arabia with its reputed 265 billion oil reserves, or 22 % of total OPEC reserves, be worried that the IOCs and other consumer countries go to Iraq and Iran for their energy supplies? While the immediate answer weighs heavily in Saudi Arabia's favor, yet as explored below, Saudi Arabia's hegemonistic role in OPEC is gradually being eroded. An entry of incremental production from cash strapped Iraq and Iran, even at discounted prices, will only add to market perceptions that the long-term shape of oil prices is indeed one that looks more like an *M* rather than a *V* or a *W*, on the basis that both Iraq and Iran will instigate a strategic price war in a world of shale oil, whether Iraq reaches its planned 6–10 million bpd or not.

The Long and Winding Road: Iran's "Historic" Nuclear Standstill Agreement

After months of tortuous negotiations, going to the wire, the major powers and Iran agreed on a set of principles in *Lausanne* on April 2, 2015, with the aim of concluding a comprehensive agreement by end of June 2015 to end the nuclear standoff and to gradually remove Iran trade sanctions (New York Times 2015; Siddiqui and Lewis 2015). After some delays, the long awaited agreement was clinched on 14 July 2015 between Iran and the big powers, capping more than a decade of negotaitions with an agreement that could potentailly transform the Middle East, despite some fierce resistance from some countries and US lawmakers (Harezi et al. 2015; Lakshmanan et al. 2015). According to the final agreements reached, full implementation and a gradual lifting of economic and other sanctions against Iran will contingent on Iran meetin its obligations to curtail its nuclear program. The reaction of the oil market to the breakthrough in diplomacy was symptomatic of what might come in the future, with the price of *Brent* falling to $54.95, a fall of $2.15 a barrel or just under 4 % on the news, even without a single additional barrel of Iranian oil put onto the market. While there are many hurdles yet to overcome which might derail a final agreement, analysts were somewhat divided on the short- and long-term impact of Iran's breakout from sanctions on its oil exports. Some noted even before the April *Lausanne* headline agreement that a deal that would phase out economic sanctions against Tehran was unlikely to flood the world markets with more oil any time soon, despite Iran's declared intention to "claw back market share lost because of the curbs" which saw that country's oil exports fall to just 1 million bpd from 2.5 million bpd in 2012 prior to the sanctions (Torbati 2015).

Most analysts focused on how quickly Iran could technically resume pumping oil to pre-sanction levels, assuming shipments could follow quickly, and overcoming concerns about a diminished customer base and potentially neglected Iranian oil fields. Iranian officials were upbeat following the *Lausanne* agreement and the final July agreement, stating that their country could raise exports by 1 million bpd within "a few months" if sanctions were lifted according to Iran's Oil Minister *Bijan Zanganeh*, who also assured that an increase in Iranian oil exports would not have much impact "given other factors in the market" (Kalantari 2015[a]). Presumably these "other factors" were shale oil production related. The Iranian oil minister also advised that developments at the *Yadavaran, Azadegan,* and *Yaran* oil fields would be expedited with additional drilling rigs and that construction of offshore phases 15–19 of *South Pars* gas field in the Gulf will be completed by year end 2015 (Kalantari 2015[b]). Further Iranian reassurances on cooperation came when Iranian Deputy Oil Minister Hossein Zamaninia was quoted as saying that Iran hoped to restore its market share within a short period through "cooperation with OPEC" (Amott 2015).

Some energy experts believed that a return of 1 million bpd Iranian exports is "at least a year away" but that smaller amounts of around 200,000–300,000 bpd are possible in the short term. According to Barclays analysts these may reach 500,000 bpd by end of Q1 2016 and that Iranian oil sales may "displace oil sales by West African and other OPEC countries that have filled the gap" (Cheong 2015). Analysts at Morgan Stanley were inclined to believe that there will be "no impact on physical oil market from the Iran deal before 2016" and that if Iran lifts output by 1 million bpd and clears its *floating* storage of 30 million barrels, it could potentially "delay any cyclical recovery in global oil prices by 6–12 months" and concludes that Iran's goal to lift production well beyond initial export relief could take many years, given recent Iranian underinvestment in the sector (Lee 2015[a]).

While there is some uncertainty on when Iranian oil production could once again come in significant amounts, there is a possibility that some increased sales could come about from Iran's current Asian customers. Iran's four main Asia customers are China, India, South Korea, and Japan who imported around 800,000 bpd in Q1 2015 illustrated in Fig. 7.8, with the potential for some additional purchases from China and India, especially from the quick release of Iran's floating storage if sanctions permit.

Fig. 7.8 Top four Iranian crude oil importers

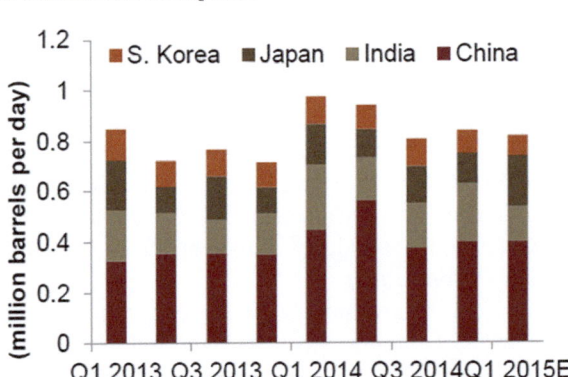

Source: Jadwa Investment (2015[b], p. 5)

Even under the best circumstances, low oil prices may limit how much Iran will want to export and may cause the country to think carefully before flooding the market, even if sanctions are totally lifted. In the final analysis this may depend on how quickly and urgently Iran is in need for foreign exchange earnings. Further discounts to Asian buyers may trigger a price war and stiff competition with fellow OPEC members, particularly Saudi Arabia, Kuwait, and Iraq. However, with the Iranian population's expectations having been raised for tangible benefits coming from quick sanction lifting, careful long-term oil price considerations and market supply-demand imbalances might give way to short-term political expediency. Iranian Oil Minister Zanganeh has said that the country's oil industry could survive prices as "low as $25 per barrel" prior to the April 2015 agreement (Torbati 2015). Whether this signals Iran's intentions to flood the market post sanctions is another matter. Despite the Iranian people's high expectations for quick sanctions relief, there are some differences expressed on when such sanctions will actually be lifted, with the EU stating that these will not be lifted automatically on June 30, 2015, but might "take time" and the EU cannot say how much time (Neugar 2015). However, according to the Energy Information Administration's latest assessment of the impact of additional Iranian oil on the market, the EIA estimates that oil prices could tumble $15 a barrel in 2016 if sanctions are completely lifted and a comprehensive agreement is reached, which according to the IEA "could change significantly the STEO forecast for oil supply, demand" (Zhou 2015^a).

Given this confluence of a gathering "perfect storm" from additional Iraqi, Iranian, and US shale oils, despite some contradictory statements on when the sanctions might be lifted, what are the options facing other major OPEC producers, especially Saudi Arabia, the largest member?

Saudi Arabia and OPEC: End of the Hegemony?

Saudi Arabia imposed its viewpoint on all other willing and reluctant OPEC members during the November 2014 conference meeting, arguing that there would be no production cuts, to help balance global supply and demand with an expected exit of high-cost non-OPEC marginal producers due to sharply reduced prices. The intention seemed flawless: *sweat out* the higher-cost producers and see who blinks first in this collective OPEC *shock-and-awe* production stance. As prices continued to remain soft and even fell on slightest updates of marginal increases from shale producers, the Saudi-led OPEC stance began to be questioned, with calls for an emergency OPEC meeting to be convened ahead of the scheduled June 2015 meeting. Nigeria's then Oil Minister and OPEC President *Diezani Alison-Madueke* pushed for one in view of oil price volatility and stated that she was talking to other OPEC members and that "almost all OPEC countries, except perhaps the Arab bloc, are very uncomfortable ….. It is hoped that the price will stabilize at no less than $60, but we cannot be sure." The former OPEC President said that she supported the November 2014 collective decision, but added that "when you cede market share continuously, you drive yourself into oblivion" and that many OPEC members "are going to suffer greatly from a drastic fall in the price" (Goodley 2015). The emergency OPEC meeting never materialized despite prices falling to $54 a barrel levels. This has not stopped more calls for such OPEC emergency meetings later on, especialy when oil prices fell below the $40 level in August 2015, but once again the organisation decided to sweat it out under pressure from Saudi Arabia that the market would eventually balance itself out.

For how long financially stressed OPEC countries such as Nigeria can grudgingly follow the Saudi-imposed production policy is difficult to judge, given that Nigeria with its population of 172 million, representing nearly 40 % of OPEC's population, produced a mere 2.44 million bpd in December 2014, or around 7.8 % of OPEC's daily output for the same period. According to latest data, oil production from Nigeria decreased by 5 % during Q1 2015 as Nigeria's production is frequently affected by theft and sabotage, especially in the Niger Delta, the country's main oil-producing region. Nigeria is also facing some severe competition from US shale production as the African country's oil exports are mainly light sweet crude, which were previously supplied to the USA, with Nigeria struggling to find new long-term customer contracts (Jadwa Investments 2015[b]).

If Nigeria and other OPEC members feel that oil prices will rebound in a sharp *V* shape to $100 levels again, they will be disappointed by the general consensus of most oil analysts that this will not happen any time soon. According to these analysts, the shock to oil prices reflected what economists would characterize as "unusually unfavorable movements both on and among supply and demand curves (and that) these development in combination caught many off guard" (El Erian 2015). OPEC members who are adopting a "stand your ground" policy are predicting that sharply lower oil prices are inducing significant supply destruction that has yet to run its course and that eventually low price levels will render many existing high-cost fields uncompetitive, curtail alternative energy sources, and even cancel long-term energy capacity expansion investments. The upshot for such an optimistic low-cost OPEC comeback is that, assuming no major geopolitical shocks, there will be some marginal high-cost producer consolidation, with a tendency for higher prices, but that there will not be a quick return to $100 price level for some while. Under this scenario, it will not take Saudi Arabia and other OPEC members years to re-establish some lost market share (El Erian 2015).

The above once again assumes that US shale production will gradually decline to enable OPEC to regain *some* of its lost market share assuming that OPEC and the rest of the world can agree on what really, constitutes OPEC's "market share." By all indications, this will be difficult to achieve due to several factors. Non-OPEC supplies grew by 2.1 million bpd in Q1 2015, year-on-year, largely as a result of increased US oil production. In Q2 2015 it is forecasted that there will be continued output increases from non-OPEC sources amounting to 1.4 million bpd, year-on-year, with output again led by the USA, as well as increased supplies from Russia and Iraq, excluding any new Iranian supply to the market. According to the latest Energy Information Agency (IEA) data, total US oil production was expected to have risen by 15 % in Q1 2015 as shale oil continues, regardless it seems of the low oil market price.

From Fig. 7.9, the IEA expects slightly lower growth in Q2 2015 with growth falling more rapidly in the second half of 2015. Overall US crude production is expected to rise to 9.32 million bpd in 2015 compared with 8.59 million bpd in 2014. What factors are accounting for the resilience of US shale production in the face of weak oil prices, a continuing fall in the number of US oil rig counts, and high capital expenditure? The answer lies in US shale *productivity* as illustrated in Fig. 7.10 setting out US rig count and production per rig at the five largest US shale plays.

Fig. 7.9 US oil production growth

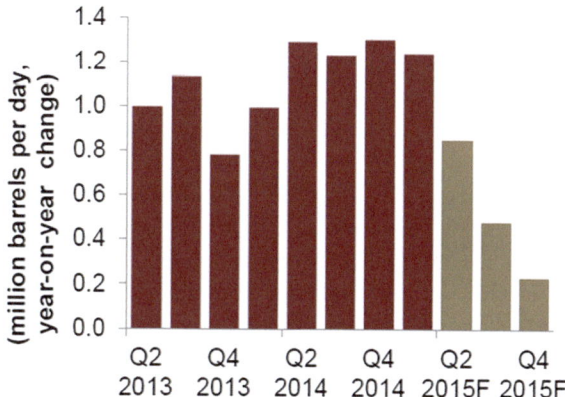

Source: Jadwa Investment (2015b, p. 4)

Fig. 7.10 US rig count and production per rig at the five largest shale plays in the USA

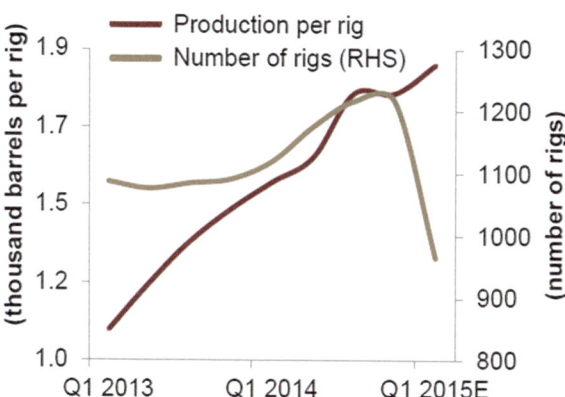

Source: Jadwa Investment (2015b, p. 4)

Against all expectations, according to the US Energy Information Administration, US shale producers have not only cut back on rig counts and capital expenditure (capex) but have actually *seen productivity per rig rise* by 22 % in Q1 2015 for the five largest shale plays, underlying the resilience of the nonconventional industry.

Shale operators are now learning to focus on low-capital efficiency areas of their shale portfolio, better use of seismic technology to ensure more precise drilling in the richest geological fields, compared with previous random drilling, coupled with focusing on new wells which have not yet reached their peak production levels, as well as requesting cost reductions from their suppliers. While the above shale producers' actions have increased productivity and reduced costs, there is another factor that OPEC producers have to take into account that could still provide nonconventional players with the means to make a rebound when market circumstances warrant it. This relates to what has been termed *drilled uncompleted wells*, or *DUCs*. Slowdown in production growth comes from increasing numbers of *DUCs,* and it has been estimated that these number around 200 for the *Bakken* shale play of North Dakota alone. These *DUCs* in effect act as a form of *oil inventory* to be reactivated when prices rise. As such, the build-up of *DUCs* in the USA are not signs of a permanent exit of shale producers, but instead act as a hedge in *maintaining a ceiling on prices going forward*, since shale operators are likely to compete these drilled uncompleted wells as soon as oil prices rebound. This will result in a surge in supply, which puts downward pressure on prices, and once again leading to the M oil price curve analyzed earlier.

While *DUCs* are a long-term shale supply reservoir, the US shale producers have reacted to the 2014–2015 oil price falls by being innovative and "re-fracking" old wells. According to mid-2015 reports, around 2 % of shale wells in North America are being "re-fracked" with numbers expected to increase in 2016 as technology improves. The use of these new technologies cost as much as 50 % of well's original costs, lowering cost of production, and helping to offset around 20–30 % drop in completion costs from peak levels (Harvey 2015). This demonstrates, once again, the powerful effect of T—technology—in the energy equation.

Saudi Arabia's Policy Options

A seemingly persistent opinion in the analysis of the 2014 price collapse is that the real driver and culprit of the market glut is Saudi Arabia's rigid stand on production policy. According to such analysts, Saudi Arabia should have saved fellow OPEC members and, once again, obediently played its expected role as a "swing producer," reducing its output as prices fell in order to put a floor under prices. Conspiracy theories then abounded that it did not do so for various political reasons, as opposed to economic reasons, such as falling prices hurt its geopolitical rivals Iran, and Bashar Al-Assad's ally Russia, much more than Saudi Arabia with its large cushion of financial reserves. As analyzed in the volume, one cannot deny that fundamental changes in the global production of energy have taken place which is reshaping market oil prices however flawed these "market" oil prices may be due to some manipulation and the activities of financial commodity players and speculators. *Saudi Arabia cannot now create a glut or shortage simply on its own.* The historical precedents discussed in earlier chapters that drive Saudi Arabian oil policymakers are not the same as those earlier hegemonistic days of the 1970s when cryptic utterances and vague statements by the Saudi oil minister could send oil market prices rocketing or tumbling, but rather the traumatic experience of Saudi Arabia of the mid-1980s period. During the latter period, the Saudis tried to arrest the fall of prices by obligingly cutting production with their OPEC partners pledging the same but rarely following through. By 1985, Saudi Arabia, the *Sultan of the Swing*, saw its production fall to just above 3 million barrels a day from a high of nearly 10 million barrels a day in 1980, illustrated earlier in Table 7.5. The Kingdom lost market share that has left a deep scar in the collective memory of Saudi oil policymakers to this day, with Saudi officials now deciding that they were done with playing the dumb role, unless everyone played by the same rules.

While Saudi Arabia is no longer willing to play the old swing producer game, it still certainly sees itself as a reliable partner to the world in times of extreme supply shortages, in effect the world's *central oil banker* when needed as a measure of last resort. This is because of the 2.5 million barrels or so that the Kingdom maintains as a spare capacity. The "call on Saudi Arabia" (as opposed to one on OPEC) has been made on many occasions: during the supply shortages of the Islamic revolution in Iran, the 1980–1988 Iran-Iraq war, the 1990–1991 Iraq occupation of Kuwait, the 2008 global financial crisis, and, most recently, during the loss of Libyan production in 2013. According to some analysts, even adopting a conservative outlook and allowing for technical problems and unscheduled maintenance work, it can be said that Saudi Arabia is capable of sustaining an average production rate of 12 million bpd over the next few years (unless Saudi Arabia fundamentally changes its reserve capacity maintenance policy), while, on a surge basis, production of 12.5 million bpd can be maintained for a period of *several months* if such a need should arise (CGES 2011). However, according to a senior Aramco veteran, to reach such a sustained production capacity, Saudi Arabia would have to drill many new wells and pipelines to bring the 2.5 million bpd on stream which is costly and takes time, despite having a supporting infrastructure in place (Al Khowaiter 2015).

Given the weight markets place on public announcements by Saudi oil ministers and other senior officials, it is important to make a brief summary of what they have *actually been saying* over the past few years to assess whether there has been continuity in policy statements over time and whether the markets are ignoring the Saudi message. Some recent key pronouncements are listed below:

- Oil Minister Ali Al Naimi (Doha 2012)
  "There hasn't been an incident in the past where oil was taken off the market for a different reason without Saudi Arabia responding to that shortage. And I can name you all of them. I think we have enough proof of that. I have a whole list, when PDVSA strike, 2 million were taken off the market, and we supplied the balance. Invasion of Iraq, again we responded. Chinese surge in 2004, we responded to the demand, Katrina, we responded to the demand. The latest one was Libya, we responded to the demand. So there hasn't been an incident, an event, where we in Saudi Arabia, actually OPEC, failed the market. So why is

there a concern today that if something were to happen in the market, that we should not be additional crude, 2.5 million barrels if need be? I don't know why there is any question in that area."
- Oil Minister Ali Al Naimi (Berlin, March 2015)
  "The oil price is not quite so clear cut. Supply and demand are key aspects, but it also takes into account a range of other factors. These include speculation, conjecture informed or otherwise—and perception about what the future holds. Also oil is increasingly used as an asset class and this also impacts the price."

  "During periods of rapid price movement, up or down, there is often a frenzy of commentary ascribing various bizarre theories and motives—about collusion or conspiracy—to OPEC and to major producers, most notably Saudi Arabia. With the recent price drop, OPEC and Saudi Arabia have yet again been maliciously—and unfairly—criticized for what is, in reality, a market reaction. Some speak of OPEC's 'war on shale', others claim 'OPEC is dead.' Theories abound. They are all wrong. It has always been the aim of OPEC nations to work together to do what they can to stabilize prices, ensure fair returns for producers and steady supplies for consumers."

  "We have a long-term view. We try to avoid knee-jerk reactions to short-term market movements. Over the past eight months, though, with the market in surplus, it is Saudi Arabia that is called upon to make swift and dramatic cuts in production. This policy was tried in the 1980s and it was not a success. We will not make the same mistake again. Today, it is not the role of Saudi Arabia, or certain other OPEC nations, to subsidize higher cost producers by ceding market share."

  "Saudi Arabia remains committed to helping balance the market but circumstances require other non-OPEC nations to cooperate. Currently, they choose not to do so. They have their reasons. But I would like it to be known that Saudi Arabia continues to seek consensus."

  "I would also like to reiterate a point that I made several times before. This new oil supply growth—much of it coming from the US—is a welcome development for world oil markets and the global economy over the past several years. These new supplies, along with Saudi Arabia's own efforts, have helped offset outages from other oil producing countries. Without them, a still vulnerable global economy could have faced much higher energy prices. Saudi Arabia has consistently welcomed new unconventional supplies, including shale."

  "We will never be able to curb volatile oil market investment cycles, but perhaps we can work to moderate them for the benefit of all producers. It is vital that all producing countries—OPEC and non-OPEC—continue to focus on long term common objectives of ensuring oil market stability and a sustainable future for both oil producers and consumers."
- Oil Minister Ali Al Naimi (Riyadh, April 2015)
  "Prices will improve in the near future. The challenge is to restore the supply-demand balance and reach price stability. This requires the cooperation of non-OPEC major producers, just as it did in the 1998–1999 crises."

"*Some non-OPEC major producing countries said they were unable or unwilling to participate in production cuts. For this reason, OPEC decided at its November 27 meeting to maintain production levels and not to give up its market share in favor of others*".

"*Saudi Arabia produced 10.3 million b/d in March and will keep pumping for now at least 10 million barrels a day*".

"*Saudi Arabia plans to produce 20 million to 50 million cubic feet of shale gas next year (2016) and raise shale gas output to 500 million cubic feet in 2018. At around 2025, and beyond, the Kingdom will produce 4 billion cubic feet of shale gas. We are very active in searching for shale oil and gas within the Kingdom*".

From the above public statements by the Saudi oil minister, one can deduce several policy options that might be followed, each with its advantages and drawbacks. These are summarized in Table 7.10.

Table 7.10 Saudi oil policy options

Policy target	Issues	Success/likely outcome
I. Defends market share at any cost	• OPEC unanimity breaks down • Hopes that non-OPEC producers give way and cooperate and supply and demand rebalance • Ramp up production on expectations of rising oil prices	• Medium to low
II. Accepts a low price regime dictated by the market	• No more announcements of what constitutes "fair price bands" • Can withstand a low price due to financial reserves • Accepts low price for time being to maintain market share • Prepares for tougher OPEC negotiations • Maintains clients in competitive markets	• Medium to high
III. Helplessness in the face of global energy market events	• Miscalculation on growth of US shale, entry of Iraq and Iran • No one knows the end game • No point in defending oil prices in the face of weak demand and supply growth • Conscious of the rise of domestic consumption and of heavy/light crude production mix imbalances	• High
IV. Develops own alternative energy sources/economic structure	• Develops shale gas for local petrochemical economy and refinery base • Become less dependent on crude exports • Hope for the best	• Low to medium

While Table 7.10 may indicate a range of possible Saudi oil policy options, in reality such options are rarely tackled or effected in isolation of others. Since 1973, Saudi Arabia has not used oil as a political tool and is not likely to do so. Defending market share *at any cost* to achieve implicit political objectives such as hurting other countries with whom Saudi Arabia has political disagreements has been explicitly ruled out. However, adopting a policy of defending market share at any cost might cause a breakdown in OPEC unanimity from fiscally stressed OPEC members, and this could be further aggravated in the face of recent sharp increases in Saudi oil production to over 10 million bpd, with plans to maintain this level in the hope of capturing a large market share when prices rise (Reuters 2015[b]).

Accepting *a low price regime* that is "dictated" by market forces indicates that making pronouncements about what constitutes a "fair" price whether it is $75 or $90 is no longer an option and that this has become an ineffective market communication tool. Accepting lower market prices helps Saudi Arabia to maintain its market share and also prepares the Kingdom for tougher upcoming OPEC negotiations to forestall pressure on Saudi Arabia to cut back on its production, unless other non-OPEC producers do likewise, as the Saudi oil minister's statements have repeatedly made clear.

In all probability there is also a high degree of helplessness in the face of fundamental geopolitical energy landscape developments, especially the rise of shale oil and its continued resilience despite low prices which has not materially dented continued supply growth. This has created constraints in Saudi policymaking in face of high levels of uncertainty. Added to this is a *seeming helplessness* about the effect of Iranian production entry to the market, assuming a successful outcome of the final nuclear deal and sanctions lifting as well as a long-term threat from Iraqi output. A possible long-term policy option is to *develop Saudi Arabia's own nonconventional energy resources,* especially gas to support the existing mega petrochemical projects and expand them in order to wean Saudi Arabia from being primarily dependent on crude oil exports. This however depends on the type of unconventional gas that is found, whether dry or wet, and its suitability for petrochemical manufacturing with the probability that it will be dry. In 2015, the Saudi oil minister underscored this invigorated commitment to shale exploration and production by announcing that the Kingdom plans to produce 20–50 million cubic feet of shale gas by 2016 and raise the country's output of shale gas to 500 million cubic feet in 2018 and *4 billion* cubic feet by around 2025 (Mahdi 2015[c]). The above announcements are underpinned by Aramco's plans to spend an additional $7 billion on nonconventional gas exploration and production, according to the company's former CEO Khalid Al Falih (Saudi Gazette 2015[b]).

The Saudi petrochemical industry has grown significantly in both size and product diversity over the past decade as illustrated in Fig. 7.11 and an increased stock of cheap gas feedstock is essential to provide the sector with a competitive international advantage.

Fig. 7.11 (a) Chemical production capacity in Saudi Arabia (Million tons) 2003–2013. (b) Saudi Arabia's chemicals capacity by product segment (total 91.5 million tons) 2013

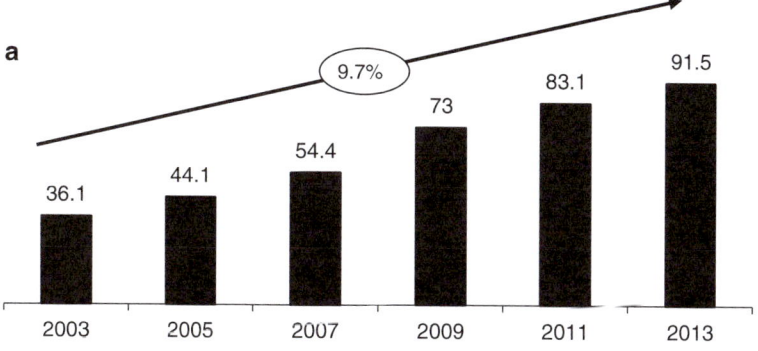

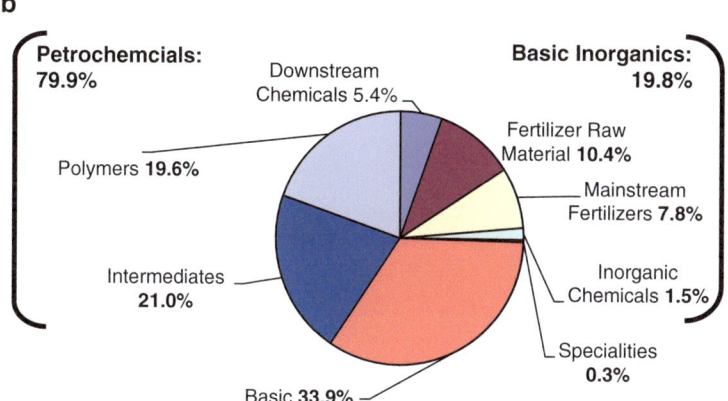

Source: Gulf Petrochemicals and Chemicals Association (GPCA) (2014), Fattouh (2014c) Oil Market Dynamics: Saudi Arabia oil policies and US shale supply responses. Oxford Institute for Energy Studies, Gulf Petrochemicals and Chemicals Association (GPCA) (2014)

Having such a high-value product line will assist the Kingdom in its economic diversification drive and make it less dependent on the vagaries of crude oil price market volatility. The change in composition of Saudi Arabia's energy exports indicates that this strategy is beginning to pay off. According to SAMA, the Kingdom's crude oil exports stood at 2783 billion barrels in 2012, 2763 billion in 2013, and 2611 billion in 2014 while the corresponding exports of refined products rose from 315 million barrels in 2012 to 360 million barrels in 2014 (SAMA 2015, p. 143).

The Saudi decision to ramp up oil production to protect its market share is also part of the Kingdom's strategy to prepare it for tougher negotiations with other OPEC members who, unlike Saudi Arabia and its Gulf allies, do not have spare capacity to lift their own production. Such a ramping of Saudi production will add to a global glut and will also send a signal to non-OPEC producers that Saudi Arabia will use its excess reserve capacity to speed OPEC-non-OPEC cooperation. However, as stated earlier, this carries the risk of breaking OPEC's unanimous position of no production cuts and which seems to have been *quietly changed to one of increased Saudi production* which other OPEC producers cannot match suit, except to offer more price discounts. Saudi Arabia is also aware that it needs to maintain its market share both in its traditional market in the USA and in the new Asian markets, particularly China. As Fig. 7.12 illustrates, the Kingdom has been losing market share in both.

Fig. 7.12 Saudi exports to China and the USA

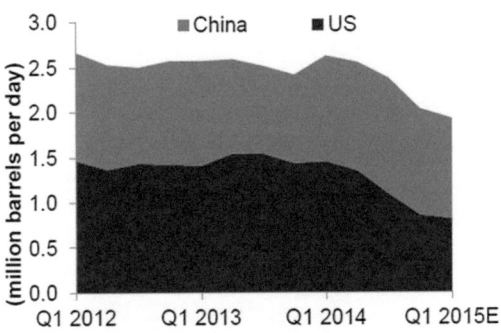

Source: Jadwa Investment (2015[b], p. 5)

However, such price discounts and added Saudi productions could have *unintended consequences* and may pose a risk for OPEC as a whole, as it may send signals to the market that OPEC approves, albeit reluctantly falls in oil prices regardless of how low they might reach.

The competitive pressure among OPEC's largest three Middle East countries for Asia's market share is illustrated in the next set of diagrams which indicates fierce export competition to China from Saudi Arabia, Iran, and Iraq, as well as sharp falls in official selling prices (OSPs) by both Saudi Arabia and Iraq to Chinese clients, something that will be aggravated when Iran re-enters the market with its post-sanctions oil exports (Fig. 7.13).

Fig. 7.13 (a) Chinese imports by country (million bpd). (b) Iraq and Saudi OSPs to Asia ($/barrel)

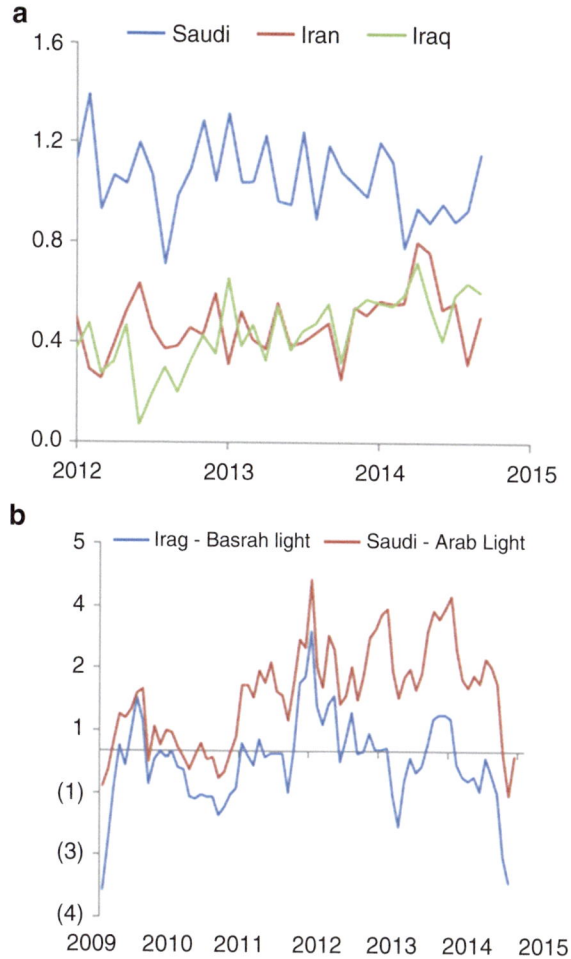

Source: Fattouh (2014ᶜ) Oil Market Dynamics: Saudi Arabia oil policies and US shale supply responses

This fierce price discounting competition could be one of the reasons for the surge in Saudi oil production so as to position the Kingdom against increased competition not only from Middle East OPEC producers but major non-OPEC competitors, especially Russia. From available data, Russian crude oil exports grew seven times faster than output in Q1 2015, making Russia the world's largest producer ahead of Saudi Arabia and the USA, with Russian production reaching 10.7 million bpd between January and March 2015 and exports registering 5.29 million bpd compared with 4.92 million bpd a year earlier (Rudnitsky 2015). Such rising Russian shipments added pressure on crude prices as global inventories swelled and the oil price slump of 2014/2015 combined with US and European Union sanctions in response to Russia's role in the Ukraine conflict seemed to have failed to reduce Russian oil production. The Russians in fact *released* more output for *exports* as domestic Russian consumption fell due to a sluggish economy hit by sanctions (Rudnitsky 2015). Despite these energy chess board movements by Russia and Saudi Arabia on who "blinks first," Russian officials met with OPEC representatives from Venezuela, Saudi Arabia, as well as non-OPEC Mexico prior to the November 27, 2014, OPEC Vienna meeting to discuss responses to falling oil prices, but no agreement was reached, although it was reported that more discussions between Russia and Saudi Arabia had taken place in Riyadh during March/April 2015 which indicates that both sides were still trying to agree on a common ground to help all parties and that Russia and OPEC will hold senior level talks in Vienna in June ahead of the planned June 5 OPEC conference meeting. There is a distinct possibility that the recent production increases in Saudi Arabia and Russia could be used as a negotiation tactic to announce production cuts to *lower levels* by both countries and send a signal to the market, hoping that a price rally could then take place but this did not materialize. However, the problem is that in a largely privatized Russian oil producer market, the Russian state has no means to cut production unless it puts some unofficial pressure on private Russian producers to cut back on their production and exports, while ensuing that Saudi Arabia did the same, so that these two largest oil producers could at least have a semblance of control over a market *price feedback mechanism* that has been eroded due to US shale and other incremental non-OPEC supply and anemic global demand. However, in the final analysis, Saudi Arabia has to continually consider the likely political consequences of its actions on other OPEC members.

Besides the above uncertainties, Saudi Arabia also faces a problem of what type of oil to produce and export. Currently, with the boom in US shale oil, there is a *surplus* in light, sweet grade crude oil rather than medium to light crude oil grades that Saudi Arabia also produces.

Table 7.11 summarizes the capacity and type of crude oil grades produced in Saudi Arabia's major fields.

Table 7.11 Major Saudi Arabian oil fields (2012)

Field	Location	Capacity/type of crude
Ghawar	Onshore	5.8 million bpd of Arab Light crude
Safaniya	Offshore	1.2 million bpd of Arab Heavy crude
Khurais	Onshore	1.2 million bpd of Arab Light crude. Plans to expand capacity by 0.30 million bpd by 2017
Manifa	Offshore	0.90 million bpd of Arab Heavy crude oil after completion at the end of 2014. Production will be used to offset declines in mature fields
Shaybah	Onshore	0.75 million bpd of Arab Extra Light. Plans to expand capacity by 0.25 million bpd by 2017
Qatf	Onshore	0.50 million bpd of Arab Light crude
Khursaniyah	Onshore	0.50 million bpd Arab Light crude
Zuluf	Offshore	0.50 million bpd of Arab Medium crude
Abqaiq	Onshore	0.40 million bpd Arab Extra Light crude

Source: EIA, Saudi Aramco (2014)

What is noticeable is that the majority of Saudi oil production capacity is Arab Light crude (8 million bpd) as well as a further 1.5 million bpd of Arab Extra Light crude, with 2.6 million bpd of Arab Heavy and Arab Medium crude. In theory, should Saudi Arabia cut back on its production of heavier crudes, then this will create an even larger imbalance between light and heavy oil grade supplies. However, the growth in Saudi Arabia's petrochemical and refining base discussed earlier ensures that the country's heavy grades are used in these petrochemical projects, as the new Saudi refineries are designed to process heavy crudes, thus freeing up Saudi light and medium grades for exports, especially to Asian refineries who can still process light grades, especially if supplies are sufficiently discounted, while preferring heavy grades (Fattouh 2014[b]).

To Invest or Not to Invest in Spare Capacity?

Given global changes in energy supplies which are undermining both OPEC and Saudi Arabia's dominant position within the organization, should Saudi Arabia continue to maintain an expensive production capacity level or even consider expanding its current capacity? Figure 7.14 illustrates the huge capacity advantage that Saudi Arabia has over its fellow OPEC members at around 2.4 million bpd and which helps to illustrate the grudging acceptance of Saudi Arabia's lead role in determining OPEC's policy.

Fig. 7.14 OPEC members spare production capacity (K bpd) 2014

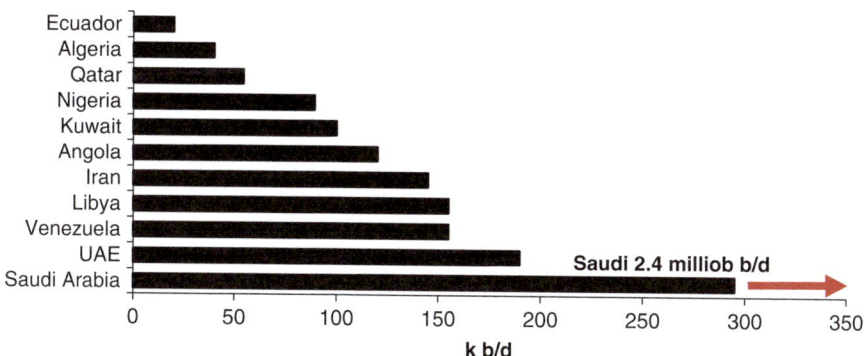

Source: Bank of America Merrill Lynch Commodity Research (2015), IEA (2015[a])

While Saudi Arabia is producing at record levels of 10.3 million bpd, this has not been reflected in terms of its share of the global market as analyzed earlier. The question remains whether Saudi Arabia wants to put its spare capacity to work in the coming years and, above all, whether it wants to increase capacity output beyond its stated 12.5 million bpd level. According to some analysts, under current price and global supply-demand uncertainties, the option for Saudi Arabia to wait becomes very valuable with no plans or incentives to increase oil productive capacity from current levels (Fattouh 2014c). Figure 7.15 illustrates Saudi capacity expansion options, with indications that new investment is taking place only to replace declining fields.

Fig. 7.15 Saudi Arabia crude capacity growth projects

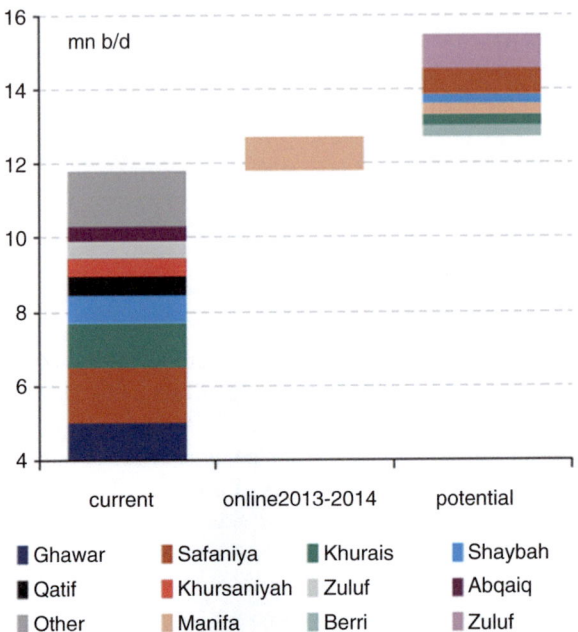

Source: Fattouh (2014c) Oil Market Dynamics: Saudi Arabia oil policies and US shale supply responses, March 2014

From Fig. 7.15, Saudi Arabia intends to maintain its crude capacity at 12.5 million bpd with the ongoing *Manifa* offshore project adding 900,000 bpd of Arab heavy crude to offset declines in more mature fields. The major potential crude capacity expansion projects are also offshore in *Zuluf* and *Safaniya*, both producing Arab Medium and Arab Heavy crude grades respectively. Other future potential expansion projects are to increase production of Arab Light crude grades, but these are not as significant as the potential expansion of heavier crude grades. If all goes to plan, the Kingdom is expected to reach around 15.0 million bpd total production capacity. It is doubtful however whether most of the longer-term projects will proceed under current price and market uncertainties, with Saudi Aramco already announcing some cuts in its capital investment plans (Carey and DiPaola 2015). Some argue that Saudi Arabia should produce around 8 million bpd as this was the Saudi "OPEC share" for many years and that it should keep "oil in the ground" to avoid speedier depletion of reserves at higher production rates, which also requires higher "water cuts" in maintaining production pressure (Al Khowaiter 2015).

The issue of what type of crude oil OPEC and non-OPEC competitors are producing is important to determine future oil prices. As most of the crude oil produced in the Arabian Gulf is sour-medium and heavy—and most of the global US surplus is sweet—any production cut by Saudi Arabia and its allies would not drive prices back up and the only way to do that would be to reduce production of light sweet crude, including by US producers which would lose market share. The US shale revolution divided OPEC countries according to the quality of its members' crude oil with exporters of light sweet crude-like Algeria, Nigeria, and Angola—losing nearly all their market share in the USA, while exporters of sour, heavier crude, including Saudi Arabia and Kuwait, have lost a little but are beginning to see their export in this crude also fall due to several factors. The first is competition from heavy crude from Canada and, second, the domestic refinery needs of countries like Saudi Arabia, with these refineries mostly using heavy crude to enable export of value-added products.

Aramco is now emphasizing more focus on exploration and development of non-conventional gas reserves to increase its share in the domestic economy and free oil for exports. This energy-switching strategy is driven by the sharp growth in domestic oil-intensive energy demand. This has capped Saudi crude exports at around the 7.1–7.3 million bpd levels, with declines during the summer months due to surges in domestic demand underpinned by the current subsidy system explored earlier. Managing such domestic energy subsidies in many of the OPEC countries is important, if not *more* important than being held hostage to the vagaries of volatile market oil prices. While the latter seems to have slipped beyond the *control*, as opposed to *influence* of OPEC members, the former—subsidies—is entirely in the control of member nations. The key decisions to be made are political ones involving a trade-off between long-term economic efficiency and short-term sociopolitical expediency. At some stage, one has to give way to the other, irrespective of the current financial advantages that some OPEC countries have over others. In an example of what might happen to some OPEC countries that do not try to rationalize their current oil subsidy programs, Nigeria found itself with the unlikely prospect of having no diesel fuel to distribute to its retail outlets, with the situation apparently aggravated by endemic corruption in the way the government manages its wholesale subsidy scheme estimated to cost the Nigerian economy $14 billion as its peak in 2011 and $5 billion in 2014 (Economist 2015[b]). Cash-strapped OPEC members can no longer afford such costly or corrupt practices.

Chapter 8
OPEC Reinvented: A New OPEC for the Twenty-First Century

> *Everything we hear is an opinion, not a fact; everything we see is a perspective, not a truth.*
>
> Marcus Aurelius

Introduction

"*OPEC is dead. Long live OPEC.*" There seems to be a bit of truth in both statements. When a king dies, the new one ushers a reign that sometimes follows what his predecessor has done and builds upon it, but most often the successors establish their own stamp with new policies and directions, whose outcome is not so assured but which promises a brighter future however difficult the new path chosen.

And so with OPEC, as it passes more than a half century of existence at 54 years old and now wonders what new role to play in a fast-changing, shale oil-led world. It is a time for reflection and not recrimination and for one to assess in fundamental and sometimes uncomfortable terms what the organization wishes to achieve to reach 100 years to continue to contribute to the world's energy needs. As one analyst commented, "if OPEC did not exist, it would be necessary to invent another organization much like it" (Kemp 2014). The key is *what type* of organization and role should OPEC play that makes it acceptable to consumers and reduce latent hostility to the organization ever since it burst into the popular imagination and became a favorite bogey figure during the first oil shock in 1973.

Let's Not Call It a Cartel

OPEC has always had fierce critics with opponents ready to forecast its eminent demise, leading a Kuwaiti Oil Minister to state in 1981, to hostile journalists, that "news of OPEC's death has been much exaggerated." The animosity stems from a popular perception that OPEC is a cartel, when our earlier analysis indicated that while it may have some *cartel-like features*, it was not a real cartel. The classic textbook definition of a cartel is one of a group of like-minded producers that control production and sometimes allocate customers among themselves, limit free competition to secure higher prices, and earn "above normal" returns. The reality is that OPEC has never really done all these things and even in its heydays, it never controlled anything that approached total market share and complete control over prices. OPEC members more often than not have competed against each other, at times offering secret deals and discounts and not abiding by agreed quotas. This is compounded by the fact that most members, especially those with the largest reserves and capacity, treat national information about their reserves, exploration, and investments as state secrets and do not disclose them to the OPEC Secretariat in a transparent manner, let alone to outsiders. As such, OPEC members in turn resent and object to their organization being labeled a cartel, and yet many people still believe that OPEC can and does manipulate the global price of oil even when the 2014 fall in oil prices and the almost helpless reaction to it by OPEC indicate the opposite.

Why then do so many still believe OPEC to be a cartel and does this really matter in the final analysis? The fact that some members of the US Congress keep trying to pass punitive *NOPEC* (*No Oil Producing and Exporting Cartels*) legislation to criminalize the organization indicates that it does matter. It is important to assess why the "OPEC is a cartel" has evolved into a self-sustaining "rational myth," with a life of its own. According to some, the fact that such a widespread belief could be wrong "sheds light on the process of ideational change and the failure to *update beliefs*" (Colgan 2012).

The birth of OPEC and its actions, or more specifically the actions of the Arab members of OPEC in 1973, created the notion that OPEC is a cartel. However, much has changed in the world of energy supply since 1973 and two of the three actions that OPEC took in 1973 cannot be repeated now: specifically, *posted prices no longer exist*, with prices now being determined by market supply and demand as OPEC Ministers repeatedly confirm, and second, *oil nationalization* has already happened in most major oil producers. The third action, an *oil embargo*, could theoretically happen in today's oil market, but in terms of significantly affecting oil prices—unless there is a complete shut down by *all* OPEC countries (which did not even happen during the 1973 embargo)—then this will not have the same effect as in 1973 on prices. Major consumer countries have not only developed their own energy resources and nonrenewable energy sources, but significant strategic petroleum reserves (SPRs) exist which will blunt the effects of an oil embargo.

To put this in perspective, in comparison with the 1973 embargo, the amount of oil production "lost" to the world from stoppages and sanctions in OPEC members, Libya and Iran, was around 1.470 million bpd in 2013 compared with 2011 levels, or around 1.7 % of global production (BP 2014). In 1973 all the countries of the Organization of Arab Petroleum Exporting Countries (OAPEC) cut their production for 3 months from October to December, averaging at 2.633 million bpd for the period, representing just fewer than 5 % of global production for that year (Johany 1980). If one takes out Iran from the 2013 "lost" production (given that Iran did not participate in the 1973 Arab oil embargo), then the loss to the markets in 2013 is even more negligible. According to Johany, the rise in oil prices during the 1973 embargo was due to the fact that there was "tightness in the oil market in the 1970–1973 period that was caused by actual demand being greater than projected demand" (Johany 1980, p. 47).

Given the above, why is there persistence among some groups that OPEC is and acts as a cartel? According to analysts, the following groups can be identified (Colgan 2012):

- *Group A*: *OPEC insiders and the "rational myth"*. This group sustains a "rational myth" about OPEC's influence over the world market for oil, and *crucially it contains members of OPEC itself*, as they are the least likely to undermine the narrative that so long as OPEC is viewed as "powerful," its leaders can claim credit for their "economic stewardship" of the global economy and the positive impact they are having on humanity's economic and social progress. Sometimes national leaders in OPEC countries take credit among their people for actions on the international stage. This is best illustrated when supporters of the late Venezuelan President *Hugo Chavez*, who was elected in 1998 as oil prices were plunging, argued that their president "revitalized OPEC and thus almost single-handedly brought about the rise in world oil prices" (Wilpert 2007, pp. 93–94).
- *Group B*: *Oil market participants*. This group consists of "informed" oil market participants outside of OPEC, such as commodity traders and international oil companies (IOCs). They have access to proprietary data sources on shipments of oil, contacts within the energy industry, and analytical skills to asses OPEC's behavior and likely impact on prices. Such participants aim to find out what are the likely actions of the perceived "strongest" OPEC member on market developments. In this scenario, a country like *Saudi Arabia* will come under intense scrutiny as to its pricing intentions, spare capacity and production plans, marginal cost of production, and "reading between the lines" of senior oil officials' statements about oil production levels that correlate with intentions to loosen or tighten global oil supply, as vividly illustrated in previous chapters about the intentions of Saudi Arabia to cut production. Such *Group B* participants believe that understanding what the most powerful OPEC member's intentions are will provide an insight into what the whole OPEC membership will do, acting like a cartel. As discussed earlier, Saudi Arabia can and does decide on production policies, such as raising its production to new high levels of over 10 million bpd in March 2015, without meaningful coordination with anyone, much as a dominant firm might do in a semi-monopolistic market, with the dominant firm believing that *it is* the market. The new, more assertive Saudi production policy and refusal to act as OPEC's swing producer on the downside may indicate that the country is signaling that *it is* OPEC for all practical purposes. However, for participants in *Group B*, they probably understand, to varying degrees, that OPEC is *not* a cartel, but they do not care, instead focusing their efforts on OPEC for signals about the present and future behavior of the dominant members, in this case Saudi Arabia. If the cartel hypothesis is then obviously incorrect, why does this group persist with it? The reasons might be mundane in that it takes a lot of

time and effort for these market participants to dispel it. There could also be some modest benefits to the general public's confusion about OPEC and the organization's somewhat dismal public relations initiatives to dispel the notion that it is the main cause of high domestic oil prices in consumer nations, when OPEC has pointed out that taxation by consumer nation's governments is *the* primary cause for higher oil prices. This is illustrated in Fig. 8.1 that sets out a composite barrel of oil and its crude price, industry margin, and tax components for the major industrial economies.

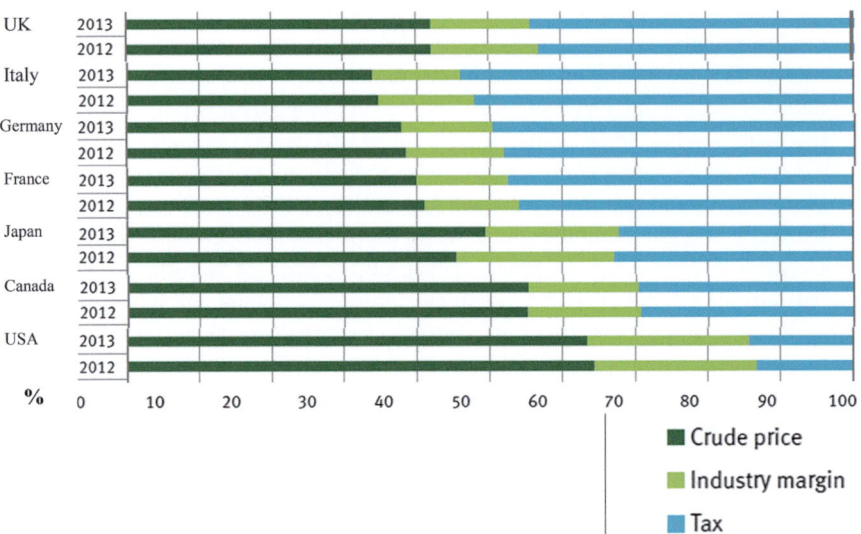

Fig. 8.1 Composite barrel and its components (%) 2012–2013. Major industrial countries

Source: OPEC Annual Statistical Bulletin (2014, p. 92)

- From the above diagram, the USA has the lowest tax rate at around 15 % compared with 50 % levels in the UK, Italy, Germany, and France. What is also interesting are industry margins ranging from around 10 to 15 % for the industrialized countries, indicating that oil company executives of the IOCs are possibly happy to join *Group B* and have OPEC as a scapegoat for high oil prices, deflecting blame away from their own companies. Despite erosion of some of their market powers, with nationalization in many OPEC countries, the IOCs are still a dominant force today with significant reserves and production capacities as illustrated in the diagram below.
- Figure 8.2a, b indicates that the world's largest five oil companies—*Chevron, Royal Dutch/Shell, Total, ExxonMobil, and BP*—hold around the same level of reserves as three OPEC members, *Algeria, Angola, and Ecuador*, of around 30 billion barrels, while their combined production in 2013 of just over 9 million bpd equaled those of *Iran, Iraq, and Kuwait* at 9.41 million bpd (OPEC 2014, p. 8).
- *Group C*: Government analysts and scholars.

Fig. 8.2 (**a**) Major oil companies' crude oil and NGL reserves (bn/b). (**b**) Major oil companies' crude oil and NGL production (mm. bpd)

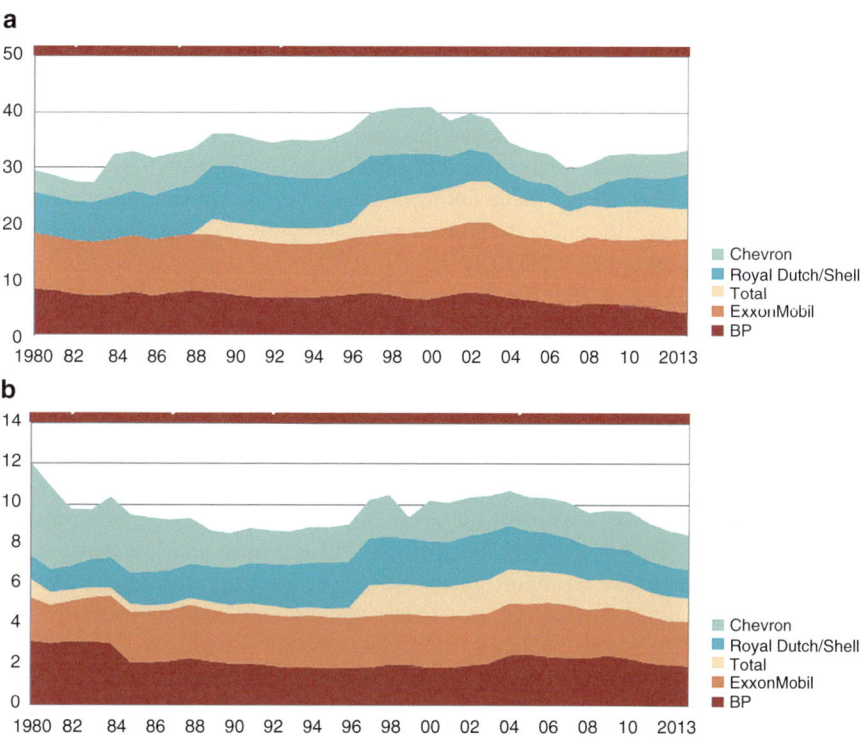

Source: OPEC Annual Statistical Bulletin (2014, p. 98)

- This group consists of government analysts and academic scholars who wish to understand how OPEC works or does not operate as a cartel. While government analysts have access to confidential data sources on production and shipment of oil destinations, academics rely on public data. Government officials often follow "perceived wisdom" and political directives, as opposing this might damage careers. It is not surprising that some Western government energy research and data collection institutions such as the US Department of Energy website still label OPEC as a cartel. The problem for government analysts is trying to *infer* OPEC's impact in the face of frequent price volatility and then further inferring from this that the group acts like a cartel. For example, if OPEC production increases as prices are falling, some will infer that OPEC is causing prices to fall, but if instead OPEC production decreases, some will infer that OPEC's impact is delayed and will cause a price increase some time in the future.

From this, one cannot directly infer OPEC's status as a cartel and even scholars are divided on the matter. Some cast doubts about OPEC's causal impact on the oil markets (Goldthau and Witte 2011; Bremond et al. 2002; Cairns and Calfucura 2012), while others insist upon OPEC's role as a cartel (Bentzen 2007; Hyndman 2008).

Given such divergent academic opinion, it is not surprising then that they have failed to persuade some key policymakers or the general public to change their view about OPEC.

Who then should take up the mantle? Some would argue that professional energy *journalists* could assume this role as they have, collectively, a sufficiently deep understanding of oil markets to realize that OPEC is weak as an organization, especially in trying to set market prices, and that any such market power that may exist is driven largely by Saudi Arabia, despite some recent or potential erosion of its market share by other OPEC members as discussed earlier. Despite this in-depth professional expertise, the enduring popular image of OPEC as a cartel still persists, and it is up to the organization to dispel it.

Consumer-Producer Cooperation and Supplier's Security to the Center Stage

The issue of meeting consumer supply concerns and the drive of some major economies, principally the USA to ensure domestic "energy independence," have already been highlighted. To some extent, the shale energy revolution has created a large degree of consumer energy import independence, causing understandable anxiety from producer nations, the majority of whose economies are single-resource based. The issue of meeting producer supply security has now taken center stage, which has to be addressed if producers are still to invest substantial amounts in maintaining current as well as future production capacity, as discussed in Saudi Arabia's case. OPEC's problem has always been its inability to restrain production, even if they could, by non-OPEC members. Sustained oil price rises, however volatile, have generally brought a boom in non-OPEC exploration and production, gradually eroding the organization's market share. The non-OPEC surge has come in different waves: in the 1970s and 1980s, the surge came from the North Sea, Alaska, the Gulf of Mexico, the Soviet Union, and even China. In the 2010s, it has come from US shale.

In face of these surges, and others yet to come from deepwater exploration in South America, OPEC seemingly has had limited success in persuading nonmembers to cooperate and restrain their output to ensure higher prices and better revenues for *all* members and non-OPEC members alike. This does not mean that OPEC does not try. As discussed in the volume, OPEC members recently have made repeated public statements that they are willing to cooperate but that they cannot and will not do it alone. A trust deficiency between the two blocs seems to have risen, and this mirrors earlier attempts at cooperation. In the mid-1980s, Saudi Arabia lobbied Britain and Norway to limit output from the North Sea and maintain prices but was rebuffed even if Norway at first tinkered with the idea (Claes 2001, p. 302). In the 2010s the problem for cooperation with surging US shale producers was even more problematical for OPEC, because it would be prohibited by US antitrust laws.

The Saudi negotiations with non-OPEC Norway in 1986 on production cuts are illustrative of the pitfalls faced even when agreements in principle seemed to have been reached, at least in the opinion of one of the parties. The then Saudi Oil Minister Zaki Yamani was quoted by *Platts* on January 24, 1986, as praising "Norwegian Oil Minister *Kaare Kristiansen's* statement yesterday indicating that his country would be willing to cut production," while another news bulletin read "Norwegian Oil Minister—Kristiansen, who won Yamani's praise for his earlier comments, took back much of the support he implied yesterday for such a pact. Yamani had no sooner set forth his statement than Kristiansen issued a counter-statement largely rescinding his remarks of the day before. Today's version of the Norwegian position defined it as "up to Saudi Arabia to do something about prices and production." Kristiansen added, "I can't see any reason for the Norwegian government to impose on companies a reduction in production" (Claes 2001, pp. 308–309).

This was an astonishing turn of events, causing embarrassment to OPEC and Saudi Arabia in particular, when the organization had believed that a firm cooperative agreement had been reached only to be publicly rebuffed. One possible reason for the Norwegian stance is whether OPEC was aware of the *breakeven price on the Norwegian oil fields* where a substantial part of production had been running on costs of less than $10 a barrel and that a price war could not have had the effect of lowering Norwegian production before prices had fallen to this level (Claes 2001, p. 307). The 1986 situation echoes the fog of confusion over estimated 2014 shale oil producer's cost of production and their breakeven pricing levels.

Why Cooperate?

In general, some countries might join oil cooperative agreements from a calculation of potential economic gains, but as pointed out by some researchers, there is no single norm of cooperation, but rather several *distinct* norms which include moral norms of "fairness" or "interest-based" selfish reasons (Elster 1989). Sometimes there is a "mix" of motives making the issue of cooperation more complex than a simple economics game theory of winners and losers. A mix of individual governments' interests and resources and their individual calculations of "own risk aversion and probability of likely success of different strategies could possibly determine the potential success of oil producers cooperation" (Claes 2001, p. 295). What might seem rational and the obvious thing to do by cooperating might not be as straightforward given national complexities in reaching a decision. This can lead to some frustration and puzzlement by senior OPEC officials, illustrated below from the seminar and special meeting by the organization held in March 2009:

– *Alvaro Silva-Calderon (former Energy and Mines Minister of Venezuela)*: "OPEC committed itself to market stability, so that the interests of producers, consumers and investors could be considered. It has not been easy for OPEC to fulfill these tasks, since in order to respond to world demand in a balanced way it has to fundamentally manage the other non-OPEC oil producers. This has required the organization to adopt cooperative approaches as an indispensable work tool … The time seems to have come to request a deeper, more formal

cooperation—one based on a more binding character-extended to non-OPEC producers, as well as consumer countries and their energy organizations. Like OPEC, they are all similarly interested in market stability and energy security, both in the present and in the future."
- *Dr. Fadhil Al-Chalabi (OPEC's first Deputy Secretary-General)*: "Now OPEC, I believe, is more of a price-taker than a price maker ... In my opinion, the world now is heading forwards a new era, where there are many sources of pressure on oil ...."
- *Dr. Ramzi Salman (OPEC's second Deputy Secretary-General)*: "Another thing that is really important is the non-OPEC issue. After all, what we are seeing happened now is not the first time. We had an energy crisis in 1975. We also had a big problem in 1986, and then non-OPEC had to suffer with us and they helped to impose a new deal. Then we had another problem in 1998. So this is not new ...."
- (OPEC 2009[a])

More recently, the same concerns about OPEC-non-OPEC cooperation were frankly stated by the *OPEC Secretary-General Abdalla El-Badri* when he said in an interview with Bloomberg, "we met in the 27th of November (2014) ... We debated the world economy. We debated supply and demand ...and we see that OPEC did not decrease its production for the last 10 years from 2004 to 2014. We have not increased it. We produce 30 million barrels a day... the whole non-OPEC supply; they increased the production by 7 million barrels a day. I mean, why could it not increase our production? So we said if going—if we are going to reduce our production, then this would be a collective decision by OPEC and non-OPEC" (Bloomberg 2015[l]).

And what is the reaction of non-OPEC producers to such cooperation calls? According to an emerging global offshore energy power, *Brazil's Minister of Mines and Energy, Edison Labao*, when asked about closer ties with OPEC, stated that his country was "studying" such a move and that "we have not determined when, but at some point, we will respond. But we certainly understand that the existence of OPEC is extremely necessary to help improve relations among producing and consumer nations" (OPEC 2009[a], p. 71).

With such seeming goodwill from a leading non-OPEC producer, and presumably the same apt choice of words also expressed for cooperation by other countries, like Mexico and Russia whose oil ministers were canvassed prior to the November 2014 OPEC meeting, what are the obstacles then for successful OPEC-non-OPEC agreements? One answer seems to lie in the *bargaining process* between the two parties. According to experts on the subject matter, a bargaining process demands some degree of agreement to reach a decision or outcome, and while there is no requirement for the counterparts to be in *complete agreement* (their highest preferences), neither can they be in *complete disagreement* (their lowest preferences) in order to achieve a cooperative outcome. Second, bargaining requires communication whereby a counterpart must be made aware that a *concrete proposal has been made* which is different from somewhat vague media statements and platitudes. Third, the parties must have both conflicting interests and common interests (Ikle 1964). One of the most serious reasons for conflicting interests is probably OPEC's inability to correctly assess so-called higher marginal costs of non-OPEC producers, which reduces their bargaining power in trying to match changes in production levels. Just waiting for non-OPEC producers to "see reality" by sweating them out during periods of sharp oil price falls is no longer a basis for a successful OPEC strategy.

The OPEC Secretariat put the matter of OPEC-non-OPEC cooperation even more bluntly when in its April 2015 *OPEC Bulletin* it stated that:

> Today, operating purely through self-interest is quite simply frowned upon. As the old adage says, a problem shared, is a problem halved. Yet, when it comes to the supply of petroleum, there is a stubborn willingness of some non-OPEC producers to adopt a go-it-alone attitude, with scant regard for the consequences. These parties consider producing to the maximum as being the norm. To them, rationalizing the development of one's precious natural resources in keeping with market demands appears to be an alien concept (OPEC 2015c).

Suppliers' Energy Security: An Unattainable Goal?

While major consumers, to a certain extent, can feel more confident that there seems to be a more abundant and variety of energy sources, the same feeling of a predictable energy supply security is not prevalent for OPEC producers. The issue of Gulf producers' energy security was specifically highlighted in a 2010 conference organized by the Abu Dhabi-based *Emirates Center for Strategic Studies and Research* entitled "Energy Security in the Gulf: Challenges and Prospects." According to some, the core conceptual understanding of energy security in the Gulf relates to the pursuit of a *secure* flow of oil and gas supplies to consuming countries, but energy security goes beyond simply securing supply and demand as it involves achieving stable energy prices and the value of the US dollar, as the impact of dollar fluctuations on real export earnings and producers' terms of trade is highly significant, highlighted during the early days of the 2008 global financial crisis (Al Sahlawi 2010).

Others argue that since energy markets are global and interdependent, then the concept of "reciprocal energy security" arises, whereby enhancing energy security in oil-producing countries will eventually enhance energy security in consuming countries and vice versa. Furthermore, the concept of energy security should be broadened to include assessing the effects of increased threats of terrorism and conflict, for example, in Libya and Iraq, as well as increased threats of technology failure and cyber terrorism disrupting output and having the same impact as any major political event in an oil-producing country (Al Hajji 2010). The mysterious cyber attacks on *Saudi Aramco* in August 2012, with conflicting reports on the origin of the virus attacks, whether external or internal, as well as a cyber attack on Qatar's *RasGas*, highlight the seriousness of such threats, although in both countries there was no stoppage of oil and gas production (Finkle 2012).

Besides the above, dependence of energy-producing countries on energy exports has become a threat to global energy security as declines in oil income may lead to social and political problems that might threaten oil supplies to the rest of the world (Al Hajji 2010).

According to some estimates, there has been significant drawdown on financial reserves held by some OPEC countries hit hard by the 50 % fall in oil prices from November 2014, with reports that Saudi Arabia's foreign exchange reserves fell by $20.2 billion in February 2015, the biggest monthly drop in the last 15 years according to data from the Saudi Arabian Monetary Agency (SAMA), and SAMA's net foreign assets dropping further by around $46 billion at the end of April 2015 (Saudi Gazette 2015[c]). Many OPEC countries are in fact disinvesting from assets held in Western institutions instead of adding to their investments, with US estimates that OPEC countries as a whole are expected to earn around $380 billion from oil revenues during 2015, representing a $350 billion drop from 2014, the largest 1-year decline in history (Blas 2015[a]).

Producer energy security has also other dimensions that incorporate an *environmental* and *technological* element. The objective of the environmental dimension of energy security is to "reduce, or eliminate if possible, the negative environmental impact of exploration, production, transportation, processing and use of energy sources" (Al Hajji 2010, p. 205). With global environmental pressure groups calling for fossil fuel divestment as analyzed earlier, OPEC producers will have to meet this challenge and come up with a novel way to reduce fossil fuel emission, which is costly and cannot be afforded by all members of the organization.

The *technological* dimension is to ensure that a low price for oil and government regulations do not choke new technologies that lead to improvements in energy efficiency, increase productivity, and lower emissions. Such technological advances should be made available to producer and consumer countries, as it benefits both and extends the life of a depletable resource that will still be an important source of global energy for many years to come. However, environmental management is going to be a slow, lengthy, and painful process of mitigation and adaptation (Khatib 2010). There is also an argument that possessing sustained high levels of spare capacity could pose security of supply challenges for OPEC, as it would test the organizations' market management cohesion. In times of low prices over a long period, this might be one method of keeping some OPEC members in check, particularly those who were serial quota busters, especially if they have the least spare capacity. The prospect of higher levels of spare capacity might lessen for those like Saudi Arabia to invest in new production capacity in case it is not needed (Al Kadiri 2010). This is not an idle matter, as Saudi Arabia, which has a policy of maintaining a current spare capacity of around 2.5 million bpd, does not want to risk over-investing and thereby repeat incurring the costs of the massive 10 million bpd capacity surplus in OPEC during 1986. As such, the subject of spare capacity's "who benefits, and who should pay, is a recurring theme in the producer-consumer dialogue" (Skinner 2006, p. 3).

In conclusion, what can one make out of supply security? Is it achieved as a *quantifiable* economic concept, or is it more a *psychological* notion of security, which is a feeling, but one that can trigger either "rational" or "irrational" decisions by government?

One country can be entirely dependent on imports yet feels "secure" because supplies are coming from "friendly" countries, for example, Canadian oil exports to the USA, while another can rely partly on imports yet feels this constitutes a major vulnerability and insecurity. Supply quantity and the degree of dependence can remain unchanged, yet the psychological feeling of insecurity can change over time depending on geopolitical circumstances and shifting political alliances. The recent worsening relationship between Russia and some countries of the European Union over events in the Ukraine illustrates shifting gas energy supply insecurities.

For OPEC countries embarking on long-term capital-intensive capacity expansion plans, the issue of "psychological" security is important. The global energy supply system is a vast complex of large, fixed capital assets that take years to plan, approve, and construct, even if governments decide to approve them in face of conflicting socio-economic priorities. Once in place, these assets are in place for decades and can either add to national wealth or turn out to be costly "white elephant" projects. As discussed in the preceding chapter, Saudi Arabia's hesitancy in expanding its current production capacity from 12.5 million bpd to 15.0 million bpd in the face of multiple "unknowns" is a rational one, given that a high degree of "psychological security" is absent.

OPEC: Holding Together?

There is no question that the shale revolution has, despite some production fluctuations due to lower oil prices, tipped the balance of power in the oil market by opening up vast new energy resources, not only in the USA but in other frontier countries with future technological advancement and environmental controls in this sector. The question then, is OPEC relevant, or more precisely, is it *as relevant today* as it has ever been since its establishment by the founding fathers? Is it less "powerful" today and where can it still add a meaningful role to ensure that it still holds together or, as a former senior Aramco energy executive asked, "does anyone now need OPEC?" (Al Husseini 2015).

By all accounts, and despite its many detractors, OPEC members still hold a significant amount of the world's proven and potential oil reserves, and their ability and willingness to produce them will continue to have an impact on world prices for many years to come. Like many other international and multilateral institutions such as the *North Atlantic Treaty Organization* (NATO), the *World Bank*, and the *European Union* (EU), these bodies have had to adapt to changed economic and political circumstances. Others fell by the way of history, like the *Central Treaty Organization* (CENTO) which was formed in 1955 by Iran, Iraq, Pakistan, Turkey, and the United States as the *Baghdad Pact*, but was dissolved in 1979. Others like the United Nations still exist but are plagued by superpower rivalries, interest blocs, and groups and have not been very successful in containing global conflict through peaceful means, akin to OPEC's failure to meet all its members' different interests as explored earlier in the volume.

As for OPEC even if it did not exist, oil-producing nations would still want to discuss oil market issues with one another, like other major commodity producers for copper or wheat, and if OPEC did *not* exist, it would be necessary to invent another organization much like it. What such an organization might best be doing is another issue which is explored later in more detail.

It is not only OPEC that may have to readjust itself to changing events but also others in the global energy market and those that are in competition with OPEC. As noted, even higher-cost shale oil producers are beginning to adapt to lower prices, and a rebound in prices will see some of those idle *DUCs* or drilled-uncompleted-wells come back on stream. Likewise, IOCs with their significant production and reserves of energy are quick, often quicker than sovereign states, to adapt to constrained economic circumstances. Energy analysts have noted that IOCs "have a knack for picking the bottom of crude prices and history may be about to repeat itself" through mergers in the oil industry like *Royal Dutch Shell's* takeover of *BG Group* for $70 billion in April 2015 (Coulter 2015). Some believe that his could herald the first wave of 2010's acquisitions as the IOCs seek to drive out costs following the sharp fall in oil prices, repeating the massive deals of the 1990s that restructured the international oil industry, when *BP PLC* bought *Amoco Corp.* for $56 billion in 1998 and *Exxon* took over *Mobil Corp.* for $80 billion in the same year, when oil prices crashed to $9.55 a barrel before beginning their rally. The oil price recovery did not deter other mega acquisition deals, with Spain's *Repsol SA* acquiring *YPF* of Argentina for $15 billion and *Total SA* of France buying *Elf Aquitaine* for $52 billion, while in the USA, *Chevron* took over *Texaco Inc.* and *Conoco* took out *Phillips* (Coulter 2015). As in 2009, from 2014, both producers and IOCs were battling for market share to force out the most inefficient.

The non-OPEC producers and the IOCs have very little incentive to behave differently so long as OPEC holds together and provides the IOCs with a "safety net" of production, but their very short-term profit-maximizing behavior strains the system under which they shield themselves. As OPEC officials have repeatedly stated, this is a game in which *all* producers stand to lose and whether the IOCs and non-OPEC producers are able to see the danger of noncooperation. While some might label this as a "conspiracy," others believe that it is more akin to "schizophrenia," whereby "non-OPEC producers want to enjoy the ability to play the market *and* the protection of an OPEC price assurance for which they pay no premium." The oil company wants both a low price for the crude it acquires and a higher general crude price level that prevents a free fall for product prices. "Nobody can have it both ways. Sooner or later, everybody will have a rude reawakening" (Mabro 1986, p. 65). What was true then during the price rout of 1986 is still valid in 2015. However, while IOCs can scoop one another at prices agreeable to the acquired and the acquirer, can OPEC do the same among its members or is there an alternative mechanism to differentiate non-efficient from efficient OPEC producers?

OPEC Members: Some More Equal Than Others

As OPEC tries to navigate through this period of uncertainty and find a role for itself in a shale oil world with a long and difficult period of adaptation for which some of its members are not prepared, questions arise on who are the fittest to survive? While all OPEC producers are losing out to the fall in oil prices, some OPEC producers with larger marginal costs of production and higher "breakeven" prices are suffering more than those with lower marginal production costs and breakeven prices. OPEC in essence is now divided into camps—those that are "efficient" low-cost producers and can sweat out the increased marginal non-OPEC producers and those "inefficient," high-cost OPEC producers who are struggling to make their voices heard to convince the former OPEC group to cut production and help raise prices. This assumes that *all* OPEC members agree on what they believe OPEC represents as its twin fundamental interests: *fairness* and *stability*, whereby "fairness" is about the equitable distribution of the global oil surplus in providing a reasonable return to investors and ensuring that neither producers nor consumers claim a disproportionate share of oil rent benefits (Aissaoui 2014[a]). "Stability," on the other hand, is something that OPEC aspires to, without the organization being sidelined to a role of "residual supplier" to ensure market stability primarily at the expense of its more efficient, high production capacity members who end up losing market share.

The huge fiscal divergence between the "richer" (Saudi Arabia, Kuwait, UAE, Qatar) and "poorer" (Venezuela, Nigeria, Ecuador, Iran, Iraq, Angola) OPEC members and those caught in civil conflict (Libya) has made it difficult to find a common ground to work on. On paper, and taken as a group, OPEC is the predominant owner of global oil reserves of around 81 % or 1206 billion barrels illustrated in Fig. 8.3.

Fig. 8.3 OPEC share of world crude oil reserves (2013)

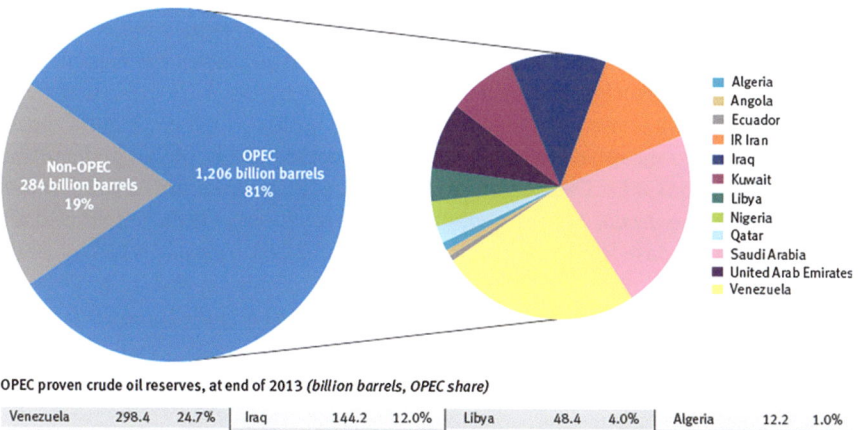

Venezuela	298.4	24.7%	Iraq	144.2	12.0%	Libya	48.4	4.0%	Algeria	12.2	1.0%
Saudi Arabia	265.8	22.0%	Kuwait	101.5	8.4%	Nigeria	37.1	3.1%	Angola	9.0	0.7%
IR Iran	157.8	13.1%	UAE	97.8	8.1%	Qatar	25.2	2.1%	Ecuador	8.8	0.7%

Source: **OPEC Annual Statistical Bulletin (2014)**

According to OPEC, while almost 81 % of the world's proven oil reserves are located in OPEC member countries, the bulk of reserves are in the Middle East, amounting to 66 % of OPEC's total. Only Venezuela, a non-Middle East country, takes the lion's share in first place with nearly 25 % of total OPEC reserves, ahead of Saudi Arabia's 22 %. Again according to OPEC, the organization as a whole continues to invest in additional upstream capacities, regardless of current challenges and oil price uncertainties, with 117 ongoing and planned projects at an overall estimated cost of $270 billion over the period 2014–2018. This is illustrated in Fig. 8.4.

Fig. 8.4 OPEC upstream investment plans (2014–2018)

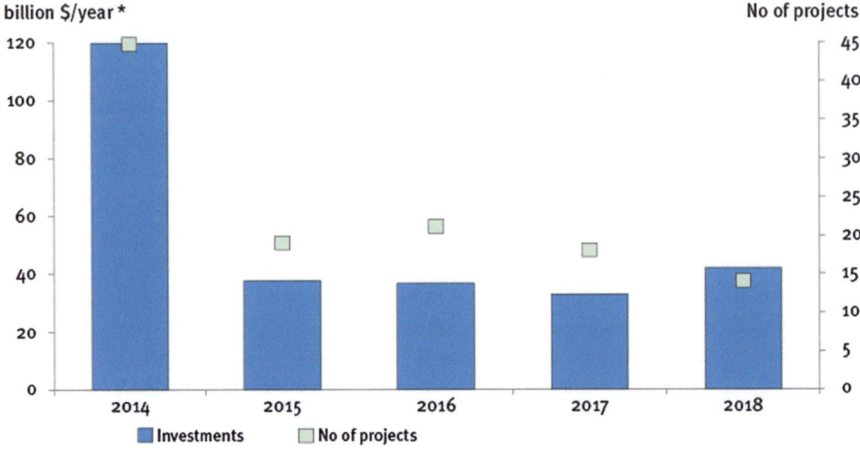

*These estimates are based on upstream project or field development requirements at field gate and do not include the infrastructure required beyond the field.

Source: http://www.opec.org/opec_web/en/data_graphs/647.htm

The impact of the sharp fall in prices in 2014 is evident from the reduced flow of projects and planned investments. It is doubtful whether all planned projects will be completed, adding to more intra-OPEC tensions on who carries the organization forward and bears the burden on behalf of other members. It is clear that those OPEC members who have the financial resources to build up their production capacity are the ones to most likely benefit from the expected rise in global demand and a renewed "call on OPEC." According to the US *Energy Information Administration* (EIA) in its *International Energy Outlook* (IEO Sept. 2014[a]), OPEC's market share is expected to increase from a low of around 39 % in 2020 to around 42 % by 2040, with a forecasted *increment* in OPEC production of 7.7 million bpd in 2030–2040, compared with an increment of around 2.2 million bpd during the forecasted period 2013–2020, as illustrated in Fig. 8.5.

Fig. 8.5 Expected OPEC output increment (2013–2040) million bpd

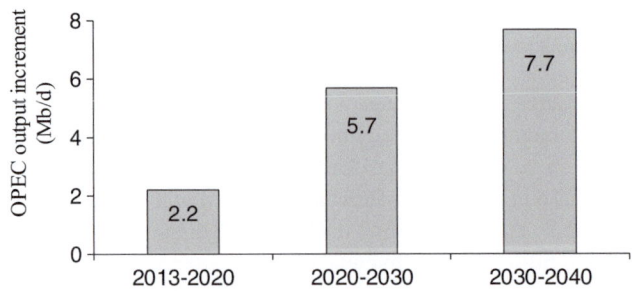

Source: Aissaoui (2014[a]) Economic Commentary Vol. 9, No. 10, Oct. 2014, EIA-IEO, Sept. 2014

While future demand for OPEC's oil is apparently assured, who in OPEC is willing to invest under current uncertainties? The answer and the reality, however unpalatable to other OPEC members, is *only Saudi Arabia*. The Kingdom possesses all the attributes necessary to meet some of the challenges, starkly illustrated in the table that follows, which differentiates OPEC members in various leadership/followers categories.

While Venezuela might hold OPEC's largest proven oil reserves, the country is dwarfed by Saudi Arabia's crude production and exports, as well as more crucially its spare capacity, making the *combined* OPEC members' spare capacity pale in comparison. To put it bluntly, *OPEC is Saudi Arabia* and *Saudi Arabia is OPEC*. From this premise, Table 8.1 differentiates various OPEC members into *Alpha* leaders (Saudi Arabia) and those members closely following the dominant Alpha leader (UAE, Kuwait, and Qatar). This does not mean that there are no aspiring OPEC *Beta* leaders waiting in the wings. Venezuela and Iraq are the most likely contenders to try to assert some OPEC leadership, direction, and say over policy or even create a parallel organization should the Saudi policy of "no production cuts" continues without a compromise in sight to assist other financially stressed OPEC members.

Table 8.1 OPEC's unequal members: leaders and followers

Category	Country	Proven oil reserves (billion bbl) 2013	Crude oil production million bpd (Feb. 2015)	Crude oil exports million bpd (2014)	Estimated spare capacity (mm bpd) Feb. 2015
• Alpha *leaders*	• Saudi Arabia	265	10.1	7.5	2.5
• Alpha *followers*	• UAE	98	2.84	2.7	0.026
	• Kuwait	101	2.80	2.1	0.020
	• Qatar	25	0.650	0.6	0.040
• *Rival*-Beta *leaders*	• Venezuela	298	2.81	1.98	0.150
	• Iraq	144	3.4	2.4	0.350
• *Go it alone*	• Iran	158	2.85	1.2	0.760
	• Libya	48	0.260	0.6	0.280
	• Nigeria	37	2.44	2.2	0.010
	• Algeria	12.2	1.2	0.75	0.060
• *Free riders*	• Angola	9	1.8	1.7	0.030
	• Ecuador	8.1	0.6	0.4	0.020

Sources: OPEC Annual Statistical Bulletin (2014), IEA, Bloomberg

Another OPEC group is classified as *go it alone*, and this includes members with modest to significant oil reserves but with a potential to develop their energy sector by striking bilateral supply relations with countries with whom they have strong ex-colonial ties or are in close geographical trading proximity (France, Italy, India, Turkey, Pakistan, the United Kingdom). Finally there are a few countries like Angola and Ecuador who are termed *free riders* who might either go it alone and opt to export to regional economies or join one or another of OPEC's rival leadership groupings depending on their perceived political and economic interests.

The above OPEC scenario might seem too overly pessimistic or even unrealistic given that human nature is often reluctant to embrace too radical a change and hope that the current *status quo* somehow improves for the better. However, even countries that are sometimes deemed to be overly cautious and conservative have recently witnessed dramatic changes. This is exemplified by the bold decision to restructure the national oil giant Saudi Aramco and separate it from the Oil Ministry in April 2015, in line with the vision of the newly appointed *Deputy Crown Prince Mohammed bin Salman*, who also created a new 10-member Supreme Board to oversee Aramco and ensure effective coordination of the main pillars of Saudi Arabia's economy (Al Mashaabi 2015; Mahdi 2015[d]). The *status quo* was not acceptable to the new Deputy Crown Prince. The June 5, 2015, OPEC meeting was a very critical one, even more so than the one held in November 2014, to determine in what direction the organization might be going. If the more recent bullish oil price news, with *Brent* reaching $58–$59 a barrel level in April 2015, continues to hold and rally, with the added evidence of slackening in US shale output, then the June meeting might not be as acrimonious as feared with possible fissures arising between competing OPEC countries on policy direction. As it turned out, the OPEC June 2015 meeting went very smoothly even from OPEC "hawks" like Venezuela and Iran, who went along with the new Saudi "market-share" protection strategy.

Venezuela in particular seems to be preparing the ground for an OPEC policy change with news that the country's Oil Minister *Asdrubal Chavez* and Foreign Minister *Delcy Rodriguez* were holding meetings with the Ambassadors from key OPEC countries in Caracas (Saudi Arabia, Algeria, Ecuador, Iraq, Kuwait, Nigeria, Qatar, and Iran) to try to agree on possible production compromises (Orozco 2015[b]). Venezuela has also proposed a novel plan for COOP members to cooperate in practical terms by blending their various grades of crude oil, using the country's heavy crude and other OPEC members' light crude that can compete with US and Canadian supplies, and expanding on a pilot scheme that Venezuela carried out with Algerian light crude during 2014 (Cawthorne 2015). The Venezuelan proposal envisions supplying refineries built for medium-grade crudes rather than the light oil that has become plentiful as a result of North American shale boom and is seen by Venezuela as a creative way to retain its US market share at a time of intensifying US competition and an expected rise in Canadian heavy crude to US refineries undermining Venezuela's market position. If agreed, the co-blending initiative could also provide a similar advantage to OPEC members whose lighter crude oil has been threatened by US shale production.

The Venezuelan diplomatic move has also been spurred by the news that, despite some slowdown in US shale oil production, the USA would become a *net energy exporter* for the first time since the 1950s, with exports exceeding imports from 2029 to 2033 and from 2037 to 2040, according to the latest estimates of the EIA in its *Annual Energy Outlook*, with "advanced technologies reshaping the US energy sector" according to the EIA (Shank 2015). The latest EIA estimates for US crude output forecasts a total 10.3 million bpd compared with earlier estimates of 9 million bpd and 7.5 million bpd for the same periods. Despite such discrepancies, the message to non-OPEC members is clear: US shale oil production is here to stay, at least for the medium term. For countries like Venezuela with its checkered history of political differences with the USA, the prospect of US oil exports at the expense of oil imports from countries like Venezuela is even more alarming in the face of the 2014/2015 global energy glut and weak oil prices.

There are however some signs that the inflexible stand of no production cuts imposed by Saudi Arabia on OPEC is beginning to bear some fruit, at least on the diplomatic non-OPEC front, with reports that Russia was holding high-level meeting with OPEC officials in June 2015, according to Russian Energy Minister *Alexander Novak* who also commented that further meetings were held in May 2015 to prepare for the June meeting and these were expected to cover shale output, refining, and tax changes (Rudnitsky and Bierman 2015). This mirrors the same sequence of meetings and events between Russia and OPEC ahead of the November 2014 organization's conference when no agreement was reached between the two parties. The deteriorating economic situation in Russia and the decline in the Russian *ruble* in the face of US-led sanctions, as well as persistent low oil prices, could just have brought about a rethink in the initial Russian refusal, possibly allowing Russia to go along with some production cuts in tandem with OPEC. The June 2015 meeting was shaping up to be critical in more ways than one for OPEC and non-OPEC relations.

Why were the Russians in particular having second thoughts on possible cooperation with OPEC besides their own economic problems? It would seem that Saudi Arabia's stance of "no production cuts" is only one aspect of the story, the other being the quiet "surge" in Saudi and other OPEC Gulf producers who have spare production capacity. According to the *International Energy Administration*, OPEC production climbed by the most in almost 4 years as Saudi Arabia, Iraq, and even Libya boosted output by as much as 890,000 barrels a day to take total OPEC production to 31.02 million bpd in March 2015, the biggest monthly gain since June 2011 (Smith 2015^b). The jump in OPEC's output in March 2015 left the organization's production about 2.5 million bpd higher than the average 28.5 million bpd needed in 2Q 2015 according to the *IEA*, led by Saudi Arabia's growth of 390,000 bpd and then by 658,000 bpd, to an average of 10.294 million bpd, the highest since September 2013 and close to record levels. By May 2015, OPEC's production climbed to 31.58 million bpd, exceeding its quota of 30 million bpd for the 12th consecutive month (Smith et al. 2015).

With Iraq also restoring output by 350,000 bpd on improved weather conditions in the Arabian Gulf, this signaled that Iraq could be a major challenge for OPEC if the country can actually begin to produce a significant amount of its planned long-term capacity expansion explored earlier. However, to put the Saudi surge in production in perspective, it was equal to *half the daily output of the US Bakken shale formation in North Dakota*, with Saudi Arabia in the space of *31 days* managing a production boost that took the *Bakken* drillers almost *3 years* to achieve (Smith 2015^c). Some analysts commented that not all of the Saudi incremental production was for ensuring market-share protection in the face of global demand pickup but was also to meet an increase in Saudi domestic needs during the hot summer months when refining stocks rise to meet demand from local utilities.

How realistic then is the likelihood of OPEC breaking up into several competing groups? The Saudi no production cuts policy and the production surge can be viewed as both a carrot and a stick, whereby as a carrot, Saudi Arabia is willing to cut back on its production *only if* other non-OPEC producers, especially Russia, will do the same. It is also a stick aimed at other OPEC members that unless they have the spare capacity and cost efficiency that Saudi Arabia possesses, then the Kingdom feels free to produce what it thinks is best for itself in the long run to maintain its market share. In a likelihood that OPEC splits into several factions, then Saudi Arabia already has a ready-made allied grouping that follows its lead, represented by the majority of the OAPEC—Kuwait, UAE, and Qatar, with Algeria and Libya probably induced to join, or go their own way as analyzed earlier.

OPEC can still play an important role and remain a cohesive force for *all* members, but the organization's role could be undermined if their policymakers fail to address the new challenges facing them compared to three decades ago and agree on new rules of leadership and collective action that takes into account the reality of differences in OPEC members—*that not all are equal* to ensure a better effectiveness in meeting global competition. This calls for reexamining the organization's founding charter and its objectives to enable for a new set of priorities be identified to underpin OPEC's continued survival.

OPEC's Founding Charter: A Time for a Revisit?

In 2005, *Ahmad Fahad Al Sabah*, then President of the OPEC Conference and Minister of Energy of Kuwait, restated OPEC's statutory objectives from the date of the organizations' establishment as being "the *coordination* and *unification* of the petroleum policies of our member countries and the best means of safeguarding their interests, individually and collectively; devising ways and means of ensuring *stability of prices* for the international oil members, with a view to eliminating harmful and unnecessary fluctuations; giving due regard to the *interest of the producing* nations and to the necessity of securing a steady income for them; an efficient, economic and *regular supply of petroleum to consuming nations*; and a *fair return on their capital* to those investing in the petroleum industry" (Mabro 2006, p. xiii).

Analyzing the words of one of OPEC's eminent officials, one is struck by the constant use of identical terms (*highlighted*) which seem to be the cornerstones of OPEC's mission, even in the face of new challenges posed by the emergence of nonconventional energy sources. A key issue is whether such well-intentioned and seemingly fair objectives, which seek the best outcomes for *all* energy stakeholders, can withstand current stresses or whether it now behooves OPEC to reexamine its founding charter and to adapt them.

The expression "mid -life crisis" is associated with that time of life when many individuals reach their 40s and ask themselves many searching questions on what they have done and, more importantly, what they should now be doing with whatever existence they think they may have left. And so with OPEC at 54 years, where does it go next? For the organization, like individuals, it is a time of crisis, reflection, and sometimes profound changes as few are blessed with total contentment with what they are doing.

Since OPEC was established in those heady days in Baghdad in 1960, petroleum crude is still viewed as a strategic and critical commodity for the world economy, highlighted earlier by the many ongoing territorial disputes over possession of this essential commodity. However, the international petroleum industry has substantially changed since 1960, with a bewildering array of alternative energy sources and a mix of public and private players, along with an added element of environmental pressure groups that was not prevalent in 1960. When OPEC came to life, production, marketing, and prices were controlled unilaterally by the multinational oil companies (IOCs), and there was little or no concern for the environmental and social impacts of the industry such as global warming and ozone depletion. Today things have fundamentally changed and, at their peril, neither OPEC nor the IOCs can ignore these issues.

When OPEC was formed, *sovereignty* was an emotive one for the recently decolonized countries. Nationalism and a desire for an alternative nonaligned movement took root to protect "national resources." In the twenty-first century, sovereignty is now more associated with a state's capability to *successfully* afford a peaceful, stable, and sustainable environment for economic development to flourish. These concepts involve issues such as energy security and financial viability and are based on mutually beneficial cooperation in a globalized world where the pursuit of economic efficiency, cost cutting, and the search for "added value" are the new mantras for multinational corporations, rather than having a political say in the affairs of resource-rich countries, although some of these accusations are still thrown at them.

The mega merger energy company deals of the 1990s and the 2010s discussed earlier were about the global upstream, exploration, and production of oil and gas around the world, with companies seeking efficiency and cost reduction and the quest for "scale"—the ability to take on larger and more complex projects as well as "their ability to mobilize the money, people, and technology to execute these projects … also the bigger and more diversified the company the less vulnerable it was to political upheavals in any country" (Yergin 2011, p. 97). The IOCs seem to have learned their earlier lessons of interference in the internal affairs of resource-rich sovereign nations as the troubled history of Iran with the west illustrates. *Muhammad Mossdadegh*, who, as Prime Minister of Iran in the early 1950s, nationalized the country's oil. This brought him into conflict with the British government, led by *Winston Churchill*, which, just before the outbreak of the First World War, had bought a majority stake in the *Anglo-Persian Oil Company* with its concession in Iran. Churchill thought that if Mossadegh's move was allowed to set a precedent, British imperial power would be under threat across the globe. The result was the MI6/CIA-led coup against him. *Anglo-Persian* went on to become *British Petroleum*. "Twenty years later, when Middle Eastern oil producers, spearheaded by Libya's Colonel Gaddafi, nationalized oil assets and withheld supply to push up the oil price, BP might have recollected Mossadegh's terms with nostalgia" (Hilsum 2012). It was not only in Iran and Libya that there exist nationalist feelings about having a bigger say in a country's resources but also in Saudi Arabia where the Kingdom's first Western educated Oil Minister *Abdullah Al Tariki* tried to curb the powers of the so-called "Seven Sisters" US oil companies owning and running Aramco and was instrumental in co-founding OPEC with Venezuela's Oil Minister *Juan Alfonso* (Ibrahim 1997).

In today's world, IOCs are more concerned with ensuring they remain on the right side of antitrust laws, as a high degree of market concentration among the few remaining "supergiants" would set off antitrust action and massive law suits. This evolved nature of the IOCs and their new international priorities could create a situation for possible joint cooperation with OPEC and will be explored later.

OPEC's founding charter is based on the traditional notion of state sovereignty. However not all countries see OPEC as a full-fledged international body enjoying legal immunity normally granted to sovereign bodies. The Swiss authorities perceived OPEC as a mere commodity producer's agreement and failed to grant it diplomatic status, and OPEC's headquarters was moved to Vienna, Austria, following an agreement between the Republic of Austria and OPEC in 1965, granting the latter diplomatic immunity. Some authors have noted that the concept of "state sovereignty" was somewhat an elastic one, for when OPEC was formed, *Kuwait* one of its founding members was not even recognized as an independent state by the international community. Similarly when both *Qatar* and *Abu Dhabi* became a member in 1961 and in 1967, respectively, they were not independent members of the United Nations (Cuervo 2008, p. 12).

Given the new challenges facing the organization, questions arise, some of which are:

1. How can OPEC's bureaucracy, and specifically its Secretariat, meet challenges of the international oil industry in a more efficient and proactive manner?
2. Do the economic and energy realities of the twenty-first century justify a rethinking of OPEC's key objectives and role?
3. Are OPEC's goals and the structure of the organization outdated and unable to meet the emergent energy trends, and which objectives and structural mechanisms need to be updated or even discarded?

Table 8.2 sets out the challenges of amending some of OPEC's key statutes to enable the organization to address issues such as improving members' domestic business climate, having access to modern technology, playing a role in the formation of market prices, building a reliable and transparent market, and international cooperation with once antagonistic IOCs.

Some of the above-suggested proposals for change might very well go against the inherent grain of national sovereignty that underpins OPEC's founding charter and the resultant adoption of rigid political stances. However, even the most seemingly ideologically opposed antagonists sometimes seek compromises and get together to do business as another "historic" political compromise took place when *US President Obama and Cuba's President Raul Castro* decided to restore relations

Table 8.2 Harmonizing key OPEC founding statutes with twenty-first-century challenges and needs

OPEC statute	Issues and possible changes
I. Stabilization of oil prices in international markets	• Joining or monitoring price-setting bodies such as *Platts* to ensure that OPEC data are reflected in daily market prices by reporting agencies
	• Independent audits of OPEC reserves and production by institutions such as *JODI*
II. Securing steady income for producing countries	• Ensuring economic replacement of exhausted reserves and spare capacity though cost-effective national and international tenders
	• Independent audits of reserves
	• Diversification of integrated energy projects across both OPEC and non-OPEC producers, as well as consumer nations across all the energy chain
III. Securing an efficient, economic, and regular supply of petroleum to consumer nations	• Coordination between OPEC, non-OPEC producers, consumer nation's data collection agencies such as *IEA*, *EIA*, *JODI* to ensure transparent and timely data is available to forecast accurate supply inventories
	• Establish joint strategic petroleum reserves in consuming nations to ensure alignment of interests
IV. Seeking a fair return on capital investments in the petroleum industry	• Open up local financial markets for international capital flows and allow long-term IOC joint ventures
V. Permanent sovereignty over natural resources	• Open up national energy sectors to privatization and joint foreign ownership
	• Ensure that disputes are settled by a combination of local and international laws while confirming the state's power of expropriation, but only under specific terms and conditions

in April 2015 after a long freeze (Spetalnick and Trotta 2015) with the two countries reopening their embassies in August 2015.

The stabilization of prices in the international oil markets remains one of OPEC's key organizational purposes and becomes particularly acute when nonindustry players such as hedge funds and financial futures players affect the price through speculation. Those entrusted to produce "fair" market prices sometimes do not shoulder this responsibility as the *LIBOR* fixing scandal revealed in the aftermath of the global financial crisis of 2007–2008. This forced regulatory bodies to oversee the process and not leave it solely in the hands of market players. Similarly, OPEC given its still significant contribution of around 30 million plus daily crude output should be in a strong position to request that its qualified representatives maintain a presence in oil price reporting settings to provide additional producer supply data if required and so avoid undue speculative pressure building up because of opaque data flows from key producers. Most OPEC members are still single commodity earners and fluctuations in oil prices are critical for the well-being of their economies.

Determining a single driver of international oil prices is *not* easy, as some of the required factors for market pricing includes information which might be insufficient or withheld, such as refining capacity, low market inventories, and "geopolitical" tensions which might be magnified by oil traders due to lack of information from the producer countries affected. Uncertainty about OPEC production levels is also another element, with the OPEC Secretariat partly relying on "secondary sources" to find out what its own members are producing. This adds to overall uncertainty and price volatility. For OPEC to be taken seriously and have a "seat" in global price fixing, some significant efforts at obtaining more transparent OPEC members' data have to be realized. Given the fact that the bulk of oil produced in the current market is non-OPEC, the organization decided to suspend its so-called "price band" targets (or *OPEC Reference Bands*) in 2005, after noting that oil prices were mostly outside these OPEC reference bands and were "unrealistic," and instead decided to use production targeting to balance global supply and demand (Mabro 2006, p. 91).

Getting OPEC to agree on a common strategy on international oil pricing policies is not as easy as it may seem. This is illustrated by the different opinions expressed by OPEC's previous senior officials, with its First Deputy Secretary-General *Dr. Fadhil Al-Chalabi* saying that "….in my opinion there has been a substantial transformation of the structure of the oil industry towards making the price. OPEC now has only one role to play. When prices are down it cuts production. That is all it can do. In the 1980s, OPEC was the real price-maker. Every producer was following OPEC. Now producers are following *WTI* and *Brent*." However, OPEC's Second Deputy Secretary-General *Dr. Ramzi Salman* felt that OPEC should set some sort of price anchor for the market by stating that "…OPEC will have to decide, sooner or later what it wants, and let the world know clearly that this is its objective, its target. It has to have two levels again like when it had $22–$28 pb for the price band…. But then we had the price collapse after that, because of other elements. Fundamentals no longer rule the market. Therefore, it is not a fundamental thing just to cut production. In declaring its target, OPEC will have to go for a minimum and maximum price. The minimum will not be set by OPEC. It will be set by others….OPEC will have to set the ceiling" (OPEC 2009a, b, pp. 68, 70).

Table 8.2 raises the issue of opening up OPEC's energy sector to foreign participation, especially for those financially distressed members who either do not have the capacity or access to global capital markets like some of the larger IOCs. In this respect, OPEC's accomplishments regarding its principal aim of coordinating the external petroleum policies of its members have been rather poor, and OPEC has not been a major source of research to produce comprehensive documents on what constitutes model petroleum codes nor oil and gas investment laws. Concerning potential foreign joint venture contracts, these have been left to national governments to draw up, instead of OPEC taking the lead and producing model granting contracts, model international treaties to govern supply and demand issues, or even contractual issues for allowing upstream foreign investments. The lack of such guiding OPEC documentation has led to important differences among OPEC members regarding their oil and gas legal systems compared with other multilateral organizations like the International Monetary Fund (IMF), World Bank, and the EU, where their Secretariat's research outputs are invaluable resources, especially to poorer members.

OPEC: Birth of a New Club

While some view OPEC as a "political club," with perceived OPEC membership as a signal of some status and prestige, especially for smaller OPEC members, others view the organization as a perpetuation of a "rational myth" of its perceived cartel power. It is not necessarily true that OPEC has no useful role to play. By assessing how the organization came into being, its structure, and the challenges it faces, a "new" OPEC can be born that can make a positive contribution not only to its members if they decide to stick together but also to many others in the world and dispel the pessimistic conclusion of some who state that "whatever OPEC's future, it has very little importance any way…. In other words, the price of oil is still being determined as it has been for at least the last thirty years by the demand for oil and by its long run supply cost" (Johany 1980, pp. 95, 97).

Concerning whether OPEC members might stay together and even benefit the larger ones, specifically in what Saudi Arabia has to gain being a member of OPEC, the same author is even more damning by stating that, "today, there is no doubt in my mind that Saudi Arabia would have been better off without OPEC. The absence of OPEC may not change Saudi oil policy, in regard to rates of oil output, but will allow it to change the 'world price' for its oil. Many Saudi officials know that OPEC is neither monopolizing the world oil market nor in any other way keeping the price of oil higher than what it would have been without it, but they still cannot imagine a world without OPEC. *The reason: they think OPEC provides stability to the world oil market*" (Johany 1980, p. 98). Whether senior Saudi oil officials or the political leadership harbors such thoughts that Saudi Arabia is better off without OPEC, they keep this private to themselves, as there have not been any public statements to support such sentiments. This is despite the frustration the Kingdom has evidenced over quota busting and cheating and the loss of its "market share" when it was called upon to act as the organization's "swing producer" to prop up prices. This has left a deep scar among Saudi oil policymakers with a vow of "never again," irrespective of the damage it might do to some higher cost and inefficient OPEC producers.

Before proposing some changes on how OPEC should operate, or its "new club rules," it is worthwhile to try and understand what the perceived benefits are of OPEC membership to both the larger and smaller members, as these could be important elements in determining whether current members stick together or not:

1. Larger members, especially Saudi Arabia, will still want OPEC to survive despite personal misgivings, as it suits the Kingdom's policy and international agenda as

a global moderating influence, by using the organization as a "collective voice" and a decision-making platform, with Saudi Arabia not being seen *acting alone* in global energy matters, thus taking all the blame.
2. Smaller OPEC members will also use OPEC for their different domestic or regional political agendas, in that they are part of a larger "collectively strong" organization and so are punching above their weight in diplomatic settings.
3. Some members are in OPEC because they have foreign IOCs operating in their countries, and these members can use whatever "official" OPEC production quotas they have agreed to as leverage to curtail production agreements with the IOCs whenever it suits them by citing OPEC as an excuse.
4. OPEC is quite important to many of the poorer member countries because of the deemed value of the OPEC Secretariat in providing the organization with *global* information and energy data trends which they might not have access to if they were not members. Such information sharing is very important in the oil industry where precise information is often hard to obtain. Smaller members also benefit whereby OPEC provides them with a forum where all members could share strategies for dealing with the IOCs as well as predictions about the oil market.
5. For all its shortcomings, OPEC is probably *the only* energy platform for some collective *decision*-making when the time for action is needed and a collective voice is to be heard, even in the face of a diminishing market share. The collective OPEC November 2014 decision not to cut production in the face of falling oil prices is a striking example of a collective decision, despite individual members' misgivings.
6. For many members, especially the smaller ones, OPEC membership is cost-effective—it costs around $1.8 million per annum in membership fees, with the perceived value of being a member of a "super club" far outweighing the monetary cost, although *Gabon* left OPEC in 1994 reputedly because it was unable to obtain a special dispensation to reduce its annual membership fee and was already in arrears for its 1994 dues (New York Times 1995).
7. Smaller OPEC members could also receive financial support from OPEC's specialized funding arm—the *OPEC Fund for International Development* (*OFID*) which was established in 1976 to promote cooperation between member states of OPEC and other developing countries. To date, OFID's total approved commitments distributed through four financing mechanisms (public sector loans, private sector loans, trade finance operations, and grants) stood at $17.54 billion as at end of September 2014 (OFID 2015). This is further illustrated in Fig. 8.6. Not only do the smaller OPEC members benefit from OFID support, but the larger OPEC members who make up the bulk of funding to OFID derive some global recognition for their *corporate social responsibility* (CSR) actions and help to dispel the notion that OPEC is simply engaged in transferring exorbitant oil rents from consumer to producer nations.

Fig. 8.6 OFID: OPEC's global corporate social responsibility in action. (**a**) Number of approved operations (total 3229). Beneficiary countries (total 134). (**b**) Sectoral distribution of public sector loans (% by value). (**c**) Number of beneficiary countries—total of 134

a Number of Approved Operations (Total 3,229)   Beneficiary countries (Total 134)

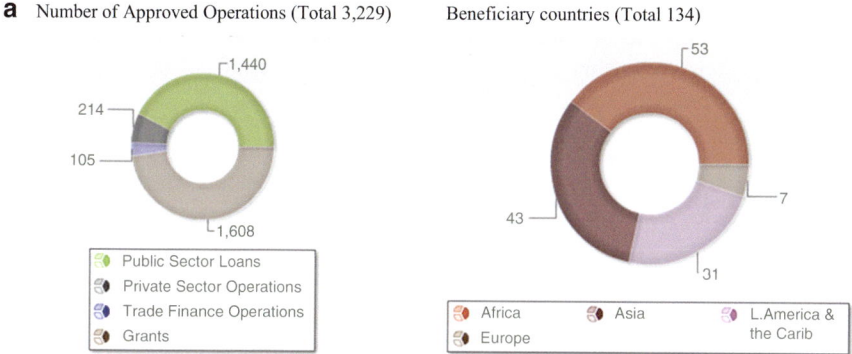

b Sectoral distribution of public sector loans (% by value)

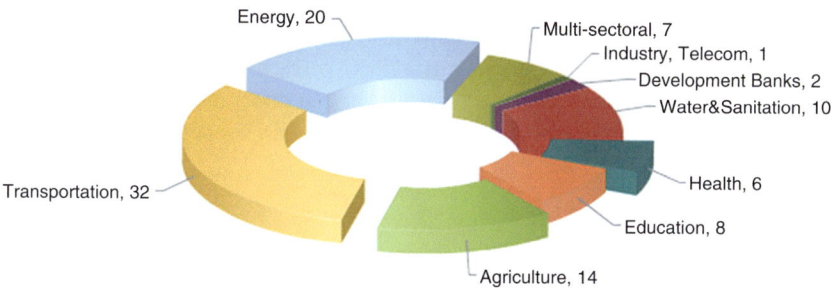

c Number of Beneficiary countries – Total of 134

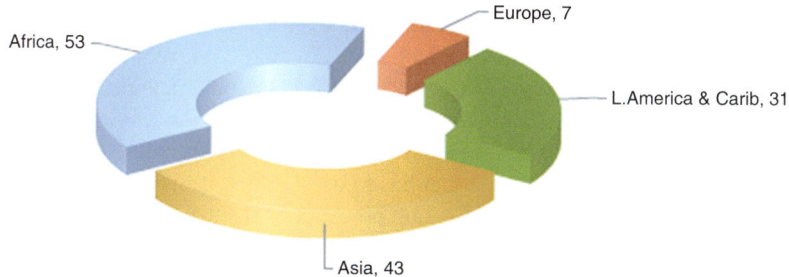

Source: http://www.ofid.org/HOME/OFID-Profile

As illustrated in Fig. 8.6, countries across the world have benefited from OFID including some in Europe, but with the majority being in Africa, Latin America, and Asia. By sector, energy and transportation, as well as agriculture, accounted for around 66 % of total disbursements. The emphasis on energy is in line with OPEC's mandate, when during the OPEC Summit held in Saudi Arabia in 2007, OFID was mandated to align its program with *energy poverty eradication*, with an *energy for the poor* initiative. In its 2012 declaration, OPEC committed a minimum revolving $1 billion facility which was announced at the Rio Summit (OFID 2015).

With such commitments to the welfare of the less advantaged nations of the world, and the tangible and intangible benefits accruing to all OPEC members irrespective of their size, what else can OPEC do to "re-brand" itself as a modern, proactive club ready to take on future challenges?

OPEC: Re-branding the Institutional Survivor

OPEC has become a part of the global international brands. Despite reduced market share and competition from new energy sources, it still facilitates dialogue and helps to educate officials and ministers from the less advanced member countries on how to deal with international issues. Its senior officials and past representatives are well respected in their field and are not afraid to speak their mind at international forums, often expressing divergent views on what the organization should be doing as illustrated earlier. This diversity is to be welcomed, but at the same time there must also be a sense of purpose on what OPEC needs to do to make it more effective, to avoid divisions between OPEC members, in effect making it the most *relevant* and *irrelevant* organization at the same time, as OPEC operates in an oil market which is the most politicized in the world.

OPEC's "Re-branding": Some Recommendations

(a) *Reconcile the differences between its members in terms of oil production, reserves, and economic needs.* The old formulas that were used to establish production quotas are no longer valid, and concepts such as existing and future spare capacity and investment spending are now more relevant to distinguish between members and their overall importance to the organization.
(b) *Abolish OPEC's conference unanimity rule of one member-one vote* or limit this to a few essential instances such as changes to the founding charter. All members are *not* equal and this fact has to be recognized by OPEC, as not all members of the IMF have equal voting powers but one that is based on contribution. Similarly, granting Saudi Arabia the same weight in voting as the smallest OPEC member might appeal to equalitarian principles, but is unjustified in practical terms.
(c) *Actively participate in environmental laws and sustainable development* as a strategic option and engage in serious international cooperation with interested stakeholders, instead of being reactive to global warming and fossil divestment pressure groups.
(d) *Change OPEC's agreed decisions to binding decisions on all members* as one of OPEC's characteristics has been the nonbinding effects of decisions taken,

with OPEC members using the organization to send "signals" to the market, while still maintaining full control over their "sovereign" production output.
(e) *Expand OPEC membership, whether on a "full" or "partial" basis to increase the diversity of producers*, while at the same time securing the involvement of non-OPEC producers in the policymaking mechanism so they feel that they are making an effective contribution. This requires creating *different membership levels* to the new OPEC club, whereby each level has different obligations and benefits.
(f) *Restructure the office of the Secretary-General* and grant it with true authority to bind member states and thus provide the organization as a whole with a degree of necessary autonomy from its members to act meaningfully and effectively, especially providing the Secretary-General with true powers that may allow him and his office to act during emergency situations without referring to the biannual OPEC Conference meetings.
(g) *OPEC's Board of Governors*: their current role might have to be reexamined from meeting twice a year to transform it to a *permanent body* conducting day-to-day work and in effect become a true legislative body that reviews and approves specific binding regulations. High-caliber national representatives with many years of experience in the upstream and downstream energy sector should be key candidates for such a changed role of the Board of Governors.
(h) *Appoint permanent OPEC "Ambassadorial" representatives to the most important consumer countries and to the IEA*, in effect to establish a new "OPEC Ambassadorial" category staffed by highly trained experts in the energy sector who will act as the organization's "eyes and ears" in the world's important consumer nations, as well as other emerging energy players in Latin America like Brazil, through organizations like the *Latin America Energy Organization (OLADE)* to ensure that OPEC is abreast on economic and political developments on a first-hand basis. This will ensure that the information flow back to OPEC is not biased or filtered through member states national representatives to suit particular countries' agendas. By having a global presence in all continents, OPEC then becomes a truly supranational organization with a global presence that will adequately address *global* problems.
(i) *Empower OPEC to enter into international binding agreements and treaties*. The organization is limited under its current statute in concluding treaties with states and other international organizations, and the statutes should be amended to enable OPEC to enter into binding international agreements and also to clarify the role of the Secretary-General to be able to do this, especially in negotiations with international conventions on climate change.
(j) *Play a more proactive role advancing the organizational "group's" interests over individual nation's interests*. Hiring qualified technocrats with specialized skills from nonmember countries, with career advancement based on merit rather than nationality, could raise the professional standards of OPEC Secretariat.
(k) *Raise OPEC's public relations and media profile* to ensure that its message is effectively heard to counter cartel-like accusations and to highlight the inequity

of energy pricing and the effect of national taxation on final consumer prices, as well as OPEC's aid and CSR to developing countries.

(l) *Create world-class centers for technological research* that could assist oil producers and institutions concerned with global warming and carbon capture storage to enable poor OPEC members "jump-start" such technologies in their own countries in a cost-effective manner and fund endowed academic chairs at leading international and member states' universities to carry out research in applied resource economics.

(m) *Cooperate in co-blending of heavy and light crudes* from different OPEC members to maintain a market share for supplying to medium crude oil grade refineries, in the face of increased Canadian heavy and US light oil crude production.

(n) *Increase support for JODI*. Provide sufficient financial and technical support to JODI (*Joint Organization Data Initiative*) based in Riyadh to enable a more proactive and credible energy center to emerge that becomes *the key reference point* not only for assessing global supply and demand data and changing conditions but with an additional focus on refining capacity expansion plans and renewable energy advances. In time JODI can become the global energy-related reference center, like the IEA and EIA today.

(o) *OPEC Fund for Joint Investments*. The organization can create inter member synergy and value added through establishing a specialized OPEC Energy Investment Fund, similar to the energy-related investment vehicle like *APICORP* established by the *OAPEC* Arab oil producers. These Arab oil producers' joint energy project investments are the only public face of inter-Arab oil producer cooperation that brings them together in a practical way. Establishing a joint OPEC investment fund can do likewise for the whole organization to participate in, whether they are big or small producers, based on their fiscal abilities.

(p) *Cross-border OPEC energy investments*. Currently there are no cross-border OPEC joint venture investments whether in upstream, midstream (transportation, pipelines), or downstream. Such inter-OPEC energy joint ventures, whether upgrading or building refineries to co-blend different types of crude oil produced by the member states, joint explorations in third-party countries, or joint acquisition of energy assets can bind OPEC countries together. This will be similar to the early days of the oil majors who cooperated tacitly in furthering their interests and jointly invested abroad—with *Aramco*, *Anglo-Persian Company*, and the *Iraq Petroleum Company* being good examples.

Conclusion

The above list of possible actions that a "new" OPEC can take is not an exhaustive one, but the challenges that the organization faces in 2015 require some immediate responses from its members. Above all, an honest dialogue between oil producers, whether OPEC or non-OPEC, and consumers is now even more important to avoid recurrent cycles of volatile oil prices that have destabilized the world energy markets several times in the past decade, with sharp overshoots and declines affecting long-term investment planning. Global energy players as a whole, irrespective on which side of the negotiating table they sit, are faced by the same pressing issues of energy security, global warming, and the eradication of poverty. Under its current structure and mistrust between members wielding different market power, OPEC is ill-suited to adequately handle such a comprehensive potential dialogue, especially with a US private sector shale energy revolution. While recognizing that politics can never be excluded from global business, there is a need for OPEC to evolve into a new platform that brings together "rationality" and reality into OPEC's decision-making process and ensures that, as much as possible, *explicit* as opposed to *implicit* interests are addressed by members. A revitalized and "re-branded" new OPEC could improve the chances of a long-term mutually beneficial dialogue with productive outcomes for generations to come. This type of strategy will be seen to be beneficial to non-OPEC members to cooperate with OPEC. It was interesting to note that even members who had left OPEC are now reconsidering joining the organization, when it was announced that Indonesia was seeking approval to rejoin OPEC in 2015, 7 years after leaving, and that the country would attend the June 2015 OPEC meeting as an observer. OPEC's initial reaction was positive, as OPEC termed Indonesia's departure as a "suspension," similar to Ecuador's which later rejoined OPEC in 2007, despite Indonesia's small oil output of under 900,000 bpd, half of the country's 1990s production peak, with very little of oil production going to exports, as Indonesia is now a net gas exporter (Reuters 2015^c). In June 2015, Indonesia gained approval from all OPEC members to rejoin the group, with Saudi Arabia providing strong support as the Kingdom and the UAE are also major suppliers of crude to Indonesia (Reuters 2015^d).

Closing Afterthought: OPEC—Has Its Strategy Succeeded?

On his first appearance in Vienna after almost 6 months since OPEC first implemented its new strategy of defending its market share in late November 2014, the Saudi Oil Minister assured reporters who were waiting for his arrival that the Saudi and OPEC policy is succeeding. Demand was increasing, supply in the market was slowing down, and things will be better over the second half of the year as they were "in the right directions," he said.

Indeed the market did stabilize in the second quarter of 2015 compared to the first quarter, but the market situation to which the minister had pointed out may seem misleading to lead to any definitive conclusion that the market-share strategy was successful. Following OPEC 167th meeting in June 2015, Barclays issued a report arguing that the available oil supply and demand data do not support Minister Naimi's statements. It added that "a more accurate verdict on the data, in our opinion, would be: demand is not picking up very much, supply growth has barely slowed at all, and things are still heading in the wrong direction for OPEC" (Norrish June 2015). A closer look is needed to understand the reality of the market situation now facing OPEC and hence any success of OPEC's new strategy.

Looking at the market conditions that prevailed over the first 8 months of 2015, there are four key issues that put a limit to the extent of the success of OPEC's strategy. *First*, demand grew unexpectedly in the second quarter of 2015 for seasonal reasons such as refinery delays and outages that cannot be sustained throughout the year to support the refinery margins which saw in 1Q15 demand surging to 1.7 mbpd, from an average 0.7 mbpd in 2014. Demand growth alone, however remarkable, could not have been the only source of oil price support, dwarfed as it was by a surge in global liquid supply to a towering 3.1 mbpd over the same period. *Second*, despite signs of a slowdown in non-OPEC supply, notably in the USA, global production growth remains exceptionally high, with Russian production reaching a record 10.7 million bpd. As a result, oil inventories have soared, but their breakdown by product and region doesn't quite match that of demand. More than the rise in demand itself, it is that mismatch between product supply and product demand that seems to have supported prices. In particular, gasoline prices have found support from robust US demand, leading to a surge in crack spreads. Clean tanker owners have been enjoying their moment in the sun amid surging product-shipping demand. Pockets of product tightness in effect seem to have helped support the oil complex, pulling crude prices along (IEA June 2015[d]). The fall in shale oil production and from other high-cost producers was not as huge as warranted by the 50 % fall in oil prices, with US shale output forecasted to decline by less than 100,000 bpd as of July 2015, according to the US EIA latest *Drilling Productivity Report* (EIA 2015[b]), and with Standard Chartered Bank forecasting total US shale production to reach 5.15 million bpd by December 2015, a fall of 191,000 bpd compared with December 2014 production levels (Horsnell et al. 2015). *Third*, the surplus in the market was still lingering despite the fall in prices and the marginal fall in production, and output is still expected to increase from shale oil and from other OPEC producers as prices stabilized at around $60–$70 toward the end of the first half of 2015.

Fourth and more important is that OPEC's countries lost around 50 % of their income over the first half of 2014 with no avail as the imbalance in the market persists. For 2014, the US EIA estimates that, excluding Iran, members of the Organization of the Petroleum Exporting Countries earned about $730 billion in net oil export revenues (unadjusted for inflation). This represents an 11 % decline from the $824 billion earned in 2013, largely because of the decline in average annual crude oil prices and to a lesser extent from decreases in the amount of OPEC net oil exports. This was the lowest earnings for the group since 2010. However, for 2015, the EIA projects that OPEC net oil export revenues (excluding Iran) could fall further to about $380 billion in 2015 (unadjusted for inflation) as a result of the much lower annual crude oil prices expected in 2015 (EIA March 2015[a]) a loss of $350 billion. These revenue losses are in stark contrast with the small fall in US and other non-OPEC production, *as in a financial comparison, the losses to shale oil producers during 2015 of around $5 billion amounts to less than 1.5 % of the estimated*

OPEC revenue losses, with the new market-led strategy feeling more like a *Pyrrrhic* victory for OPEC in the short term.

The only parties that seemed to have benefited from the entire new OPEC strategy are countries that stepped up their purchases to fill their strategic oil reserves such as India and China, who were able to secure more volumes at half of the oil cost compared to first half of 2014. In other words, cheap OPEC crude was going to fill storage tanks in many countries until the tanks were becoming full. The amount of oil going into strategic reserves in many developing countries is harder to assess and that complicates the real oil demand picture for 2015. Another interesting point is that oil stock levels remained high in the first half of the year contrary to the fact that the *contango* was getting narrower, indicating a future price fall.

Many will agree that OPEC can still play a useful role but stress that the organization has lost its effectiveness as an oil producer group and that it will continue to exist, but not with the same effectiveness as in the past. It will lose production to competitors because of depletion rates and lack of investment and an inability in maintaining current production capacity despite the fact that key members of OPEC are *the* low-cost oil producers who can out survive high-cost rivals. In the final analysis, as a respected energy executive stated "OPEC must look *inside* OPEC to set itself a high standard for the whole world on technical, economic and political factors affecting all, as without it the energy world is going to remain chaotic" (Al Husseini 2015). The alternative is for OPEC to wither away in acrimony, with several competing bodies emerging, weakening the whole body.

For the organization to effectively compete, the new Saudi market-led strategy has to succeed and succeed quickly given the fiscal stress faced by many OPEC members. However, all indications are that it has produced some mixed results. From all indications, the Saudi policy is not expected to easily change and reverse direction after reaffirming that it will not, and will not OPEC act alone in any production cuts. Given that the market share strategy has been announced and endorsed by the Saudi cabinet, any reversal in this policy would need political pressure and a high degree of coordination with other producers. In the final analysis, the events of 2014–2015 has demonstrated that OPEC has no clear ideas about the direction of global oil prices and they were not prepared for such an eventuality. Their internal studies and world demand-supply forecasts were not thorough enough to understand the complex world of non-OPEC production. OPEC was now left to manage a moving target of when world demand and supply would actually balance, and in 2015 the target date for such a balance was pushed back further to 2016 and even 2017. In order to be able to assess the success of the new OPEC strategy, at least one and possibly two full years has to elapse since the start of the implementation as a self-correction mechanism for the market is a long-term process. The problem with such long-term corrections is found in the apt words of Lord John Maynard Keynes in which he said: "but this long run is a misleading guide to current affairs. In the long run we are all dead. Economists set themselves too easy, too useless a task if in tempestuous seasons they can only tell us that when the storm is past, the ocean is flat again."

References

Abdel-Rouf, M. (2015) "Environmental Integration is a Prerequisite for GCC Unity". *Gulf Research Centre*. 27 February 2015. http://www.grc.net

Abdulaziz bin Salman, HRH. (2014) "Keynote Speech" Presented at the *Arab Gulf and Regional Challenges Forum Organized by Institute of Diplomatic Studies*. Riyadh, 17 September 2014.

Abdulaziz bin Salman, HRH. (2015) "Keynote Speech Climate Change" Presented at the *International Experts Workshop on Carbon Management and its Implications*". Al Khobar, Meridian Hotel, 27 April 2015.

Abdullah, A.K. (2012) "Repercussions of the Arab Spring on the GCC States". *Arab Center for Research and Policy Studies*. Qatar.

Adams, C. (2015) "North Sea Oil and Gas Drains Cash at fastest Rate Since 1970's". *Financial Times*, London. 24 February 2015.

Aissaoui, A. (2010) "GCC Oil Price Preferences: At the Confluence of Global Energy Security and Local Fiscal Sustainability", in Energy Security in the Gulf: Challenges and Prospects". *The Emirates Center for Strategic Studies and Research*. Abu Dhabi. pp. 107-133.

Aissaoui, A. (2013). "Modelling OPEC Fiscal Break-even Oil Prices: New Findings and Policy Insights". *APICORP Research*, Economic Commentary. 8 (9/10). September/October 2013.

Aissaoui, A. (2014[a]) "OPEC in the Future: Will it Continue to Play a Pivotal Role?". *APICORP Research*, Economic Commentary. 9 (10). October 2014.

Aissaoui, A. (2014[b]) "Revisiting the Dynamics of the Forward Crude Oil Price Curve". *APICORP Research,* Economic Commentary. 9(11). 11 November 2014.

Aissaoui, A. (2014[c]) "The Arab Energy Investment Outlook: Opportunities, Constraints and Policies". *Tenth Arab Energy Conference*. Abu Dhabi, UAE. 21-23 December 2014.

Al Ansary, K. and Ajrash, K. (2015) "Iraq Crude Exports Rise 15% in March to Highest in 35 Years". *Bloomberg*. 04 April 2015.

Al Dossari, N. (2015) In speech at the 18th Annual Meeting of the *Saudi Economic Association*, Riyadh on April 8, 2015.

Al Fathi, S. (1990) "Relations Between OPEC and Non-OPEC Oil Producers". *OPEC Review*. 14 (1). Spring.

Al Hajji, A. (2010) "Dimensions of Energy Security: Competition, Interaction and Maximization" in Energy Security in the Gulf: Challenges and Prospects". *The Emirates Centre for Strategies Studies and Research*. Abu Dhabi, UAE. pp. 191-218.

Al Hajji, A. (2014[a]) "Shale Revolution has Direct and Indirect Impacts on OPEC" *World Oil.* January 2014.

Al Hajji, A. (2014[b]) "OPEC's Newest Production Cut: A Surprise or Signal?" *Middle East Economic Survey*. 1st March 2014.

Al Harthi, Awad. (2015) *"Personal Interview"*. Dr. Awaad Al Harthi, Advisor, Ministry of Petroleum on Climate Change', Khobar, 28 April 2015.
Al Hayat Newspaper. (2013) An interview in Arabic under the title "Naimi: Protecting Oil Prices is the Responsibility of all OPEC Countries and Not Only Gulf". *Al Hayat*. 6 December 2013.
Al Husseini, Sadad. (2015) *"Personal Interview"*. Former Senior Aramco Executive. Al Khobar, Saudi Arabia, 27 April 2015.
Al Kadiri, R. (2010) "In the eye of the Beholders: Global Consumption, Uncertainties and Energy Security in the Gulf States". In "Energy Security in the Gulf: Challenges and Prospects". *The Emirates Centre for Strategic Studies and Research. Abu Dhabi, UAE*. pp 71-92.
Al Khowaiter, Othman. (2015) *"Personal Interview"*. Former senior Aramco Manager. Dhahran, Saudi Arabia, 26 April 2015.
Al Madi, M. (2015) *"Speech at GCC Petroleum Media Summit"* Riyadh, Saudi Arabia. March 4, 2015.
Al Mashaabi, D. (2015) "Saudi Arabia Separates Aramco from Oil Ministry: *Arabiya*". *Bloomberg*. 30 April 2015.
Al Moneef, M. (2011) In paper under the title *"Saudi Economy and the Future of Energy"* presented at Junadriyah National Festival, Riyadh on April 18, 2011.
Al Moneef, M. (2012) *"Personal interview with Majid Al Moneef"*. Riyadh, Saudi Arabia. 20 April 2012.
Al Muhanna, I. (2015) In speech at the 18th Annual Meeting of the *Saudi Economic Association*, Riyadh on April 9, 2015.
Al Naimi, A. (2008) In speech at Jeddah Energy Meeting, *Jeddah*, Saudi Arabia on June 22, 2008.
Al Naimi, A. (2012) *Correspondents Press Meeting*, Doha, Qatar,2012.
Al Naimi, A. (2013) Transcript of "A Conversation with His Excellency Ali al-Naimi, Minister of Petroleum and Mineral Resources, Kingdom of Saudi Arabia". *Centre for Strategic and International Studies, Washington DC*, on April 30, 2013.
Al Naimi, A. (2014) In speech at the *10th Arab Energy Forum*, Abu Dhabi, UAE on December 21, 2014.
Al Naimi, A. (2015) Speech at a Conference at the *Federal Academy for Security Policy* in Berlin on 4 March 2015.
Al Nasrawi, A. (1985) *"OPEC in a Changing World Economy"*. John Hopkins University, Baltimore, USA.
Al Sahlawi, M. (2010) "Energy Security in the Gulf and the US Dollar: An Overview" in "Energy Security in the Gulf: Challenges and Prospects". *The Emirates Centre for Strategies Studies and Research*. Abu Dhabi, UAE. Pp. 93-105.
Al Saif, Waleed. (1996) *"OPEC Quota System"*. Ph.D. Thesis, Claremont Graduate School, USA.
Alike, E and Okafor, C. (2015) "Algeria in talks with Nigeria, Angola, over oil price slump". *Bloomberg,* 18 March 2015.
Amott, J. (2015) "Final Accord to Pave way for Iran to Restore Oil Market Share: IRNA". *Bloomberg*. 4 April 2015.
Amuzegar, J. (1999) *"Managing the Oil Wealth OPEC's Windfalls and Pitfalls"*. I.B. Tauris, London, UK.
Anderson, H. (1988) *"Aramco, the United States, and Saudi Arabia: A Study of the Dynamics of Foreign Oil Policy, 1933-1950"*. Princeton University Press, USA.
Anderson, I. (1981) *"Aramco, the United States, and Saudi Arabia: A Study of the Dynamics of Foreign Oil Policy, 1933-1950"*. Princeton, Princeton University Press, USA.
Andrews Speed, A. (2010) "Asia's Energy Demand and Implications for the Oil-Producing Countries of the Middle East", in *"Energy Security in the Gulf: Challenges and Prospects"*. *The Emirates Centre for Strategic Studies and Research. Abu Dhabi, UAE*. pp 137-169.
Arab News. (2013) "Saudi Arabia Unconcerned by US Shale Output". *Arab News*, Jeddah. 21 November 2013.
Arab News. (2014[a]) "Al Naimi: OPEC 'Must Combat US Shale Boom'". *Arab News*, Jeddah. 29 November 2014.

Arab News. (2014^b) "No Conspiracy behind Oil Price Fall: Al Naimi". *Arab News*, Jeddah, 22 December 2014.

Arab News. (2014^c) "OPEC will not cut Output even at $20". *Arab News,* Jeddah, 24 December 2014.

Arab News. (2014^d) "Riyadh: Zero Emission Goal not 'realistic'". *Arab News*. Jeddah. 10 December 2014.

Arab News. (2015) "Gulf Economies have failed to Diversify". *Arab News*, Jeddah. 25 February 2015.

Armental, M. (2015) "Fitch changes Saudi Arabia's IDRs Outlook to Negative". The Wall Street Journal. New York. 21 August 2015.

Asharq Al-Awsat. (2014^a) Interview in Arabic with former Qatar oil minister Hamad Al-Attiyah. *Asharq Al-Awsat*. 22 November 2014.

Asharq Al-Awsat. (2014^b) Story in Arabic under the title "The Saudi Dragon in OEPC". *Asharq Al-Awsat*. 12 November 2014.

Baffes, J, Kose, M, Ohnsorge, F. and Stocker, M. (2015) "The Great Plunge in Oil Prices: Causes, Consequences, and Policy Responses". World Bank. *Policy Research Note*. March 2015.

Baker Hughes. (2015) "Rotary Rig Counts". http://www.bakerhughes.com/rig-count.

Baker Institute Policy Report. (2014) "Navigating the Perils of Energy Subsidy Reform in Exporting Countries". *James A. Baker III Institute for Public Policy*, Rice University, USA. 58. May 2014.

Bank of America Merrill Lynch. (2014^a) "Saudi Puts No More". Bank of America Merrill Lynch Research, *Global Energy Weekly*. 4 December 2014.

Bank of America Merrill Lynch. (2014^b) "The End of OPEC". Bank of America Merrill Lynch Research, *Global Energy Weekly*. 27 November 2014.

Bank of America Merrill Lynch. (2015) "How Low can Oil Go?" *Global Energy Weekly*. 6 January 2015.

Barclays. (2015) "OPEC: the 'happy meeting'". *Barclays Commodities Research*. Oil Market Outlook. 8 June 2015.

Barnett, J, Dessai, S, and Webber, M. (2004) "Will OPEC Lose from the Kyoto Protocol?" *Energy Policy*. 32. Pp. 2077-2088.

BBC. (2015^a) "US House passes Controversial Keystone Pipeline Bill". *BBC News*. USA and Canada. London. 9 January 2015.

BBC. (2015^b) "Climate Change: China Official Warns of 'huge impact'". *BBC*. London. 22 March 2015.

BBC. (2015^c). "Pope Francis encyclical calls for end to fossil fuel." BBC, London. 18 June 2015.

Beblawi, H. and Luciani, G. (ed.) (1987) "*The Rentier State*". Croom Helm. London.

Bentzen, J. (2007) "Does OPEC Influence crude Oil Prices? Testing for Co-movements and Causality Between Regional Crude Oil Prices". *Applied Economics*. Vol. 39(11). pp. 1375-1385.

Bernstein Research. (2014) "Bernstein Energy: A Good Sweating—2015 Global Upstream Capex to Collapse". *Bernstein Research*, USA. 12 December 2014.

Bershidsky, L. (2015) "OPEC's Production Limit just Ceased to Exist: Leonid Bershidsky". *Bloomberg*. 5 June 2015.

Bierman, S. (2015) "Lukoil Says Aramco in talks to boost JV Stakes to 50 %." Bloomberg, 24 June 2015.

Blas, J. (2015^a) "Oil-Rich Nations burn through Petrodollar Assets at Record Pace". *Bloomberg*. 13 April 2015.

Blas, J. (2015^b) "Shale's Resilience Vindicates Conoco Boss on Return to OPEC". *Bloomberg*. 4 June 2015.

Bloomberg Businessweek. (2014) "Why Russia Said 'No Deal' to OPEC on Cutting Oil Production". *Bloomberg Businessweek*. 27 November. 2014.

Bloomberg. (2009) "Putin Blinking on Exports Signals Lower Oil for OPEC". *Bloomberg*. 8 September 2009.

Bloomberg. (2011) "Saudi Arabia Sees 'No Shortage' of Oil, Says $80 'Fair'". *Bloomberg*, Riyadh. 22 February 2011.
Bloomberg. (2013) "Naimi Says $100 Crude Oil is Reasonable Price". *Bloomberg*. 18 March 2013.
Bloomberg. (2014^a) "OPEC Takes No Action to Ease Supply Glut as Crude Oil Slumps". *Bloomberg*. 27 November 2014.
Bloomberg. (2014^b) "UAE Sees OPEC Output Unchanged Even of Oil Drops to $40". *Bloomberg*. 15 December 2014.
Bloomberg. (2014^c) "Iraqi-Kurdish Accord Lifts Exports, Blunts OPEC Quota: Bear Case". *Bloomberg*, 3 December 2014.
Bloomberg. (2014^d) "Oil Rebounds as Libyan Conflict Offers Relieve From Global Glut". *Bloomberg*, 29 December 2014.
Bloomberg. (2014^e) "Oil at $40 Possible as Market Transforms Caracas to Tehran". *Bloomberg*, 30 November 2014.
Bloomberg. (2015^a) "Iran Can Add 600k B/D to Capacity If Sanctions Lifted: Husseini". *Bloomberg*. 31 March 2015.
Bloomberg. (2015^b) "Iran's Oil Output Capacity Will Rise to 3.96m B/D by March 2016". *Bloomberg*. 7 May 2015.
Bloomberg. (2015^c) "Venezuela Working With OPEC to Stabilize Oil at $100/bbl". *Bloomberg*. 15 May 2015.
Bloomberg. (2015^d) "U.S. Reduces 2015 Oil-Output Forecast in Short-Term Outlook". *Bloomberg*. 12 May 2015.
Bloomberg. (2015^e) "BP CEO Says Consolidation May Happen If Oil Prices Stay Low". *Bloomberg*. 21 April 2015.
Bloomberg. (2015^f) "IEA Sees China, India Filling Strategic Reserves with Cheap Oil". *Bloomberg*. 13 March 2015.
Bloomberg. (2015^g) "Iran Joins OPEC Member Libya Calling on Group to Cut Oil Output". *Bloomberg*. 14 April 2015.
Bloomberg. (2015^h) "Oil Buyer's Guide". *Bloomberg Briefs*. 17 January 2015.
Bloomberg. (2015^i) "Oil Industry May Cancel $1T in Investment on Price Fall: Aramco". *Bloomberg*. 9 March 2015.
Bloomberg. (2015^j) "Oil Is Now a Free Market for First Time: U.S. Energy Envoy". *Bloomberg*. 7 May 2015.
Bloomberg. (2015^k) "Bakken County Breakevens Illuminate Oasis Rig Allocation Plans". *Bloomberg*. 8 January 2015.
Bloomberg. (2015^l) "OPEC's El-Badri Speaks in Bloomberg TV Interview". *Bloomberg*. 22 January 2015.
Bloomberg. (2015^m) "Why OPEC is Talking Oil Down, Not Up, after 48% Sell Off". *Bloomberg*, 12 January 2015.
Bloomberg. (2015^n) "Bakken Break-Evens, Output Cast Doubts on High Cost of Drilling". *Bloomberg*, 21 January 2015.
Bloomberg. (2015°) "Oil's Plunge Leaves $27 Billion of Energy Bonds Priced for Junk". *Bloomberg*, 8 January 2015.
Bochem, S. (2004) "Cartel Formation and Oligopoly Structure: A New Assessment of the World Crude Oil Market". *Applied Economics*. 36. pp. 1355-69.
Bolin, N. (2008) "A History of the Science and Politics of Climate Change: The Role of the Inter-Governmental Panel on Climate Change". *Cambridge University Press*. Cambridge. 2008.
BP. (2014) "*Statistical Review of World Energy*". June 2014. www.bp.com.
BP. (2015) "*Statistical Review of World Energy*". June 2015. www.bp.com.
Bremond, V, Hache, E. and Mignon, V. (2002) "Does OPEC Still Exist as a Cartel? An Empirical Estimation". *Energy Economics*. Vol. 34 (1), pp. 125-131. January 2012.
Business Monitor International. (2013) "*Latin America Monitor—Brazil Business Fundamentals*". October 2013 Analysis.
Cairns, R. and Calfucura, E. (2012) "OPEC: Market Failure or Power Failure?". *Energy Policy*. 50. pp. 570-580. November 2012.

Carey, G. and DiPaola, A. (2015) "Saudi Aramco Capital Investment to Drop more than Expected". *Bloomberg.* 27 January 2015.

Caroom, E. (2015) "EIA Over-estimated April Oil Output by 1.1 mbpd, Verlegar Says". *Bloomberg.* 11 May 2015.

Carpenter, C. (2015) "Oil Companies Slowing Capex Seen by Citi Cutting 1M b/d Surplus". *Bloomberg.* 14 January 2015.

Carpenter, C and Khan, S. (2015) "U.A.E. removes fuel subsidy as oil drop hurts Arab economies." Bloomberg. 22 July 2015.

Carroll, J. (2015) "Riskiest Oil Fields Crushed by Plunge—Cripple Future Growth". *Bloomberg*, 8 January 2015.

Cawthorne, A. (2015) "Venezuela Proposes Novel OPEC Blending Deal to Fight Market Share". *Reuters.* 20 April 2015.

CGES. (2011) "Saudi Arabia's Oil Capacity: Can it Deliver?" *CGES-Centre for Global Energy Studies.* London.

Chalabi, F. (1980) *"OPEC and the International Oil Industry: A Changing Structure"*. Oxford University Press.

Chalabi, F. (1989) *"OPEC at the Crossroads"*. Oxford: Pergamon Press.

Chalabi, F. (2010) *"Oil Policies, Oil Myths: Analysis and Memoir of an OPEC Insider"*. Tauris Press, UK.

Cheong, S. (2015) "Iran Oil exports may Rise 200K-300K B/D on Nuke Deal: Barclays". *Bloomberg.* 6 April 2015.

China Customs General Administration. (2015) *Monthly oil imports data.* January 2015. http://www.chinacustomsstat.com/

Chmaytelli, M. (2015) "Saudi's Naimi Says Oil Production is Sovereign Right". *Bloomberg.* 5 June 2015.

Cho, S. and Zhu, W. (2015) "Saudi Arabia seen giving bigger Oil Discounts to Asian Customers". *Bloomberg.* 3 February 2015.

Citi Research (2014). "Catching the Knife—Finding Oil's Short-Term Equilibrium". *Citi Research Commodities.* 18 December 2014.

Citi Research. (2015) "Alphabet Soup—L, U,V, or W? What's the Likely Shape of the Oil Price Recovery?" *Citi Research Commodities.* 9 February 2015.

Claes, D. H. (2001) *"The Politics of Oil-Producer Cooperation: The Political Economy of Global Interdependence"*. Westview Press, USA.

Colgan, J. (2012) "The Emperor has no Clothes: The Limits of OPEC in the Global Oil Market". Retrieved from www.uni-heidelberg.de/md/awi/peio/colgan_13.09.2012.pdf.

Colgan, J. (2014) "OPEC: The Phantom Menace". *Bloomberg.* 16 June 2014.

Corden, W. (1984) "Booming sector and Dutch disease economics Survey and consolidation ". *Oxford Economic Papers.* No. 36. pp. 359-380.

Corrigan, P. (2010) "Editorial Cartoon". *The Toronto Star,* Toronto, Canada. 1 May.

Coulter, T. (2015) "If History is any Guide, Big Deals Signal Oil Market Bottoming". *Bloomberg.* 8 April 2015.

Coulter, T. and Chilcote, R. (2015) "At OPEC the Saudi Oil Minister wants to discuss Solar Power". *Bloomberg.* 3rd June 2015.

Cremer, J. and Isfahani, D. (1991) *"Models of the Oil Market"*. Chur: Harwood Academic Publishers.

Critchlow, A. (2014) "Should the US Create its own North American OPEC Oil Cartel?" *Bloomberg.* 9 December 2014.

Critchlow, A. (2015) " Saudi official warns BOE's Carney oil slump due to speculators". *The Telegraph*, London, 16 March 2016.

CSIS. (2013) "A Conversation with His Excellency Ali Al Naimi, Minister of Petroleum and Mineral Resources, Kingdom of Saudi Arabia". Centre for Strategic and International Studies, Washington DC, on April 30, 2013.

Cuervo, L. (2008) "OPEC from Myth to Reality". *Houston Journal of International Law.* 30 (2). pp. 433-541.

Currie,J, Natuan, A, Creedy, D. and Courralin, D. (2010) "Commodity Prices and Volatility: Old Answers to New Questions". Goldman Sachs. *Global Economics Paper*. 194. pp. 1-44. 30 March.

Dagher, A. (2014)"Unconventional Control: Impacts of Unconventional Oil and Gas in the GCC", in Luciani, G. and Ferroukhi, *"The Political Economy of Foreign Reform: The Clean Energy-Fossil Balance in the Gulf States"*. Gerlach Press, Germany. pp. 59-83.

Dahl, C. and Yucel, M. (1991) "Testing Alternative Hypothesis of Oil Producer Behavior". The *Energy Journal*, 12, pp. 117-37.

Davenport, C. (2014) "Deal on Carbon Emissions by Obama and Xi Jinping raises Hopes for Upcoming Paris Climate Talks". *The New York Times*, New York. 12 November 2014.

Davis, J, Ossowski, R, and Fedelino, A. (2003) "Fiscal Policy Formulation and Implementation in Oil Producing Countries". *International Monetary Fund*. Washington D.C.

Daya A. (2011) "Saudis Seek to Ensure Climate Talks won't Hurt OPEC Oil Income". *Bloomberg*. 24 November 2011.

Deffeyes, K. (2001) *"Hubbert's Peak: The Impending World Oil Shortage"*. Princeton University Press, Princeton, USA.

Department of Energy. (2013) "Onshore Oil and Gas Exploration in the UK: Regulation and Best Practice". *Department of Energy and Climate Change*. England. December 2013.

Deutsche Bank Research. (2014^a) "The Morning Call: 25-Nov—OPEC Output Cuts Expected". *Deutsche Bank Research*. 25 November 2014.

Deutsche Bank Research. (2014^b) "EM Oil Producers: Breakeven Pain Thresholds". *Deutsche Bank Research*. London. 16 October.

Deutsche Bank. (2015) "Adjusting to Lower Oil Prices: Budget Breakeven Thresholds". *Deutsche Bank Research*. London. 14 May 2015.

Dietz, J. and Stern, N. (2014) "Endogenous Growth, Convexity of Damages and Climate risk: How Nordhaus's framework supports deep cuts in carbon emissions". *Centre for Climate Change, economics and Policy*. Working Paper No. 180. London. June 2014. http://www.cccep.ac.uk

DiPaola, A. (2015^a) "US Won't Intervene in Oil Market Amid Price Declines on Supply". *Bloomberg*. 19 January 2015.

DiPaola, A. (2015^b) "UAE Sticks to Oil Output Boost even as Glut Pushes Prices down". *Bloomberg*. 12 January 2015.

DiPaola, A. (2015^c) " Saudi Aramco aims for oil recovery rate twice the global average". Bloomberg, 18 March 2015.

Doan, L. (2015) "Shale Producers in US Cutting Rigs Loose Early Amid Oil Slump". *Bloomberg*. 8 January 2015.

Donilon, T. (2013) "Energy and American Power: Farewell to Declinism". *Foreign Affairs*. 15 June 2013.

Downey, M. (2009) *"Oil 101"*. Wooden Table Press. 2009.

Dunn, D.H. and McClelland, M. (2013) "Shale Gas and Revival of American Power: Debunking Decline?". *International Affairs*. 89 (6). pp. 1411-28.

Economist. (2015^a) "The Economist Explains Why China Economy is Slowing". *The Economist*. London. 11 March 2015.

Economist. (2015^b) "Problems at the Pump". *The Economist*. London. 30th May 2015.

EIU. (2015) *"Renewable Energy Demand to Significantly Outgrow Fossil Fuels in 2015"*. *Economist Intelligence Unit*. London. 14 January 2015.

El Badri, A. (2015) *"Personal Interview"*. Secretary General of the Organization of Petroleum Exporting Countries. Doha, Qatar, 9 April 2014.

El Badri, A. (2015) In speech at *Middle East Oil & Gas Show*, Manama, Bahrain on March 8, 2015.

EL Erian, M. (2015) "Oil won't Swing back to $100 a Barrel Soon: Mohammed EL-Arian". *Bloomberg*. 5 March 2015.

El Katiri, L and Fattouh, B. (2015) "A Brief Political Economy of Energy Subsidies in the Middle East and North Africa". *The Oxford Institute for Energy Studies*, Oxford. February 2015.

Elster, J. (1989) *"The Cement of Society"*. Cambridge University Press, Cambridge. UK.

References

Emerson, S. and Winner, A. (2014) "The Myth of Petroleum Independence and Foreign Policy Isolation". *Washington Quarterly.* 37 (1), Pp. 21-34.

Energy Information Administration. (2011) "Annual Energy Review 2011". US Energy Information Administration, Washington D.C., September 2012.

Energy Information Administration. (2013^a) "Technically Recoverable Shale Oil and Shale Gas Resources: An Assessment of 137 Shale Formations in 41 Countries outside the United States". *US Energy Information Administration*, Washington, D.C., June 2013.

Energy Information Administration. (2013^b) "International Energy Outlook 2013". US Energy Information Administration, Washington, July 2013.

Energy Information Administration. (2014^a) "U.S. Liquids Fuels Production Growth More Than Offsets Unplanned Supply Disruptions". *US Energy Information Administration*, Washington. August 27 2014.

Energy Information Administration. (2014^b) "Petroleum and Other Liquids Data". *US Energy Information Administration*, Washington, USA. http://www.eia.gov/dnav/pet/PET/-CRD-CRPDN-ADC-MBBLPD-M.htm.

Energy Information Administration. (2015^a) "OPEC Revenues Fact Sheet". *US Energy Information Administration*, Washington, USA. 31 March 2015. http://www.eia.gov/about/new/index.cfm?r=60

Energy Information Administration. (2015^b) "Drilling Productivity Report". *US Energy Information Administration*, Washington, USA. 8 June 2015. http://www.eia.gov/petroleum/drilling/

Etebari, M. (2013) "Iran Press Report: The Quest to cut Cash Subsidies". *Markaz- Middle East Politics and Policy.* 1 November 2013.

European Commission. (2013) "Anti-Trust: Commission confirms unannounced Inspections on Oil and Bio-fuels Sector". *European Commission,* Brussels. 14 May 2013.

Fagan, M. (2000) "Sheikh Yamani Predicts Price Crash as Age of Oil Ends". *Daily Telegraph.* London. 25 June 2000.

Fattouh, B. (2006) "The Origins and Evolution of the Current International Oil Pricing System: A Critical Assessment" in Mabro, R,(ed) *"Oil in the 21st Century: Issues, Challenges and Opportunities".* Oxford University Press.

Fattouh, B. (2008) "To Cut or Not To Cut: The dilemma Facing OPEC". *Oxford Institute for Energy Studies, Oxford Energy Comment.* Oxford, U.K. October 2008.

Fattouh, B. (2011^a) "Inter-linkages and Regulation of Oil Derivatives". *Oxford Institute for Energy Studies.* Oxford, U.K. 28 January 2011.

Fattouh, B. (2011^b) "Oil Market and OPEC Behavior: Looking ahead". *Oxford Institute for Energy Studies.* Oxford, U.K. 2 Feb. 2011.

Fattouh, B. (2011^c) "Saudi Arabia's Oil Policy in Uncertain Times: A Shift in Paradigm?". *Oxford Institute for Energy Studies.* Oxford, U.K. 21 October 2014.

Fattouh, B. (2014^a) "Current Developments in the Oil and Natural Gas Markets and their Implications for the Energy Sector in the Arab World". *Tenth Arab Energy Conference*, Abu Dhabi, UAE. 21-23 December 2014.

Fattouh, B. (2014^b) "The US Shale Revolution and the Changes in LPG Trade Dynamics: A Threat to the GCC?" *Oxford Energy Comment,* Oxford. July, 2014.

Fattouh, B. (2014^c) "Oil Market Dynamics: Saudi Arabia Oil Policies and US Shale Supply Responses". *Oxford Institute for Energy Studies.* Oxford, U.K. 18 March 2014.

Fattouh, B. and Sen, A. (2014) "China's Rebalancing and Oil Consumption Patterns". *Oxford Energy Forum.* February, Issue 95.

Faucon, B. and Said, S. (2015) "OPEC Sees Oil Price Below $100 a Barrel in the Next Decade". *Bloomberg.* 11 May 2015.

Feng, C. (2015) " Embracing Interdependence : The dynamics of China and the Middle East". *Brookings Doha Centre,* Doha, Qatar.

Financial Times. (2015) "Obama sets Stage for Debate over US Oil Export Ban". *Financial Times,* London. 4 January.

Finkle, J. (2012) "Exclusive: Insiders Suspected in Saudi Cyber Attack". *Reuters*. 7 September 2012.
Finon, D. (1991) "The Prospects for a New International Petroleum Order". *Energy Studies Review*. 3(3), pp. 260-276.
Galpern, S. and Keefer, E. (2013) "Energy Crisis, 1974-1980". *US State Dept. Office of the History Government Printing Office*. 3 January 2013.
Gately, D, and Al Yousef, N. (2012) "The Rapid Growth of Domestic Oil Consumption in Saudi Arabia and the Opportunity Cost of Oil Exports Foregone". *Energy Policy*. 47. pp 57-68.
Gately, D. (1979) "OPEC Pricing and Output Decisions". *Applied Game Theory*. Physica-Verlag, Germany.
Gause, G.F. (2015) "Sultans of Swing? The Geopolitics of Falling Oil Prices". *Brookings Doha Centre*, Doha, Qatar.
Ghanem, S, Lounnas, R, and Brennand, G. (1999) "The Impact of Emissions Trading on OPEC". *OPEC Review*. June 23, 1999.
Ghanem, S. (1986) "*OPEC: The Rise and Fall of an Exclusive Club*". KPI Ltd. London.
Goldenberg, S. (2015) "Stanford Professor's urge Withdrawal from Fossil Fuel Investments". *The Guardian*, London. 11 January 2015.
Goldman Sachs Research. (2014) "*OPEC Loses Pricing Power, Shale Shifts to the Margin*" *Goldman Sachs Research, the New Oil Order*. 26 October 2014.
Goldthau, A. and Witte, J. (2011) "Assessing OPEC's Performance in Global Energy". *Global Policy*. Vol. 2. pp. 31-39. September 2011.
Goodley, S. (2015) "OPEC Prepares to Call Emergency Meeting over Oil Price Slump". *The Guardian*. London. 23 February 2015.
Gorelick, S. (2010) "Oil panic and the global crisis: Predictions and Myths". Wiley-Blackwell, USA.
Gould, A. (2006) "Technologies to Extend Oil Production", in Mabro, R. (ed) (2006) "*Oil in the 21st Century: Issues, Challenges and Opportunities*". Oxford University Press, UK. pp. 178-202.
GPCA. (2014) "Gulf petrochemcial trends." Gulf Petrochemicals and Chemicals Association, Bahrain, March 2014.
Griffin, J.M. (1985) "OPEC behavior: A Test of Alternative Hypothesis". *The American Economic Review*. 75, pp. 954-63.
Guardian. (2011) "Canada Pulls out of Kyoto Protocol". *The Guardian*, London. 13 December 2011.
Guardian. (2014) "Lima Climate Deal: What was Agreed- and What Wasn't".*The Guardian*, London.http://www.theguardian.com/environment/2014/dec/15/lima-climate-deal-what-was-agreed-and-what-wasnt
Guardian. (2015) "Russia's Debt Downgraded to Junk by Moody's". *The Guardian*. London. http://www.theguardian.com/world/2015/fev/21/Russias-debt-downgraded-junk-moodys
Haber, S, and Menaldo, V. (2011) "Do Natural Resources Fuel Authoritarianism? A Reappraisal of the Resource Curse". *American Political Science Review*. 105(1).pp. 1-26.
Hafezi, P, Charbonneau, L, Irish, J. and Mohammed. A. (2015) "Iran, big powers clinch historic nuclear deal." Rueters. 14 July 2015.
Hansen, P.V, and Lindhollt, L. (2008) "The Market Power of OPEC: 1973-2001". *Applied Economics*. 40, pp. 2939-2959.
Hanware, K. (2015). "The Kingdom issues SR 15 bn bonds. "Arab News, Jedah, 11 July , 2015. Reuters. (2015 e). "SAMA's net foreign assets fall 1 % to SR 2.521 trillion". Reuters, London, 01 July 2015. Carpenter, C. and Khan, S. (2015). "UAE removes fuel subsidiy as oil drop hurts Arab economies. "Bloomberg. 22 July 2015.
Hanware, K. (2015) "Kingdom issues SR15bn bonds." Arab News, Jeddah. 11 July 2015.
Harris, K. (2010) "The Politics of Subsidy Reform in Iran". *Middle East Research and Information Project*. http://www.merip.org/mer/mer254/politics-subsidy-reform-iran.
Harvey, C. (2015^a) "Refracking Seen as 'Act Two' for N. American Shale Revolution". *Bloomberg*. 9 June 2015.
Harvey, C. (2015^b) "New rules on disposal could effect oil producers in 2016: Bloomberg." Bloomberg. 23 June 2015.
Herb, M. (2005) "No Representation without Taxation? Rents, Development and Democracy". *Comparative Politics*. 37 (3). Pp 297-316.

Hilsum, L. (2012) "Patriot of Persia: Muhammad Mossadegh and a very British Coup by Christopher de Bellaigue—Review". *The Observer*. London, 5 February 2012.

Hinton, I. (2010) "In India: A Clear Victor on the Climate Front". *Yale Environment*, 260. 1 March 2010.

Horn, M.K. (2006) "Giant Oil Fields 1868-2003" in Halbouty, M, (ed) "Giant Oil and Gas Fields of the Decade 1990-1999". *AAPG Memoir*, 78, 2003, Modified November 2006.

Horn, M.K. (2009) www.sourcetoreservoir.com

Horsnell, P. Lim, S. and Snowdon, N. (2015) "Focus: The decline in US shale oil output deepens". *Standard Chartered Commodities Research, Commodities Roadmap*. 9 June 2015.

Hubbert, M.K. (1971) "The Energy Resources of the Earth". *Scientific American*. 225. pp. 31-41. September 2971.

Hughes, David. (2013) "Drill, Baby, Drill: Can Unconventional Fuels usher a new Era in Energy Abundance?", *Post Carbon Institute*, Santa Rosa, CA.

Hughes, Siobhan. (2008) "Steele gives GOP Delegates New Cheer: "Drill, baby, drill" *Wall Street Journal,* New York, USA. 3 Sept.

Hyndman, K. (2008) "Disagreement in Bargaining: An Empirical Analysis of OPEC". *International Journal of Industrial Organization.* Vol. 26. pp. 811-828.

Ibrahim, Y. (1997) "Sheikh Abdullah Al Tariki, 80, First Saudi Arabian Oil Minister" *New York Times*. 16 September 1997.

IEA. (2007) "Energy Technology at the Cutting Edge". International Energy Technology Collaboration. IEA Implementing Agreements, *International Energy Agency*, OECD, Paris.

IEA. (2011[a]) "Redeploying Renewable: Best and Future Policy Practice". Renewable Energy Markets and Policies. *International Energy Agency*, OECD, Paris.

IEA. (2011[b]) "Joint Report by the IEA, OPEC, OECD and World Bank on Fossil-Fuel and Other Energy Subsidies: An Update of the G20 Pittsburgh and Toronto Commitments". IEA, Paris.

IEA. (2012[a]) "Golden Rules for a Golden Age of Gas: World Energy Outlook—Special Report on Unconventional Gas". *International Energy Agency*, OECD, Paris. 12 November 2012.

IEA. (2012[b]) "World Energy Outlook 2012". *International Energy Agency*, OECD, Paris. 12 November 2012.

IEA. (2013) "World Energy Outlook 2013". *International Energy Agency*, OECD, Paris. 12 November 2013. www.iea.org.

IEA. (2014) "World Energy Outlook 2014". *International Energy Agency*, OECD, Paris. 12 November 2014. www.iea.org.

IEA. (2015[a]) "Oil Market Report". *International Energy Agency*, Paris. 13 May 2015. www.iea.org.

IEA. (2015[b]) "Oil Medium-Term Market Report 2015". *International Energy Agency*, Paris. 10 February 2015. www.iea.org.

IEA. (2015[c]) "Oil Market Report". *International Energy Agency*, Paris. 13 March 2015. www.iea.org.

IEA. (2015[d]) "Oil Market Report". *International Energy Agency*, Paris. 11 June 2015. www.iea.org.

IHS. (2015) "US Crude Oil Export Decision: Assessing the Impact of the Export Ban and Free Trade on the US Economy". *IHS Inc*. https://www.ihs.com/info/0514/crude-oil.html.

Ikle, F. (1964) "*How Nations Negotiate*". Harper & Row, New York.

Intergovernmental Panel on Climate Change (2007) "*Climate Change 2007: The Physical Science Basis*". New York: Cambridge University Press.

Iran Daily. (2015) "Oil, Gas Condensates to Exceed 5.7m bpd". *Iran Daily*. Tehran, 13 January 2015.

Iran Daily. (2015) "Oil, Gas Condensates to Exceed 5.7m bpd". *Iran Daily*. Tehran, 13 January 2015.

Ismail, I. (1995) "Raising Oil Output in Major Producing Regions—The Financial Implications". *OPEC Bulletin*. Vienna. 26 (10) pp. 14-19.

Jadwa Investment. (2013) "The Outlook for Unconventional Oil and Gas Production: Focus on Tight Oil and Shale Gas Production Impact on Saudi Arabia". *Jadwa Investment*, Riyadh. December.

Jadwa Investment. (2014[a]) "Oil Market Dynamics and Saudi Fiscal Challenges". *Jadwa Investment*, Riyadh. December 2014.

Jadwa Investment. (2014[b]) "Saudi Arabia's 2015 fiscal budget" *Jadwa Investment*. Riyadh, 28 December.

Jadwa Investment. (2015[a]) "Quarterly Oil Market Update (Q4 2014)". *Jadwa Investment*. Riyadh.

Jadwa Investment. (2015[b]) "Quarterly Oil Market Update (Q1 2015)". *Jadwa Investment*. Riyadh. April 2015.

Johany, A. (1980) "The Myth of the OPEC Cartel: The Role of Saudi Arabia." John Wiley, New York.

Kalantari, H. (2015[a]) " Iran says could raise oil exports by 1 M B/d if sanctions lifted". *Bloomberg*. 16 March 2015.

Kalantari, H. (2015[b]) "Iran to speed up Development of Oil Fields Shared with Iraq: INN". *Bloomberg*. 4 April 2015.

Kaplow, D. and Kretzmann. (2010) "G20 Fossil Fuel Subsidy Phase Out: A Review of Current Gaps and Needed Changes to Achieve Success". *Earth Track Inc. and Oil Change International*. 13.

KAPSARC. (2013) "Measuring the Energy Intensity of Nations: Towards a Framework for Transparent Comparisons". *KAPSARC*, Riyadh. Workshop Policy Brief. December 2013.

KAPSARC. (2014) "A Framework for Fuel and Technology Transitions in Energy: Evaluating Policy Effectiveness". *KAPSARC*, Riyadh. Workshop Policy Brief. March, 2014.

Karl , T. (1997) " *The paradox of plenty : Oil booms and petro states* ". University of California Press. Berkeley, USA.

Kemp, J. (2014) "News of OPEC's Demise has been much Exaggerated: Kemp". *Reuters*. 3 December 2014.

Kemp, J. (2015) Column in Reuters under the title "Who Wants to be a Swing Producer? No One". *Reuters*. 14 May 2015.

Kennedy, John. (1962) "Address at Rice University". Rice University President John. F. Kennedy Moon Speech. 12 December 1962. *hppt://er.jsc.nasa.gov/seh/ricetalk/htm*.

Khan, M. (2009) "The 2008 Oil Price 'Bubble'". Peterson Institute for International Economics, *Policy Brief*. August 2009.

Khatib, H. (2010) "Global Energy Security: Implications for the Gulf", in "Energy Security in the Gulf: Challenges and Prospects", *The Emirates Center for Strategic Studies and Research*, Abu Dhabi, UAE. pp. 19-70.

Klare, M. (2002) " *Resource Wars : The new landscape for Global Conflicts"*. Holt Paperbacks. USA.

Klare, M. (2008) " *Rising powers , shrinking planets: The new geopolitics of energy"*. Metropolitan Books , USA.

Kwiatkowski, A, and Zhu, W. (2013) "EU Oil Manipulation Probe Shines Light on Platts Pricing". *Bloomberg*. 15 May 2013.

Lakshmanan, I, Tirone, J. and Froohar, K. (2015) "Iran reaches nuclear deal with world powers to end sancitons." Bloomberg. 14 July 2015.

Landis, R, and Klass, M. (1980) "*OPEC: Policy Implications for the United States*". Preager, New York.

Learsy, R. (2012) "NOPEC (No Oil Producing and Exporting Cartels Act) A Presidential Issues and a Test of Political Integrity". *Huffington Post*, USA. 10 Sept.

Lee, H. (2015) "No Impact on Physical Oil Market from Iran Deal before 2016: Morgan Stanley". *Bloomberg*. 4 April 2015.

Lee, J. (2015) "Beware Year of the Goat in Middle East Oil Fortune". *Bloomberg*. 20 February 2015.

Lewis, J. (2008) "China's Strategic Priorities in International Climate Change Negotiations". *Washington Quarterly*. 31(1): Winter 2007-8. Pp. 155-174.

Litvan, L. (2015) "Senate Fails to Override Obama's Veto of Keystone XL Legislation". *Bloomberg*. 4 March 2015.

Luciani, G. (1987) "Allocation Vs. Production States: A Theoretical Framework" in Bablawi, H. and Luciani, G. *"The Rentier State"*, New York, Groom Helm.

Luciani, G. (1995) "The Dynamics of Reintegration in the International Petroleum Industry", in *Oil in the New World Order*, edited by Gillespie, K. and Moore-Henry, C., Gainesville, FL: University Press of Florida.

Lugo, L. (1985) *"OPEC: The Inside Story"* English Version. Zed Books, London, UK.

Mabey, N, Hall, S, Smith C, and Gupta, S. (1997) *"Argument in the Greenhouse: The International Economics of Controlling Global Warming"*. Routledge, London.

Mabro, R. (1989) "OPEC's Production Policies". *O.I.E.S. Paper*. Oxford: Aldgate Press.

Mabro, R. (ed) (1986) *"OPEC and the World Oil Market: The Genesis of the 1986 Price Crisis"*. Oxford University Press, UK.

Mabro, R. (ed) (2006) *"Oil in the 21st Century: Issues, Challenges and Opportunities"*. Oxford University Press/OPEC.

Mahdi, W. (2014) "We Support Saudi OPEC Policy: Iraqi Oil Minister". *Asharq Al Awsat*. Saudi Arabia. 30 December 2014.

Mahdi, W. (2015^a) "Saudi Oil Exports Falls as China Cuts Back Amid Price Rout". *Bloomberg*. 18 February 2015.

Mahdi, W. (2015^b) "Return to $100 Oil Seen Unlikely by Saudis amid Shale Surge". *Bloomberg*. 22 March 2015.

Mahdi, W. (2015^c) "Saudi Arabia to Start Producing Shale gas in 2016, Naimi Says". *Bloomberg*. 7 April 2015.

Mahdi, W. (2015^d) "Saudi Aramco Creates Supreme Board as Oil Exporter Reorganizes". *Bloomberg*. 01 May 2015.

Mahdi, W. and MacDonald, F. (2015) "Kuwait sees Crude Price Recovering, set to add more Rigs". *Bloomberg*. 16 February 2015.

Mahesh, M, Russell, W, and Cohen, M. (2015) "The Race to Store". *Barclays Commodities Research, Oil Market Outlook*. 26 January 2015.

Mahesh, M, Russell, W, and Cohen, M. (2015) "The Race to Store". Barclays Commodities Research, Oil Market Outlook. 26 January 2015.

Majid, A.H. (2014) "Technological Development in Utilizing Unconventional Resources of Gas and Oil". *Tenth Arab Energy Conference*, Abu Dhabi, UAE. 21-23 December 2014.

Martin, M. and Blanchi, S. (2015) "Saudi Aramco said to Plan as Much as $80 bn of Foreign Investment". *Bloomberg*. 11 May 2015.

Mattera, P. (2007) "Is Big Business Buying out the Environment Movement?". *Corporate Research E-Letter* No. 65. May-June 2007. http://www.corp-research.org.

McGrath, M. (2015) "Islamic call on rich countries to end fossil fuel use." BBC News. London. 18 August 2015.

MEES. (2013) "Naimi: Saudi Arabia to Stay Above Iran, Iraq Fray". *MEES*. 56 (49) 6 December 2013.

MEES. (2013) "Naimi: Saudi Arabia to Stay Above Iran, Iraq Fray". *MEES*. 56 (49) 6 December 2013.

MEES. (2014) "MEES Interview with Al Naimi: "OPEC will Never Plan to Cut". *Middle East Petroleum and Economic Publications*, Cyprus. 22 December 2014.

Meric, A. and Hacaoglu, S. (2014) "Iraqi Kurds, Turkey to Double Oil Export Pipeline Capacity". *Bloomberg*. 20 August 2014.

Metz B. (Ed). (2001) *"Climate Change 2001: Mitigation: Contribution of Working Group III to the Third Assessment Report of the Intergovernmental Panel on Climate Change"*. Cambridge University Press, UK.

Michaels, P, Singer, S, and Doughlas, D. (2004) "Settling Global Warming Science". *Tech. Central Station—Where Free Markets Meet Technology*. 8 August 2004. www.techcentralstation.com

Moguera, P, Douglas, C, and Herrera, A. (2011) "Testing for Cartel in OPEC: Non-Cooperative Collusion or Just Non-Cooperative?". *Oxford Review of Economic Policy*. 27(1). pp. 144-168.

Morales, A. (2015) "Poor Nations need $400 billion a Year in Climate Aid, Study Says". *Bloomberg*. 16 March 2015.

Motevalli, G. (2014) "Iran Oil Minister Doubts Drop in Shale Output After OPEC Meeting" *Bloomberg*. 29 November 2014.

Muller, B.(2006) "Some Aspects of the Climate Change Issue", in Mabro, R (ed). *Oil in the 21st Century: Issues, Challenges and Opportunities"*. Oxford University Press. Pp 203-240.

Nair, D. (2015) "Kuwait's kufpec said to consider bid for Eon's north sea assets." Bloomberg. 29 June 2015.

Nasser, A. H, and Sabri, N.G. (2004) "Reserves and Sustainable Oil Supplies: Role of Technology and Management". *5th International Oil Summit*. Paris, France. 29 April 2004.

Natarajan, S. (2015) "Oil's Plunge Leaves $27 billion of Energy Bonds Priced for Junk". *Bloomberg*, 8 January.

Navarro, A, and Murtaugh, D. (2015) "Mexico Ready to take Crude 'Tomorrow' if US Approves Swap". *Bloomberg*. 31 March 2015.

Neugar, J. (2015) "EU won't Lift Iran Sanctions on June 30, will Take More Time" *Bloomberg*. 7 April 2014.

New York Times. (1982) "The World Agrees to Limits on Production". *The New York Times*. New York. 21 March 1982.

New York Times. (1983^a) "Nigerian Oil Price Cut Spurs Urgent Talks Within OPEC". *The New York Times*. New York. 21 February 1983.

New York Times. (1983^b) "Why OPEC is not a Cartel". *The New York Times*. New York. 11 March 1983.

New York Times. (1983^c) "OPEC: Trying to be a Cartel". *The New York Times*. New York. 16 March 1982.

New York Times. (1986^a) "Economic Scene: Negative Effect of Cheap Oil". *The New York Times*. New York. 25 July 1986.

New York Times. (1986^b) "Price Blur at End of Oil Accord". *The New York Times*. New York. 6 August 1986.

New York Times. (1986^c) "US Criticizes OPEC Agreement". *The New York Times*. New York. 6 August 1986.

New York Times. (1986^d) "OPEC Sets Extension of Accord". *The New York Times*. New York. 22 October 1986.

New York Times. (1986^e) "The World: OPEC Reaches an Agreement". *The New York Times*. New York. 21 December 1986.

New York Times. (1986^f) "Fahd Sees $18 Price as Floor". *The New York Times*. New York. 27 December 1986.

New York Times. (1987) "Business Forum: Saudi Arabian Gulf War Strategy; World Oil Prices Are on the Way Down". *The New York Times*. New York. 18 October 1987.

New York Times. (1995) "Gabon Plans to Quit OPEC". *The New York Times*. New York. 9 January 1995.

New York Times. (2015) "A Promising Nuclear Deal with Iran". *The New York Times*. New York. 2nd April 2015.

Nixon, Richard (1974) "State of the Union Address 30 Jan. 1974". http://print.infoplease.com/t/hist/state-of-the-union/187.html.

Nordhaus, W.D. (1991) "To slow or not to slow: the economics of the greenhouse gases". *Economic Journal*. 101 (407), pp. 920-937.

Noreng, O. (1978) "*Oil Politics in the 1980's: Patterns of International Cooperation*". McGraw Hill, New York.

Noreng, O. (1982) "Friends or Fellow Travelers? The Relationship of Non-OPEC Exporters with OPEC", in Mallakh, R. (ed) "*OPEC: Twenty Years and Beyond*". Westview Croom Helm.

Norrish, K. (2015) "Oil: Not adding up". *Barclays Commodities Research, Cross Commodities Focus*. 8 June 2015.

O'Connor, P. (2012) "Doha Climate Summit Concludes without Agreement on Emissions Reduction". http://www.wsws.org/en/article2012/12/11/doha-d11.htmtl.

O'Sullivan, E. (2015) "Restructuring Saudi Aramco". *MEED: Middle East Economic Digest*. London. 2 May 2015.

OECD. (2004) "OECD Economic Outlook". No. 76, *OECD*, Paris

OECD. (2014) OECD Stat Extracts GDP. *OECD*, Paris. http://stats.oecd.org/index.aspx?queryid=60702

OFID. (2015) *"OPEC Fund for International Development"* http://www.ofid.org/HOME/OFID-Profile.
Okada, Y, and Lee, H. (2015) "Oil Bargains from Latin America Erode Saudi Market Clout in Asia". *Bloomberg*. 01 April 2015.
Olson, B. (2015) "Global Layoffs Exceed 100,000". *Bloomberg*. 12 February 2015.
OPEC. (1990) "Official Resolutions and Press Releases 1960-1990". *OPEC*, Vienna. The Secretariat of the Organization of Petroleum Exporting Countries.
OPEC. (1997) "OPEC's Views on Climate Change". http://www.opec.org The Secretariat of the Organization of Petroleum Exporting Countries
OPEC. (1998) "Statement by Dr. Rilwanu Lukman, Secretary General OPEC". http://www.opec.org The Secretariat of the Organization of Petroleum Exporting Countries
OPEC. (2001) "Frequently Asked Questions". http://www.opec.org The Secretariat of the Organization of Petroleum Exporting Countries
OPEC. (2002^a) "119th Meeting of the OPEC Conference". *OPEC*, Vienna, Press Release No. 3/2002. 15 March 2002. The Secretariat of the Organization of Petroleum Exporting Countries.
OPEC. (2002^b) "OPEC and Russia in Talks on Second-Quarter Market-Stabilization Measures". *OPEC*, Vienna, Press Release No. 1/2002. 1 March 2002. The Secretariat of the Organization of Petroleum Exporting Countries.
OPEC. (2005^a) "OPEC Bulletin No. 3/15". *OPEC*, Vienna, April 2015.
OPEC. (2005^b) "OPEC Bulletin". *OPEC*, Vienna. XXXVI (2) February 2005.
OPEC. (2006) *"The Clean Development Mechanism: Is it Meeting the Expectations?"* Riyadh, Saudi Arabia. The Secretariat of the Organization of Petroleum Exporting Countries. 19-20 September 2006.
OPEC. (2008) "151st (Extraordinary) Meeting of the OPEC Conference". *OPEC*, Vienna, Press Release No. 17/2008. 17 December 2008. The Secretariat of the Organization of Petroleum Exporting Countries.
OPEC. (2009^a) "OPEC Bulletin—Seminar Special". *OPEC*, Vienna, 40(4). May 2009. The Secretariat of the Organization of Petroleum Exporting Countries.
OPEC. (2009^b) "OPEC Monthly Oil Market Report". *OPEC,* Vienna, January 2009. The Secretariat of the Organization of Petroleum Exporting Countries.
OPEC. (2014) "Annual Statistical Bulletin 2014". *OPEC*, Vienna. 2014.
OPEC. (2015^a) "OPEC 166th Meeting Concludes". *OPEC*, Vienna, Press Release. 27 November 2015. The Secretariat of the Organization of Petroleum Exporting Countries.
OPEC. (2015^b) "OPEC Monthly Oil Market Report". *OPEC,* Vienna, February 2015. The Secretariat of the Organization of Petroleum Exporting Countries.
OPEC. (2015^c) "OPEC Monthly Oil Market Report". *OPEC,* Vienna, April 2015. The Secretariat of the Organization of Petroleum Exporting Countries.
Orozco, J. (2015^a) " Venezuela prepared for US action on oil , minister Chavez says". *Bloomberg*, 18 March 2015.
Orozco, J. (2015^b) "Venezuela holds meeting with OPEC country Ambassadors: Oil Ministry". *Bloomberg*, 15 April 2015.
Pals, F. (2015) "Shell Plans to Drill in Alaska's Chukchi Sea this year, 2016: CEO" *Bloomberg*. 19 May 2015.
Parra, F. (2004) *"Oil Politics A Modern History of Petroleum"*. I.B. Tauris, New York, USA.
Paton, J. (2015) "Chevron's Exit Signals Delays in Unlocking Australia's Shale". *Bloomberg*. 31 March 2015.
Platts. (2012) "Majority in OPEC, including Saudis, favour $100/barrel oil". *Platts,* Dubai. 18 September 2012.
Prokop, J. (1999) "Process of Dominant-Cartel Formation". *International Journal of Industrial Organization*. 17. pp. 241-57.
Quinn, D. (2015) "US Mulls Crude Exports to Mexico, Commerce Secretary Says". *Bloomberg*. 7 January.
Ramady, M. (2013) *"Political, Economic and Financial Country Risk: Analysis of the Gulf Cooperation Council"*. Springer, New York.

Razzouk, N. (2015) "Saudi's Naimi, Russian Envoy Discuss OPEC, Non-OPEC Cooperation". *Bloomberg*. 11 January 2015.

Reuters. (1983) "Communiqué by OPEC". *Reuters*. New York Times. 15 March 1983.

Reuters. (2014) "Push to Relax US Oil Export Ban ends 2014 with Break Through". *Reuters*. 31 December.

Reuters. (2015^a) "Kuwait considering imposing corporate tax- KUNA". *Reuters,* 17 March 2015.

Reuters. (2015^b) "Saudi Arabia Boosts Crude Oil Production to Highest Level on Record". *Reuters*. London. 8 April 2015.

Reuters. (2015^c) "Indonesia Seeks to Rejoin OPEC". *Reuters*. 8 May 2015.

Reuters. (2015^d) "OPEC Members approve Indonesia's bid to Region: Indonesian Energy Minister". *Reuters*. 5 June 2015.

Rueters. (2015^e) "SAMA's net foreign assets fall 1 % to SR2.521 trillion." Rueters. 01 July 2015.

Robinson, D. (2015) "Paris 2015: Just a first Step". *The Oxford Institute for Energy Studies,* Oxford. February 2015.

Roeber, J. (1994) "Oil Industry Structure and Evolving Markets". *Energy Journal*. 15 (Special Issue). pp. 253-276.

Rose, M. (2011) "Does Oil Hinder Democracy" *World Politics*. 53(3). pp 325-361.

Rosenberg, E. (2014) "Energy Rush: Shale Production and U.S. National Security". *Centre for New American Security*, Washington, D.C. February 2014.

Rosneft. (2015) "Oil Markets and their Transformation". *International Petroleum Week*. London. February 2015.

Rowling, R. and Arnsdrof, I. (2015) "Whatever OPEC Decides Oil Supplies are Rising from all Sides". *Bloomberg Business*. 5 June 2015.

Rudnitsky, J. (2015) "Russian Oil Floods Export Market as Teapot Plants Lose Money". *Bloomberg*. 9 April 2015.

Rudnitsky, J. and Bierman, S. (2015) "Russia, OPEC Plan high-level June Meeting: Energy Minister Novak". *Bloomberg*. 15 April 2015.

Rudnitsky, J. and Chmaytelli, M. (2015) "OPEC must Accept New Reality of Oil Price Below $100: EL Badri". *Bloomberg*. 9 June 2015.

Saif, O, Mezher, T, and Arafat, H. (2014) "Water Security in the GCC Countries: Challenges and Opportunities". *Journal Environmental Studies and Sciences*. 4(4), pp. 329-346. August.

Salameh, M. (2012) "Changing Oil Fundamentals: Impacts on Energy Security and Global Energy Market" in "Global Energy Markets: Changes in the Strategic Landscape". *The Emirates Center for Strategic Studies and Research*, Abu Dhabi, pp. 97-115.

SAMA. (2012) "Forth Eight Annual Report". *SAMA*, Riyadh.

SAMA. (2013) "Forty Ninth Annual Report". *SAMA*, Riyadh.

SAMA. (2015) "Fifty First Annual Report 2015". *SAMA*, Riyadh. June 2015.

SAMBA. (2015) *"Saudi Arabia Chart Book"*. SAMBA Riyadh. May 2015.

Saudi Aramco. (2014) "2014 Annual Report" http://saudiaramco.com/en/home/our-business/worlds-leading-supplier-of-energy

Saudi Aramco. (2015) "Annual Review 2014". *Saudi Aramco*. 11 May 2015.

Saudi Gazette. (2014) "Zero Carbon Emission Goal not Realistic: Saudi Arabia". *Saudi Gazette*, Jeddah. 10 December 2014.

Saudi Gazette. (2015^a) "Experts risk losing *Iqamas* with 14 Days Left for Mandatory Health Cover for Dependents". *Saudi Gazette*, Jeddah. 6 January 2015.

Saudi Gazette. (2015^b) Saudi Arabia: Next 'Frontier' in Shale Revolution: Al Falih". *Saudi Gazette*, Jeddah. 28 January 2015.

Saudi Gazette. (2015^c) "Saudi Reserves Drop". *Saudi Gazette*, Jeddah. 2 June 2015.

Sen, A. (2014). "The Known Unknowns". *Energy Aspects Research*, Perspectives. 8 September 2014.

Sfakianakis, J. (2014) "Oil Prices and the GCC: The resilient and the Less So". *Ashmore- The Emerging View*. London. December.

References

Shank, M. (2015) "US becoming Net Energy Exporter for First Time since 1950's". *Bloomberg*, 15 April 2015.

Siddiqui, S, and Lewis, P. (2015) "Barack Obama says Historic Agreement with Iran Meets come Objectives". *The Guardian*. London. 3 April 2015.

Simmons, M.R. (2006) *"Twilight in the Desert: The Coming Saudi Oil Shock and the World Economy"*. John Wiley, USA.

Simoniya, N. (2010) "The OPEC-Russia Relationship: Current Status and Future Outlook" In *"Energy Security in the Gulf: Challenges and Prospects"*. The Emirates Centre for Strategic Studies and Research. Abu Dhabi, UAE. pp 171-188.

Sinn, H.W. (2008) "Public Policies Against Global Warming: A Supply Side Approach". *International Tax and Public Finance*. 15. pp. 360-394.

Skeet, I. (1988) *"OPEC: Twenty Five Years of Press and Politics"*. Cambridge University Press, Cambridge.

Skinner, R. (2006) "Strategies for Greater Energy Security and Resource Security". *Oxford Institute for Energy Studies*, Oxford. UK. June 2006.

Slimani, S. (2015) "Algeria Replaces Energy, Finance Ministers, AP Says" *Bloomberg*. 19 May 2015.

Smith, G. (2015^a) "OPEC Cuts Forecast for US Oil Supply Growth after Price Rout". *Bloomberg*. 9 February 2015.

Smith, G. (2015^b) "IEA Sees OPEC Supply Jumping most in Four Years on Saudi Surge". *Bloomberg*. 15 April 2015.

Smith, G. (2015^c) "Saudi Arabia Adds Half a Bakken to Oil Market in a Month". *Bloomberg*. 16 April 2015.

Smith, G. and Chmaytelli, M. (2015) "Saudi Arabia Oil Minister Sees Day when Nation Exports Gigawatts" *Bloomberg*. 21 May 2015.

Smith, G. and Chmaytelli, M. and Razzouk, N. (2015) "OPEC to Maintain Current Oil Output" *Bloomberg Business*. 5 June 2015.

Smith, G. and Mahdi, W. (2015) "Saudi Oil Strategy Seen Working by Naimi as OPEC Set to Meet". *Bloomberg*. 1 June 2015.

Smith, J. (2005) "Inscrutable OPEC: Behavioral Test of Cartel Hypothesis". *Energy Journal*. 26. pp 51-82.

Spetalnick, M. and Trotta, D. (2015) "Obama meets Raul Castro in Highest-Level U.S.—Cuba talks in Decades". *Reuters*. 11 April 2015.

Stainforth, D. (2005) "Modeling Climate Change: Known Unknowns" published by *OpenDemocracy.net* 3rd June 2005. http://www.opendemocracy.net

Standard Chartered Research (2014). "The Morning Call: 25-Nov—OPEC Output Cuts Expected". *Standard Chartered Research*, UK. 25 November 2014.

Stern, N. (2013) "The structure of economic modeling of the potential impacts of climate change: grafting gross underestimation of risk onto already narrow science models". *Journal of Economic Literature*. 51(3). pp. 838-859.

Sunday Times. (2015) "Indian PM Narendra Modi says no Climate Change Pressure on India after US-China pact". *The Sunday Times*. 25 January 2015. http://www.straitstimes.com/news/asia/south-asia/story/indian-pm-narendra.

Sweetnam, G. (2008) "Long Term Global Oil Scenarios: Looking Beyond 2030". *Energy Information Administration*, 7 April 2008. Energy Conference Washington, DC.

Sykuta, M. (1994) "Real Effects of Futures Market on Firm and Industry Behavior: A Study of Institutions and Contracting in the Crude Oil Industry". Ph,D, *Washington University*. Department of Economics. St Louis. MO.

Tamimi, N. (2013) "Brazil: The Future Energy Superpower". 19 Nov. 2013. http://english.alarabiya.net/en/views/business/economy/2013/11/15/Brazil

Tamimi, N. (2015) "Navigating uncertainty : Qatar's response to the Global Gas Boom". *Brookings Doha Centre* , Doha , Qatar.

Terzian, P. (1985) *"OPEC: The Inside Story"*. English Version. Zed Books, London, UK.

Torbati, Y. (2015) "Iran nuclear deal may open oil taps in months, not weeks". *Reuters*, Washington, 17 March 2015.
Tuttle, R. (2015) "Saudi Oil Minister Discuses Market Stability with Russia, Norway". *Bloomberg*. 27 January 2015.
Umishek, G. (2003) "Petroleum Geology and Resources of the West Siberian Basin, Russia". *United Sates Geological Survey*, Washington, Bulletin No. 2201-G.
UNFCC. (2011) "Report of the Individual Review of the Annual Submission of Canada submitted in 2010". *United Nations Framework Convention on Climate Change*. New York. 19 December 2011.
United Nations. (2014) "Framework Convention on Climate Change". Lima, Peru. *Conference of the Parties. 20th Session*. 1-12 December 2014.
Valko, P, and Lee, W. (2010) "A Better Way to Forecast Production from Unconventional Gas Wells". Paper Presented at the SPE Annual Technical Conference. Florence, Italy, *Society of Petroleum Engineers*. 134231-MS.
Vietor, R. (1984) *"Energy Policy in America since 1945: A Study of Business—Government Relations"*. Cambridge University Press. Cambridge. UK.
Waldman, P. (2015) "Saudi Arabia's Plan to Extend the Age of Oil" *Bloomberg Business*. 13 April 2015.
Wall Street Journal. (2011) "Saudis See No Reason to Raise Oil Output Capacity". *Wall Street Journal*. 10 October 2011.
Waskow, D, and Bapna, M. (2015) "US-India Climate Partnership can Benefit Environment and Economy". *World Resource Institute*. 27 January 2015. http://www.wri.org
Weber, H. (2015) "Oil Will Recover Once Producers Quit Spending Money, Hamm Says". *Bloomberg*, 28 January 2015.
Westphal, K, Overhaus, M, and Steinberg, G. (2014) "The US Shale Revolution and the Arab Gulf States: The Economic and Political Impact of Changing Energy Markets". *SWP Research Paper*. German Institute for International and Security Affairs. Berlin., November.
White House. (2015[a]) "Fact Sheet: US and India Climate and Clean Energy Cooperation". *The White House*, Washington, Office of the Press Secretary. 25 January 2015.
White House. (2015[b]) "National Security Strategy". *The White House,* Washington, 7 February 2015.
Williams, P. (2015) "The $80 bn Spend that puts Shale on Shaky Ground". *Bloomberg*. 11 May 2015.
Wilpert, G. (2007) *"Changing Venezuela: The History and Politics of the Chavez Government"*. Verso, London.
Wittner, M. (2014) "OPEC meeting: A new era dawns for oil markets. Price itself will balance supply". *Societe Generale Research*, Oil Special. 27 November 2014.
World Bank. (2009) "Global Economic Prospects" *World Bank*. January 2009.
World Bank. (2014) World Bank Data GDP Growth (Annual %). World Bank Group, Washington DC, http://data.worldbank.org/indicator/NY.GDP.MKTP.KD.ZG
WTRG Economic. (2011) *"Oil Price History and Analysis"*, http://www.wtrg.com/prices.htm
Xie, Y and Popina, E. (2015) "Saudi Arabia's Credit Rating Outlook cut to Negative at S&P". *Bloomberg*. 9 February 2015.
Yamani, A. (1994) "OPEC in crisis: Oil Price Implications in the 1990's". *The Centre for Global Energy Studies*, London, UK.
Yamani, A. (1996) "Can OPEC Survive in Face of Mounting Competition?" *The Centre for Global Energy Studies,* London, UK.
Yergin, D. (2011) *"The Quest: Energy, Security and the Remaking of the Modern World"*. Penguin Press, New York.
Yergin, D. (2015) "Who Will Rule the Oil Market?". *The New York Times*. New York. 23 January 2015.
Zhou, M. (2015[a]) "Iran Nuke Deal may cut Oil Prices by $15 a Barrel, EIA Says". *Bloomberg*. 8 April 2015.
Zhou, M. (2015[b]) "Oil ETF Holders Cash out First Time in 7 months on Rebound". *Bloomberg*. 14 April 2015.

Index

A
Abdulaziz bin Salman, HRH, 26, 116, 119, 173
Abdulmahdi, Adel, 23
Abu Dhabi, 36, 44, 71, 75, 120, 148, 194, 223, 234
Abu Leif, Khaled, 158, 177
Acid rain, 156
Adaption Fund, 157
Africa, 5, 93, 94, 96, 144, 166, 239
Ahmadinejad, Mahmoud, 132
Al Dossari, Nasser, 13, 26
Al Falih, Khaled, 25, 84, 116, 206
Al Husseini, Sadad, 31, 225, 245
Al Khowaiter, Ahmed, 186
Al Khowaiter, Othman, 212
Al Mahdi, Mohammed, 26
Al Moneef, Majid, 26, 50
Al Muhanna, Ibrahim, 11, 26
Al Naimi, Ali, 11, 18, 19, 24, 25, 27, 28, 32, 50, 58, 59, 76, 115, 119, 129, 145, 158, 169, 175, 186, 203–204
Al Tariki, Abdullah, 234
Alaska, 220
Algeria, 16, 27, 36, 37, 45, 46, 48, 53, 65, 71, 74, 75, 93, 134, 135, 170, 184, 187, 188, 194, 213, 218, 229, 230, 232
Angola, 25, 60, 65, 68, 75, 126, 134, 135, 184, 187, 213, 218, 227, 229
Antrim, 103, 106
Arab embargo, 30, 40, 43, 73, 216
Arab Spring, 132, 144, 149
Arctic, 9, 91, 127, 129
Asia, 25, 26, 31, 50, 125, 126, 194, 198, 208, 209, 239
Australia, 60, 93, 160, 185
Ayatollah Khomeini, 44

Azadegan, 198
Azerbaijan, 28, 60, 135, 181

B
Backwardation, 77
Baghdad Pact, 225
Baker Hughes rig count, 111
Barnett, 103–108, 156, 170, 176
Basrah oilfield, 196
Berlin, 19, 24, 25, 27, 154, 204
Biden, Joe, 95
Biofuels, 6, 123, 160
Bitumen, 89
Brazil, 9, 25, 60, 62–64, 93, 112, 117, 155, 156, 168, 182, 194, 222, 241
Brent prices, 8
Brunei, 60, 181
Budget deficits, 136, 180
Bush, George Jr., 30, 80, 97, 147, 154
BWAVE, 75

C
Call on OPEC, 15, 21, 143, 182, 183, 188, 193, 194, 228
Call on shale, 21, 193, 194
Canada, 9, 60, 79, 89, 93, 97, 98, 112, 117, 154, 155, 157, 162, 165, 168, 176, 182, 213
Cantarell, 84
Carbon capture, 173–175, 242
Cartel
  OPEC, 57, 59, 97, 131, 180, 215, 219
  power, 58, 98, 180, 237
  pricing, 41, 42, 57, 180, 217, 242
  theory, 59

Carter, Jimmy, 80
Caspian Sea, 181
CENTO, 225
Ceyhan, 196
Chevaz, Hugo, 68, 217
China
    coal, 123, 153, 167
    oil demand, 9, 29, 31, 81, 245
    oil imports, 25, 26, 30, 122, 126, 167
    oil production, 121, 148, 206
    shale gas, 92, 117, 121, 124
Churchill, Winston, 233
CIA, 233
Clean Development Mechanism, 156, 174
Climate change, 81, 101, 134, 147–177, 241
Clinton, Bill, 154
CO_2, 89, 149–156, 164–166, 171, 172, 174, 177
Coal, 45, 88, 123, 150, 151, 153, 161, 162, 167
Coal resources, 88
Consumer–concerns, 162, 179, 222
Contango, 7, 10, 12, 77, 78, 245
Copenhagen agreement, 148, 154, 157
Crude
    heavy, 98, 125, 126, 210–213, 230
    light, 74, 98, 117, 119, 205, 210–212, 230, 242
    medium, 210, 211, 242
Cuba
    US relations, 234
Cyberattack, 223
Cyprus, 79

D

Demand shock, 9, 45, 199
Democracy, 133
Developing world, 163
Diversification
    energy security, 97, 180
Drill, Baby, Drill, 94–97, 109
Drilling rigs, 16, 95, 112, 115, 198
Dry gas, 206

E

Eagle Ford, 102–104, 107, 112, 183
Economic Commission Board, 13, 26, 67, 69
Economic growth, 15, 24, 28–31, 65, 79, 81, 116, 121–123, 147–149, 154, 155, 160, 161, 165–167

Ecuador, 48, 65, 68, 74, 134, 135, 170, 184, 187, 218, 227, 229, 230, 243
Efficiency
    consumption, 172, 188
    energy, 31, 45, 81, 156, 160, 169, 171–173, 175, 179, 188, 224
El Badri, Abdulla, 8, 9, 11, 27–30, 58, 119, 120, 129, 222
Emission, 65, 134, 147–161, 163–170, 172–177, 180, 224
Energy density, 98
Energy independence, 29, 79, 80, 95, 148, 220
Energy Policy and Conservation Act, 80
Enhanced Oil Recovery (EOR), 86, 88, 90, 173, 174
Environment, 8, 17, 85, 99–101, 114, 144, 154, 158–161, 163, 174, 175, 177, 233
Environmental debt, 164
Europe, 29, 75, 79, 124, 154, 156, 190, 239
European Union (EU), 24, 155, 156, 199, 225, 237

F

Fattouh, Bassam, 10, 75, 76, 94–96, 100, 123, 136, 190, 191, 195, 196, 207, 209, 211, 212
Fayetteville, 104, 106, 107, 111, 112
Financial crisis, 29, 128, 132, 142, 156, 175, 190, 203, 223, 235
First World War, 34, 233
Fiscal stress, 131–145
Ford, Gerald, 80, 103, 107
Forward price, 76
Fossil fuels, 65, 81, 86, 95, 133, 147–151, 153, 156, 158, 161–163, 165, 166, 169, 174, 175, 177, 224
Fracking, 89, 94, 101, 103, 108, 114, 115, 202
France, 101, 112, 176, 218, 226, 229

G

Gabon, 48, 60, 68, 238
Gaddafi, Muammar, 233
Gas
    natural, 61, 62, 64, 65, 90, 92, 104, 107, 118, 124, 150, 151, 167, 174, 179
    shale, 88, 89, 92–96, 100–110, 116, 117, 120, 123–124, 185, 205, 206
Geopolitics, 33–53

Index

Germany, 52, 101, 154, 172, 176, 218
Ghawar oilfield, 84
Giant oil fields, 83, 230
Globalization, 51, 233
Global warming, 99, 148, 149, 153, 159, 161, 163, 170, 172, 233, 240–242
Gore, Al, 156
Greater Burgan, 84
Green Climate Fund (GCF), 148, 155
Greenhouse Gases, 149, 153, 154, 163
Green paradox, 170, 171
Greenpeace, 147, 162
Gulf Cooperation Council, 94, 165
Gulf of Mexico, 9, 95, 98, 183, 220

H
Harvard University, 162
Haynesville, 103, 104, 106–111
Hedge funds, 76, 114, 235
Horizontal drilling, 89, 99
Hubbert, Henry, 81–84
Hydraulic fracturing, 24, 86, 89, 93, 99–101, 117, 132

I
India, 28, 30, 31, 60, 79, 81, 121, 151, 152, 155–158, 160–161, 166–168, 172, 180, 198, 229, 245
Indonesia, 36, 38, 48, 83, 93, 135, 170, 243
In-situ recovery, 86
Intergovernmental Panel on Climate Change (IPCC), 149, 153, 156, 163
International Energy Agency (IEA), 13, 14, 16, 29–31, 63, 64, 89, 96, 97, 99, 120, 132, 137, 144, 163, 174, 188, 189, 199–201, 231, 235, 241, 242, 244
International Monetary Fund (IMF), 180, 237, 240
International oil companies (IOCs), 3, 17, 33, 34, 37, 39, 40, 51–53, 71, 85, 112, 143, 161, 175, 185, 190, 193, 197, 217, 218, 226, 233–236, 238
Investment, 8, 15, 16, 34, 51, 53, 58, 74, 76, 77, 84, 95, 98, 99, 103, 108, 109, 113–116, 119, 127, 129, 138, 140–144, 150, 160–162, 170, 173–175, 180, 185, 187–189, 192–194, 200, 204, 212, 216, 224, 228, 235, 236, 240, 242, 245
IPCC.Intergonmental Panel on Climate Change (IPCC)

Iran, 5, 13, 15, 17, 22, 24, 25, 31, 32, 35–38, 40, 41, 43–45, 48, 49, 53, 59, 71, 75, 85, 120, 122, 124, 125, 126, 131–135, 138, 139, 142, 170, 175, 179, 181, 184, 185, 189, 193–199, 202, 203, 205, 208, 216, 218, 225, 227, 229, 230, 233, 234, 244
Iranian embargo, 40
Iranian sanctions, 32, 122, 197–199, 206
Iraq, 5, 7, 13, 15, 17, 18, 21–25, 31, 32, 34–38, 40, 41, 48, 49, 53, 65, 70–74, 84, 85, 96, 122, 124, 125, 129, 131, 135, 138, 142, 144, 170, 179, 181, 184, 185, 189, 193–197, 199, 200, 203, 205, 206, 208, 209, 218, 223, 225, 227, 229–231, 242
Iraq-Iran War, 194
ISDAFIX, 190
Israel, 79, 122

J
Japan, 79, 121, 153, 154, 156–158, 172, 176, 198
Jeddah, 24
Jinping, Xi, 159

K
Kazakhstan, 60, 135, 181
Kennedy, John, 147
Keynes, John Maynard, 245
Khurais, 24, 210
King Fahd bin Abdulaziz, 18, 49, 50
King Salman bin Abdulaziz, 116
Kirkuk oilfield, 84, 184, 196
Kuwait, 18, 20, 21, 23, 25, 32, 35, 36, 38, 41, 48, 49, 52, 53, 57, 58, 65, 71, 74, 75, 84, 85, 115, 122, 124, 125, 131, 134, 135, 138, 140, 144, 145, 148, 165, 170, 180, 181, 184, 185, 188, 189, 194, 195, 199, 203, 213, 215, 218, 227, 229, 230, 232, 234
Kyoto Conference, 155, 167
Kyoto Protocol, 154–157, 160, 162, 165, 168–170, 174

L
Lausanne, 197, 198
Lebanon, 79
LIBOR, 190, 235

Libya, 5, 13, 15, 21, 36–38, 45, 46, 48, 52, 53, 65, 71, 74, 78, 93, 94, 96, 135, 144, 170, 175, 181, 184, 185, 188, 194, 203, 216, 223, 227, 229, 231–234
Lima, 154–158
London Agreement, 47
Lukman, Rilwanu, 50, 177
Lukoil, 129

M

Maduro, Nicholas, 139
Majors, 7, 9, 11, 16–18, 22, 24–30, 33–37, 39, 40, 43, 45, 50–53, 57, 59–61, 64, 65, 71, 73–75, 79, 81, 86–89, 91, 94, 95, 99, 102, 108, 112, 114, 115, 117, 121, 124–126, 132–136, 138, 142, 147, 149, 152, 154, 156–165, 168, 170, 179–182, 185, 187, 189, 190, 197, 199, 200, 204, 205, 209, 210, 212, 216, 218–220, 223–225, 231, 236, 242, 243
Malaysia, 60, 135, 181
Marcellus, 103, 104, 106, 107, 108, 112
Marginal producer, 21, 24, 27, 44, 77, 119, 188, 193, 199
Market share, 3–32, 34, 39, 48, 49, 58, 61–63, 68, 71, 73, 95, 115, 116, 118, 125–126, 142, 143, 167, 170, 176, 184, 188, 194, 195, 197–200, 203–208, 213, 216, 220, 226–228, 230–232, 237, 238, 240, 242, 243, 245
Mexico, 9, 17, 60, 66, 84, 93, 95, 97, 98, 135, 152, 153, 155, 165, 183, 210, 220, 222
Middle East, 5, 34, 35, 40, 41, 44, 47, 63, 64, 80, 90, 91, 94, 96, 122, 132, 138, 139, 165, 166, 189, 194, 197, 208, 209, 227, 233
Modi, Narendra, 160, 161
Mohammad bin Salman, HRH, 230
Monopoly, 58–59
Mossadegh, Mohammed, 233, 234
Mufti, Yasser, 26, 27

N

Nasser, Amin, 13, 16, 26, 84, 185
National Iranian Oil Company, 31
National Security Strategy, 97, 148
Nazer, Hisham, 18, 50
Netback, 47, 48, 73
Niger Delta, 200
Nigeria, 5, 13, 36, 44, 46–48, 50, 65, 71, 74, 134, 135, 138, 144, 170, 179, 181, 184, 187, 199, 200, 213, 227, 229, 230

Nixon, Richard, 80
Non-renewables, 216
No Oil Producing and Exporting Cartels (NOPEC), 97, 216
North Sea, 9, 11, 12, 17, 39, 44, 46, 49, 53, 74, 190, 220
North Sea Dated Brent, 190
North Sudan, 181
Norway, 17, 46, 49, 52, 53, 60, 83, 112, 115, 133, 134, 220
Nuclear, 45, 49, 68, 185, 197–199, 206
NYMEX, 75, 76

O

OAPEC.Organization of Arab Petroleum Exporting Countries (OAPEC)
Obama, Barak, 80, 97, 98, 148, 159, 160, 234
OECD.Organization for Economic Cooperation and Development (OECD)
Official Sales Prices (OSP), 73, 124, 125, 126
OFID.OPEC Fund for International Development (OFID)
Oil depletion, 82, 83
Oil discoveries, 62, 70, 83, 84, 86, 126
Oil endowment, 82, 179
Oil extraction, 101
Oil fields, 210
Oil imports, 25, 26, 30, 80, 122, 125, 126, 167, 231
Oil quality, 70
Oil reserves, 29, 31, 34, 53, 59, 62, 64, 65, 84, 85, 90, 91, 105, 122, 126, 127, 149, 174, 189, 197, 225, 227, 229, 244
Oil revenues, 41, 133, 139, 143, 144, 149, 170, 175, 180, 224
Oil sands, 9, 88–91, 112, 142
Oman, 25, 60, 125
OPEC
    breakeven prices, 4–6, 21, 23, 103, 120, 221, 226, 227
    charter, 232–237, 240, 244
    competition, 9, 12, 17, 25, 27, 32, 40, 89, 124, 179, 186, 199, 200, 208, 209, 213, 216, 226, 230, 232, 240
    formation, 75, 234
    market share, 3–32, 39, 48, 49, 58, 61–63, 68, 71, 73, 95, 115, 116, 118, 125–126, 142, 143, 170, 176, 184, 188, 194, 195, 198–200, 203–208, 213, 216, 220, 226–228, 230–232, 237, 238, 240, 242, 243, 245

members, 3, 4, 6, 10, 12–14, 19, 21, 22, 31, 33, 40, 41, 44, 45, 47, 50, 53, 57–62, 64, 65, 67–72, 74, 75, 78, 85, 91, 93, 94, 97, 110, 114, 115, 117, 119, 126, 133–140, 142, 148, 165, 166, 170, 172–177, 182, 184, 186–189, 191, 194, 195, 199, 200, 202, 206, 207, 210, 211, 213, 216–218, 220, 224–232, 235–245
pricing, 41, 46, 58, 226
production, 21, 23, 28, 69–71, 138, 184, 188, 194, 219, 228, 231, 236, 238
re-branding, 240–242
reference bands, 236
reform, 94, 132, 133, 136, 149, 188
Secretariat, 13, 19, 66, 68, 69, 175, 216, 223, 234, 237, 238, 241
swing producer, 3, 12, 13, 16–22, 24, 39, 46, 47, 50, 71, 117, 145, 202, 203, 217, 237
OPEC Fund for International Development (OFID), 175, 238, 239
Organization for Economic Cooperation and Development (OECD), 9, 14, 19, 24, 28, 29, 44, 45, 132, 150, 152, 153, 164–166, 171, 172, 177, 182
Organization of Arab Petroleum Exporting Countries (OAPEC), 216, 232, 242
OSP.Official Sales Prices (OSP)
Ozone, 233

P

Palestine, 79
Palin, Sarah, 95
Paris, 154, 155, 157, 158, 160, 162, 167–169, 173
Peak oil, 10, 24, 79–129, 179
PEMEX, 98
Petrobras, 64, 112
Petro-states, 179, 180
Philippines, 181
Platts, 4, 189–191, 221, 235
Plays, 9, 11, 18, 21, 37, 39, 46, 50, 67, 71, 74, 91, 92, 94, 100, 102–113, 115, 142, 145, 148, 169, 191, 201–203, 215, 226, 232, 234, 236, 237, 241, 245
Pollution, 124, 148
Pre-salt, 63, 64
Primary energy production, 88, 89
Primary recovery, 89
Privatization, 129, 235

Producers, 3, 4, 6, 8–13, 15–25, 27, 28, 30–32, 34, 37, 39, 42–53, 58–65, 71–77, 79–129, 132, 136, 137, 139, 143–145, 147–150, 152, 156, 158, 160, 165, 168–173, 175–177, 179, 180, 182, 184–189, 193, 199,–200–206, 208–210, 213, 216, 217, 220–227, 231–233, 235–238, 240, 242, 244, 245
supply concerns, 220
Production cost, 70, 92, 227
Productivity, 100, 103–110, 172, 179, 201, 202, 224
Putin, Vladimir, 61

Q

Quito, 43
Quotas, 15, 22, 31, 43, 47–50, 58, 59, 64–73, 85, 120, 129, 131, 180, 187, 216, 224, 231, 237, 238, 240

R

Ramesh, Jairam, 158
Rational myth, 68, 216, 217, 237
Recession, 11
Refining, 51
Remaining reserves, 65, 89, 106
Rentier economies, 149
Rentier social contract, 133
Rentier state, 135
Research and development (R&D), 169, 174
Reserves, 13, 23, 29–31, 34, 51–53, 57, 59, 60, 62–65, 69, 70, 74, 80, 84–86, 89–91, 94, 105, 106, 116, 122–126, 131, 139–142, 144, 149, 174, 180, 187–189, 194, 197, 202, 203, 205, 208, 212, 213, 216, 218, 219, 224–227, 229, 235, 240, 245
Reserves–Commercial–Strategic, 30
Resource curse, 179
Resource nationalism, 52
Riyadh, 23, 26, 173, 174, 204, 210, 242
Rockefeller, David, 33, 34, 128
Rockefeller, John, 128
Rosneft, 127, 129, 187, 189, 192
R/P ratio, 65
Rub Al Khali, 74
Russia
OPEC cooperation, 60
production, 28, 60, 115
ruble, 20, 61, 127

S

Sabban, Mohamed, 174
SABIC, 117, 118
SADARA, 118
San Remo Agreement, 34
SATORP, 118
Saudi Arabia
    breakeven, 23, 136, 138, 140
    market led strategy, 245
    market share, 12, 17, 20, 23, 25–28, 31, 32, 48, 49, 71, 73, 95, 116, 118, 124, 125, 184, 200, 204, 206, 208, 220
    non-OPEC cooperation, 208
    production, 13, 15, 20, 31, 32, 48, 84, 117, 138, 140–142, 171, 184–186, 194, 199, 202, 204, 208, 217
    strategy, 12, 62, 114, 116, 145, 207
    swing producer, 12, 13, 17, 18, 24, 25, 46, 47, 71, 145, 202, 203, 217
Saudi Aramco, 16, 25, 26, 53, 84, 116, 118, 129, 148, 173, 186, 210, 212, 223, 230
Secondary recovery, 89
Second World War, 35, 40, 41
Seismic technology, 105, 202
Seven Sisters, 33, 35, 40–42, 46, 234
Shah of Iran, 43, 44
Shaibah, 74
Shale gas, 82, 88, 89, 93–96, 100–104, 106–110, 116, 117, 118, 121, 123–125, 185, 205, 206
Shale oil, 3, 6, 9–11, 13, 15, 16, 21, 22, 32, 44, 57, 62, 64, 81, 86–94, 101–103, 106, 107, 110–114, 116, 119–121, 126, 129, 145, 185, 189, 193, 197–200, 205, 206, 210, 221, 226, 230, 231, 244
Siberia, 129
Silva-Calderon, Alvaro, 169, 221
Solar energy, 158, 160, 173
South America, 89, 220
South China Sea, 181, 182
South Pars, 198
South Sudan, 5, 25, 181
Sovereign Wealth Fund (SWF), 140, 141, 180
Soviet Union, 9, 28, 220
Spare capacity, 18, 21, 22, 25, 57, 95, 116, 117, 121, 148, 189, 192, 194, 203, 211–213, 217, 224, 229, 232, 235, 240
Speculators, 11, 75–76, 187, 202
Spill, Baby, Spill, 95
Spot price, 44, 49
Standard Oil Trust, 33
Stanford University, 162
State Department, 49, 114

Statoil, 112, 185
Strategic petroleum reserve, 30, 80, 216, 235
Subsidies, 124, 132–136, 165, 169, 179, 188, 213
Sudan, 5, 25, 60, 181
Sulfur, 74, 75
Sweet spots, 100, 103, 105, 120
Swing producer, 3, 12, 13, 16–22, 24, 25, 39, 46, 47, 50, 51, 71, 117, 145, 202, 203, 217, 237
Syria, 5, 60

T

Taiwan, 181
Tar sands, 81, 88, 89, 98, 162
Taxes, 35, 37, 43, 61, 137, 169, 176, 177, 180
Technology transfer, 174
Tehran, 120, 197
Tertiary recovery, 89
Texas, 14, 22, 33, 35, 36, 74, 99, 102, 107, 226
Texas Railroad Commission (TRC), 14, 22, 43
Three Gorges Dam, 159
Tight oil, 24, 63, 65, 89, 94, 95, 102, 103, 108–110, 116, 120, 183
Turkey, 79, 225, 229
Turkish Petroleum Company, 34
Turkmenistan, 60, 135, 181

U

UAE, 20, 21, 23, 25, 38, 57, 58, 65, 71, 74, 85, 115, 120, 126, 131, 135, 136, 138, 145, 165, 170, 184, 185, 189, 227, 229, 232, 243
Ukraine, 94, 127, 132, 209, 225
Unemployment, 132
United Kingdom (UK), 46, 52, 53, 60, 79, 83, 101, 172, 176, 218, 229
United Nations, 154, 155, 157, 158, 174, 194, 225, 234
US
    Congress, 154, 216
    Exports, 99
    oil exports, 231
    oil swaps, 98
    OPEC relations, 231
    plays, 106, 107, 108, 111
    production, 21, 82, 89, 182, 183, 185, 186
    shale gas, 93, 95, 96, 104, 108, 109, 117, 125
    shale oil, 21, 22, 62, 87, 114, 115, 119, 120, 145, 199, 210, 230, 231
USSR, 129

V

Venezuela
 co-blending, 230
 fiscal stress, 131
 reserves, 64, 85, 89, 131, 139, 229
Vietnam, 30, 60, 79, 121, 181

W

Water cut, 84, 212
Water desalination, 94, 116
West Texas Intermediate (WTI), 74, 75, 99, 236
Wet gas, 206
White House, 49, 97, 148, 160

World Bank, 6, 9, 12, 29, 132, 135, 225, 237
WW II, 35, 40, 41

Y

Yadavaran, 198
Yamani, Zaki, 12, 17, 43, 47, 49, 81, 221
Yaran, 198
Yergin, Daniel, 32, 34, 46, 62, 76, 80, 82, 84, 121, 123, 127, 155, 156, 158, 161, 180, 233

Z

Zanganeh, Bijan, 120, 198, 199

MIX
Papier aus verantwortungsvollen Quellen
Paper from responsible sources
FSC® C105338

If you have any concerns about our products,
you can contact us on
ProductSafety@springernature.com

In case Publisher is established outside the EU,
the EU authorized representative is:
**Springer Nature Customer Service Center GmbH
Europaplatz 3, 69115 Heidelberg, Germany**

Printed by Libri Plureos GmbH
in Hamburg, Germany